CASTING THE NET WIDE

PAPERS IN HONOR OF GLYNN ISAAC AND
HIS APPROACH TO HUMAN ORIGINS RESEARCH

AMERICAN SCHOOL OF PREHISTORIC RESEARCH MONOGRAPH SERIES

The American School of Prehistoric Research (ASPR) Monographs in Archaeology and Paleoanthropology present a series of documents covering a variety of subjects in the archaeology of the Old World (Eurasia, Africa, Australia, and Oceania). This series encompasses a broad range of subjects - from the early prehistory to the Neolithic Revolution in the Old World, and beyond including: hunter-gatherers to complex societies; the rise of agriculture; the emergence of urban societies; human physical morphology, evolution and adaptation, as well as; various technologies such as metallurgy, pottery production, tool making, and shelter construction. Additionally, the subjects of symbolism, religion, and art will be presented within the context of archaeological studies including mortuary practices and rock art. Volumes may be authored by one investigator, a team of investigators, or may be an edited collection of shorter articles by a number of different specialists working on related topics.

American School of Prehistoric Research, Peabody Museum, Harvard University,
11 Divinity Avenue, Cambridge, MA 02138, USA

To Barbara, Ceri, and Gwyneira,

and future generations of scholars who we hope

will be inspired by Glynn's legacy

CASTING THE NET WIDE

PAPERS IN HONOR OF GLYNN ISAAC AND
HIS APPROACH TO HUMAN ORIGINS RESEARCH

Edited by Jeanne Sept and David Pilbeam

Oxbow Books
Oxford and Oakville

Published by Oxbow Books on behalf of the American School of Prehistoric Research.

Library of Congress Cataloging-in-Publication Data

Casting the net wide : papers in honor of Glynn Isaac and his approach to human origins research / edited by Jeanne Sept and David Pilbeam. -- 1st ed.
    p. cm. -- (American school of prehistoric research monograph series)
    Includes bibliographical references.
    ISBN 978-1-84217-454-8
    1. Human beings--Origin. 2. Human evolution. 3. Isaac, Glynn Llywelyn, 1937-1985. I. Sept, Jeanne. II. Pilbeam, David R.
    GN281.C377 2011
    599.9--dc23
                     2011041965

TYPESET AND PRINTED IN THE UNITED STATES OF AMERICA

Contents

Preface

David Pilbeam

This collection of essays and tributes to Glynn Isaac will mark the 26th anniversary of Glynn's premature death on October 5th, 1985. It has been a pleasure to learn from the contributions, reflecting as they do the work of many of Glynn's colleagues - students and collaborators. Reading the essays emphasizes again the extent of our loss.

Glynn and I first met in the early autumn of 1959 in the Birdwood at Peterhouse. "The Birdwood" (named after General Birdwood, Boer War and Great War veteran, Master of Peterhouse in the 1930s, and the building's donor) housed the college's showers and toilets - at that time sadly the only such resource in the college. We were both rather average cross-country runners, and saw each other mostly before during and after races (we were reading different triposes, Glynn Archaeology, me Natural Sciences, and moved in rather different circles). Occasionally when we passed each other in the courtyard I'd say hello and get a blank stare in response; as I slowly learned, I'd greeted Glynn's (very) identical twin brother Rhys, visiting from Oxford where he was reading history. The brothers were extremely close, remaining so until Glynn's death, both students of history - one of "deep time" the other of "shallow time". I got to know Rhys well later, sadly only after Glynn's death, and now Rhys is dead as well.

We next bumped into each other in Nairobi in very early 1965 when I was working on Miocene apes at the then Coryndon Museum; Glynn was serving as Warden at Olorgesailie. On a visit to the site I got to see what must have been one of the very earliest taphonomy experiments - Glynn was even then studying the effects of predators and nature on bone (mostly goat as I recall).

Glynn and I made quite regular contact at various workshops and conferences - Wenner-Gren in both Burg-Wartenstein and New York come to mind, one of the more memorable being the 1978 Nobel symposium held in Karlskoga. Glynn gave a superb presentation, which materialized in the 1980 symposium volume as a true classic, "Casting the net wide". This would have been one of the earliest (if not the earliest) articulations by Glynn of the unique features of human ranging patterns compared with those of non-human primates, of the importance of "home" to humans, and of the challenges in interpreting the archeological and fossil records. Characteristically, Glynn noted that, "This essay is more concerned with what we do not know than with what we do know", and was marked by the section, "Guesses about the factors which made the early-hominid pattern adaptive." I cannot recall seeing in the paleoanthropological literature of the past decades such an honest description of the way the game should be played. The essay showed how comfortably Glynn could move through the various flavors of paleoanthropology - primatology, archeology, the hominin, ape, and animal fossil records, even genetics, and integrate them comfortably and sensibly.

Glynn left Berkeley and moved to Harvard in 1983 where we were, too briefly, colleagues. In the following year Glynn, Barbara, my wife Maryellen Ruvolo, and I co-led a Peabody Museum sponsored safari to east Africa: gorillas in Rwanda, game and archeology Tanzania (the back cover photo is of Glynn at Olduvai), and sites and fossils at Koobi Fora in Kenya. There could have been no clearer demonstration from the many talks and discussions that Glynn was

Photograph by Maryellen Ruvolo. Used with permission.

an evolutionary biologist, focused of course on humans, but who just happened to be an archeologist. The photo shown here was taken early one morning at Ndutu Lodge in Tanzania, the safari party waiting impatiently to head off for the day's adventure, while Glynn sat calmly keeping us all waiting, oblivious to the impatience, doing a bit of flint knapping.

And then he died.

These essays are testaments to the splendid scholarly qualities of the man, and all in one way or another touch also on the other essential qualities of his personality: his calm wisdom, his generosity and kindness, his gentleness. He is still missed.

Reference

Isaac, G. L.

1980 Casting the net wide: A review of archaeological evidence for early hominid land-use and ecological relations. In *Current Argument on Early Man: Proceedings of a Nobel Symposium Organized by the Royal Swedish Academy of Sciences (21-27 May, 1978: Commemorating the 200th Anniversary of the Death of Carolus Linnaeus)*, edited by L-K. Königsson. Pergamon Press, Oxford, for The Royal Swedish Academy of Sciences.

Acknowledgments

We would like to acknowledge especially Wren Fournier, who has produced this volume and who has worked creatively and collaboratively to make the process almost painless. Bridget Alex was of enormous help in proofreading, reference checking, and more. Glynn was the George Grant MacCurdy Professor at Harvard and it is fitting therefore that the volume is funded by the American School of Prehistoric Research through endowments made possible by the generosity of Mr. and Mrs. MacCurdy, for which we are most grateful.

Contributor List

Editors

Jeanne Sept
Department of Anthropology
Indiana University
701 East Kirkwood Ave.
Bloomington IN 47405-7100, USA
E-mail: sept@indiana.edu

David Pilbeam
Department of Human Evolutionary Biology
Peabody Museum
Harvard University
Cambridge, MA 02138, USA
E-mail: pilbeam@fas.harvard.edu

Contributors

Ofer Bar-Yosef
Department of Anthropology
Peabody Museum
Harvard University
Cambridge, MA 02138, USA
E-mail: obaryos@fas.harvard.edu

Anna K. Behrensmeyer
Department of Paleobiology
National Museum of Natural History, MRC 121
Smithsonian Institution, Washington, DC
20013 USA

John W. Fisher, Jr.
Department of Sociology and Anthropology
Montana State University
Bozeman, Montana 59717, USA
Email: jfisher@montana.edu

Diane Gifford-Gonzalez
Department of Anthropology
University of California, Santa Cruz
1156 High Street, Santa Cruz, CA 95064, USA
E-mail: dianegg@ucsc.edu

John A. J. Gowlett
School of Archaeology, Classics and
Egyptology (SACE)
University of Liverpool
Liverpool L69 3GS, UK

Francis B. Musonda
Senior Lecturer, Department of History
School of Humanities and Social Sciences
University of Zambia
P.O. Box 32379
Lusaka, Zambia
fbmusonda@iconnect.zm

John Parkington
Department of Archaeology
University of Cape Town
Private Bag
Rondebosch 7701, South Africa
Email: john.parkington@uct.ac.za

Merrick Posnansky
Professor Emeritus, Anthropology and History
University of California, Los Angeles
Los Angeles, CA 90095, USA
E-mail: merrick@history.ucla.edu

Richard Potts
Human Origins Program
National Museum of Natural History,
Smithsonian Institution
Washington, DC 20013-7012, USA; and
Department of Earth Sciences, National
Museums of Kenya
P.O. Box 40658, Nairobi 00100, Kenya
E-mail: pottsr@si.edu

Hélène Roche
Directeur de Recherche au CNRS
UMR 7055 Préhistoire et Technologie
CNRS – Université Paris Ouest Nanterre
Maison René Ginouvès, 21 allée de l'Université
92023 Nanterre Cedex, France

Kathy Schick
Co-Director, The Stone Age Institute
1392 W. Dittemore Road
Gosport, IN 47433-9582 USA
www.stoneageinstitute.org
and
Department of Anthropology
Indiana University
Bloomington, IN USA
E-mail: kaschick@indiana.edu

John J. Shea
Anthropology Department and
Turkana Basin Institute
Stony Brook University, Stony Brook, NY
11794-4364, USA
E-mail: John.Shea@stonybrook.edu

John D. Speth
Museum of Anthropology
4013 Museums Building
University of Michigan, Ann Arbor
Ann Arbor, Michigan 48109-1079, USA
E-mail: jdspeth@umich.edu

Brian A. Stewart
McDonald Institute for Archaeological
Research
University of Cambridge
Downing Street
Cambridge CB2 3ER, UK
Email: bas29@cam.ac.uk

Nicholas Toth
Co-Director, The Stone Age Institute
1392 W. Dittemore Road
Gosport, IN 47433-9582 USA
www.stoneageinstitute.org;
and
Department of Anthropology
Indiana University
Bloomington, IN USA
E-mail: toth@indiana.edu

Bernard Wood
University Professor of Human Origins
Center for the Advanced Study of Hominid
Paleobiology
Department of Anthropology, George
Washington University
2110 G St. NW, Washington, DC 20052, USA

Richard W. Wrangham
Department of Human Evolutionary Biology
Peabody Museum
Harvard University
Cambridge, MA 02138, USA
E-mail: wrangham@fas.harvard.edu

List of Figures and Tables

Tables

1

Olorgesailie – Retrospective and Current Synthesis: Contribution to the Commemorative Volume in Honor of Glynn Isaac

Richard Potts

Introduction

Glynn Isaac's Ph.D. research was based at the Olorgesailie Prehistoric Site from 1961 to 1965. There, Glynn began to establish the extraordinary breadth of his thinking, and his publications immediately thereafter hinted at the overarching vision he would develop over the next two decades about how archaeological research could further and potentially transform the study of human evolution. The body of paleoanthropological analysis and interpretive synthesis that Isaac produced at Olorgesailie reached across the fields of geology, archaeology, taphonomy, ecology, and the evolution of behavior. These five areas provide a way of organizing and examining the impact of Isaac's early research, particularly how his work influenced later paleoanthropological thinking in east Africa and became a springboard for the latest research in the southern Kenya Rift Valley, where Olorgesailie is located.

The aim of this chapter is to review Isaac's investigations at Olorgesailie and, in relation to the five key areas of study, to summarize the most recent phases of scientific work and how these have extended upon Isaac's perspective on human origins research. An abbreviated history of the phases of research (Table 1.1) shows that Isaac's efforts were built upon the initial identification and excavation of Acheulean archaeological sites by Louis and Mary Leakey and also Robert M. Shackleton's painstaking mapping of

the Olorgesailie Formation and its original fourteen geological members. After Isaac departed Olorgesailie in September 1965 to write his dissertation (Isaac 1968a) at Cambridge University and the University of California, Berkeley, no systematic excavations were conducted at Olorgesailie until 1985. Since 1985, through the cooperation of the National Museums of Kenya and the Smithsonian Institution's Human Origins Program, our team's efforts have built upon: 1) Isaac's intensive excavations of Acheulean sites and his later idea of broadening archaeological studies beyond dense concentrations of artifacts; 2) his determination to improve the geochronological resolution of Pleistocene archaeological and paleoenvironmental evidence; and, 3) his fascination with the adaptive processes that led to the major behavioral transitions in human evolution.

Geology: Building Archaeological Research from the Ground Up

Located in the southern Kenya Rift Valley, the sedimentary sequence of the Olorgesailie basin preserves evidence of lacustrine, wetlands, and fluvial, and floodplain paleoenvironments. Together with a series of Acheulean archaeological sites, Olorgesailie is one of the best known and most important in east Africa, largely due to Isaac's research. The excavated Acheulean record is concentrated in the Site Museum (1°34'40"S,

Table 1.1 Phases of Olorgesailie research

Dates	Researchers/Team	Goals
1942-44	L. S. B. and M. D. Leakey	Identification of handaxe sites; initial excavation and documentation of Acheulean sites in the basin
1946-48	Shackleton	Detailed mapping and geology of the Olorgesailie Formation
1961-65	Isaac	Excavation of Acheulean sites; geological study; taphonomic experiments
1965-68	Isaac	Analysis and synthesis of Acheulean behavior and its context
1985-99	Smithsonian	Paleolandscape excavation; geochronological study; lithic sourcing; paleoecological analysis of mammalian fauna
1992-present	Smithsonian	Basin-wide correlation, comparative paleolandscape analysis; environmental dynamics framework for human evolution
2001-11	Smithsonian	Excavation of Middle Stone Age sites; documentation of a coordinated transition in fauna and hominin behavior

36°26'45"E), an area of approximately 0.12 km², and maintained year round by the National Museums of Kenya. The Olorgesailie basin, an area of ~300 km², is divided into two main sub-basins, referred to as the Legemunge Plain, to the west where sedimentary exposures are the most extensive, and the Oltepesi Plain, to the east where outcrops are mainly confined to a narrow band along the Ol Keju Nyiro River (Figure 1.1). This seasonal drainage runs east to west across both sub-basins before turning south through the Koora Graben. The river and certain north/south-oriented faults conveniently divide the basin into localities, defined by our team's research in order to organize geological sampling and the naming of excavations and surface collections. The Acheulean record is confined to the oldest geological formation, named the Olorgesailie Fm (Isaac 1978a). Recent research has also recognized a younger formation, the Oltulelei Fm, which consists of the Olkesiteti Member and the Oltepesi Member, dated between ~493 ka and ~49 ka (Deino and Potts 1990; Behrensmeyer et al. 2002; research in progress). This younger formation contains none of the standard evidence of Acheulean artifacts, but instead preserves a rich sequence of Middle Stone Age sites.

Isaac placed considerable importance on geological history as a framework for understanding the archaeology of the Olorgesailie region and for reconstructing the activities of Acheulean toolmakers. His geological research interests included three main areas. The first was stratigraphic history, which involved the definition of the Olorgesailie Formation and description of its principal units, referred to as members. The second focused on geochronology, and the third entailed study of the sedimentary context and paleogeographic reconstruction.

Isaac's formal definition of the Olorgesailie Fm followed Shackleton's astute mapping and identification of 14 members (Isaac 1978a; Shackleton 1978). These members (designated M1–M14 in our current work) represent the main lithological units of the lacustrine-wetlands-fluvial-floodplain record of the Olorgesailie Fm. The sequence of these deposits allows one to distinguish the main diatomaceous units (e.g., M2, M9, M11) and other lithological units from one another. Due to Shackleton's mapping and Isaac's description of the members, any subsequent researcher can locate where the specific members can be found in the basin and can rapidly gain an understanding of the overall geological history of the Olorgesailie Fm.

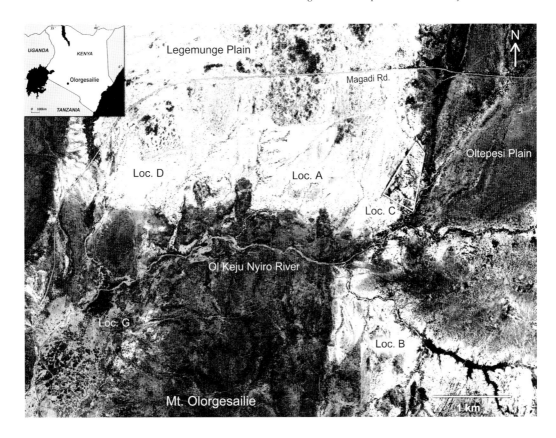

Figure 1.1 Aerial photograph of the Olorgesailie basin, southern Kenya rift valley, including the Legemunge and Oltepesi Plains, which comprise the basin's western and eastern halves, respectively. Drainages into the Ol Keju Nyiro River dissect the present land-scape and create gullies that expose artifact- and fossil-rich strata. The drainages and faults provide convenient subdivisions of the basin into localities (Locality A, B, C, D, and G, with the latter referring to the Koora Graben in the southwest).

The second area of Isaac's geological inter-ests - geochronology - most likely presented the most frustrating aspect of Acheulean research in the early 1960s. Isaac emphasized that chrono-logical resolution is a prerequisite for the validity of any narrative of early human evolution (Isaac 1972a); however, little progress could be made prior to the mid-1980s in determining age con-straints within the Olorgesailie sequence. Although tephra deposits, consisting of ash or pumice, are abundant throughout the sequence, whole-rock K-Ar analyses offered only tentative age estimates for just two members of the Olorgesailie Fm: M4 at ~486 ka, and M10 at ~425 ka, which were not concordant with the stratigraphy. While these values were considered potentially valid, caution was urged (Isaac 1977).

The third element of Isaac's geological stud-ies concerned the sediments at each excavation site and, on a broader scale, the lithologies across the basin in an effort to understand the location of hominin activities in relation to lake and land-scape features. The sedimentary context of every rich handaxe concentration that Isaac excavated,

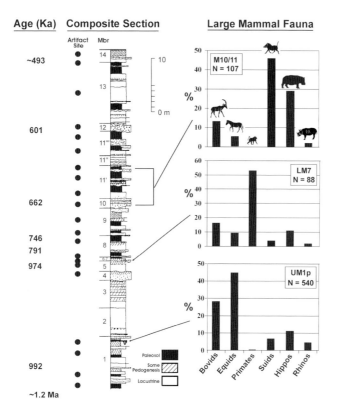

Figure 1.2 Composite geological section of the Olorgesailie Formation (based on Behrensmeyer et al. 2002), with key tephra dated by single-crystal 40Ar/39Ar analysis (Deino and Potts 1990). Acheulean artifacts have been recorded in many strata throughout the formation. Three of the most fossiliferous intervals have yielded an unusual fauna dominated at different times by Equus (Member 1), Theropithecus (M7), and suids/hippopotamids (M10/M11), while bovids are more poorly represented than expected compared with Plio-Pleistocene and present east African habitats (Potts 1998a, 2007).

such as site DE/89B, consisted of fluvial sand, and on this basis he considered sandy, seasonally-flooded channels to have provided the favored habitat of Acheulean activities. On a basin-wide scale, these fluvial loci of hominin activity were thought to have been situated typically several km away from a persistent lake that fluctuated in size. Significantly, Isaac's detailed study of the sedimentary sequence at several places across the region demonstrated that faulting occurred both within the Olorgesailie Fm and after its deposition was completed. This observation added a much needed correction to the views of L. S. B. Leakey, who considered the physical geography of the present-day basin to have persisted since the mid-Pleistocene, and of Shackleton, who considered the basin geomorphology to have been

fundamentally changed by faulting only after the deposition of the Olorgesailie Fm.

Recent Research

All three areas of Isaac's geological investigations provided a strong impetus for research initiated in 1985 by the Smithsonian-NMK team. Isaac's study of sedimentary history offered a sound geological synthesis from the start (Isaac 1977, 1978a), which, in combination with Shackleton's map (Shackleton 1978), continues to be a primary reference for the field research.

Notable advances in geochronology have, however, made a significant difference in understanding the implications of the Olorgesailie sedimentary and archaeological records. On a personal note, I originally envisioned that substantive new research at Olorgesailie would be completed in three to five years, given Isaac's interpretation that a relatively short interval, no more than a couple hundred thousand years, was available for study, confined to the middle Pleistocene. The research is now in its third decade. In 1985, a new technique – single-crystal laser-fusion 40Ar/39Ar analysis – came into play in east African research. Its application at Olorgesailie, led by Alan Deino of the Berkeley Geochronology Center, enabled distinct populations of crystals, erupted at different times yet combined in each tephra layer, to be isolated. By identifying the progressively youngest samples of crystals in each tephra, compared to the tephra below it in the sequence, the contributions from older contaminant crystals in each layer could be eliminated. The statistical treatment of the youngest-aged samples of crystals thus provided well-resolved chronological estimates for many tephra layers (Figure 1.2).

By the end of the third year of study, we knew that the Olorgesailie sediments spanned more than 1 million years. The sequence is currently the most precisely resolved record that combines archaeological remains, animal fossils, and paleoenvironmental evidence for the past 1.2 million years (Figure 1.2). An age constraint of ~1.2 Ma occurs at the base of the sequence, a published age of 493 + 1 ka occurs at the top (M14) of the Olorgesailie Fm, and age constraints of ~64 ka, 49 ka, and 10 ka have been determined near the top of the entire basin sequence (Deino and Potts 1990, 1991, and unpublished dates; Behrensmeyer et al. 2002; A. Deino pers. comm.). Research at Olorgesailie since 1990 was transformed by this relatively long chronological sequence in that it became feasible to study the detailed nature and tempo of environmental change from early to late Pleistocene, to examine hominin and overall faunal responses to

environmental dynamics, to recover evidence of post-Acheulean hominin behavior, and to test other dating methods against what is now commonly regarded as the single-crystal 40Ar /39Ar 'gold standard' of chronological resolution in east Africa. Isaac's caution about the first round of dates and his belief that the narrative of human evolution requires precise age constraints were correct. In particular, the Olorgesailie contributions to explanatory hypotheses of hominin evolution have been built on the recent advances in chronological resolution (e.g., Potts 1996a, b, 1998a, b; Trauth et al. 2005, 2007; Behrensmeyer 2006; Owen et al. 2008).

Recent research has also furthered Isaac and Shackleton's understandings of the geological sequence. To offer some examples, first, Member 6 of the Olorgesailie Fm is now understood to represent a thin overbank sand lateral to the lower M7 channels, in which the richest handaxe sites occur. Although M6 is thus removed from the sequence of members, our team has retained the original numbering of M7 through M14 (Potts et al. 1999). Second, The composite thickness of the Olorgesailie Fm is ~80 m out of a total sedimentary thickness of ~120 m in the basin (Behrensmeyer et al. 2002). The lithologies and lithofacies of the Olorgesailie Fm have now been studied in detail, and they indicate many subtleties in the paleoenvironmental history of the basin between 1.2 and 0.49 Ma (Owen et al. 2008; Potts et al. 1999; Potts 2001). It turns out that ancient lake sediments are not a persistent feature in the exposed outcrop. Instead, long periods are known when the lake gave way to wetlands or, more abruptly, when the lake potentially disappeared altogether in response to the intersection of climatic and tectonic events. The considerable thickness of the diatomaceous deposits, for which Olorgesailie is well known, resulted from far more rapid rates of deposition of lake diatoms compared to other sediment types; thus a mistaken impression had arisen that the lake persisted throughout all or much of the 700,000 years of the Olorgesailie Fm. Finally, Isaac referred to the post-Olorgesailie Fm sediments by the informal names Shanamu and Oltepesi beds; this rather extensive packet of sediments has now been studied and grouped into a new formation, the Oltulelei Fm, which also contains a rich sequence of non-Acheulean archaeological remains.

Archaeology: Digging at the Source of the Data

The space devoted in this chapter to geological studies echoes Isaac's interests and is central to our current team's outlook that paleoanthropological evidence all comes from the ground. Geological observations, therefore, occupy the vast majority of our current field research efforts just as they represented a substantial part of Isaac's. Still, the mandate of Isaac's tenure at Olorgesailie during this early part of his career was to excavate Acheulean artifacts and to learn about Acheulean archaeological sites (Isaac 1977). It is fascinating, therefore, that Glynn's huge vision for paleoanthropology flourished under such a seemingly confined directive. His 1968 dissertation and 1977 monograph on Olorgesailie stay on course in describing and seeking to explain the excavated handaxe-rich concentrations and in presenting detailed analysis of Acheulean biface morphologies. Yet in the midst of this research duty, Isaac perceived a far larger objective: "To an increasing degree, scholars are turning away from a narrow concern with artifact morphology and are making determined attempts to investigate the behavior patterns of early man" (Isaac 1967:31).

In his artifact analyses, the description of quantitative variation at Olorgesailie and comparison to other mid-Pleistocene assemblages were meant to offer insights into the cultural norms of mid-Pleistocene toolmaking. He showed that the Acheulean assemblages manifested complex variations in the presence/absence and percentages of typological categories and differences in the modal morphology of the same artifact type across a set of assemblages. For example, at Olorgesailie alone, the percentage of large cutting tools (= handaxes, cleavers, picks, and knives) ranged from near 0 to 95%. In seeking to account for such variations, Isaac cited the two standards of the time, i.e., that the artifact assemblages reflected either variations in cultural systems or different tool kits for different activities – e.g., small tools for butchery and Acheulean bifaces for non-butchery activities, as suggested by J. D. Clark (e.g., Clark and Haynes 1970). Isaac added a third possibility, what he termed the 'non-explanation' of stochastic or random walk variation, in which artifact assemblages varied with no consistent relationship to time, space, distinct cultural practices, or different activities. He also felt that this idea helped to explain the absence of broad trends in the refinement of handaxes and other tool types over time (Isaac 1975a, 1977).

In his Olorgesailie monograph, Isaac stated directly what he thus considered the most significant contribution of his early research:

> I find the clear pattern of idiosyncrasy [i.e., subtle differences in tool morphology across artifact samples] for so many of the sites extremely provocative. It seems very possible that this will prove to be an important clue to the nature of the sociocultural milieu that prevailed in east Africa during the part of the Middle Pleistocene represented at Olorgesailie. I have argued... that this phenomenon of site idiosyncrasy may reflect a pattern of microcultural differentiation of

individual bands or social groups, which in turn may indicate a cultural transmission system that combined high inertia to large-scale change, with kaleidoscopic variability of material culture habits at a local level....I regard this insight as potentially the most important outcome of my part in the Olorgesailie research (Isaac 1977:211).

Isaac linked his findings about assemblage variation, furthermore, to the great unknown of artifact function: 'the critical functions of stone tools during the Middle Pleistocene could be fulfilled by a wide range of forms and that therefore the characteristics of assemblages were free to vary within broad limits in response to influences such as raw material characteristics, stylistic idiosyncrasy, and random walk oscillation of norms' (Isaac 1977:207). With regard to the 'random walk' concept of local cultural distinctions, however, he did not see the matter settled; using a phrase he found in his later publications, he nonetheless considered it 'the best working hypothesis' (Isaac 1972a, 1977).

Recent Research

The framework established by Isaac's archaeological research has provided an important impetus in our recent field and laboratory studies. Two examples will suffice here: 1) development of the paleolandscape approach, which addresses concerns over whether Acheulean behavior is adequately sampled in handaxe concentrations (Potts 1989, 1994; Potts et al. 1999; Sikes et al. 1999); and, 2) lithic artifact analyses that have explored the effects of reduction strategies and raw material properties on handaxe morphology (Noll 2000; Figure 1.3).

In the initial phase of our team's research strategy at Olorgesailie, we sought to carry out widespread lateral sampling via excavation of the narrowest stratigraphic intervals that can be

Figure 1.3 Examples of excavations into the Olorgesailie Formation: (top) Isaac's excavation in lower Member 7 at DE/89 focused on handaxe-rich concentrations in a shallow channel-fill complex; (bottom) Smithsonian paleolandscape excavations in upper Member 1 in an area (Hyena Hill) where rich concentrations and background scatters of artifacts and fossils occur; these excavations took place as part of a broad sampling protocol of the 10- to 40-cm-thick paleosol across an outcrop length of ~4 km.

traced across outcrop distances of several hundred to thousands of meters (Figure 1.3). The goal was to substantially expand the window on the behavioral and ecological evidence preserved in the areas beyond the dense concentrations of Acheulean artifacts on which Isaac and others had focused. The paleolandscape became the unit of study, and this perspective required detailed taphonomically-oriented recovery of artifacts, fossils, and sedimentary contexts in dozens to more than 100 excavations within each narrowly defined stratum, along with the analysis of isotopic and other indicators of paleohabitat variation across the excavated areas (Potts et al. 1999). The term 'paleolandscape' emphasizes the time-averaged archive in each targeted paleosol or distinct stratum of silt or sand, quite distinct from an instantaneous snapshot of any present-day landscape.

This approach was informed by Isaac's interest, developed at Koobi Fora, concerning 'scatters and patches analysis' (Isaac 1981a), and also by the fact that Isaac's previous research at Olorgesailie sought to investigate handaxes and similar implements recovered in tightly delimited clusters. While a detailed monograph is in progress, Potts et al. (1999) presented the initial results from a dozen years of excavating three target paleolandscapes - a paleosol in upper M1 (UM1p), the handaxe-containing sand/silt unit at the base of M7 (M6/7s), and a diatomaceous silt unit in lower M7 (LM7ds). This work provided the first quantitative distinction between 'scatters' and 'patches' within excavated target intervals, and documented the considerable variation that exists in the expression of lithic artifact behaviors in the Acheulean levels at Olorgesailie. In UM1p, for example, nearly every excavation out of more than 100, ranging in area from 4 to 23 m², contained artifacts and fossil bones, with spatial densities of 3.0 (plotted artifacts) and 2.3 (plotted fossil specimens) per 0.1 m³ within the scatters, versus 15.7 and 19.3 per 0.1 m³ within the dense concentrations. By contrast, in M6/7s where Isaac uncovered the richest concentrations of Acheulean bifaces, artifacts and fossils were hardly noticeable in the scatters, with densities of only 0.08 plotted artifacts and 0.23 plotted fossils per 0.1 m³, versus 23.43 artifacts and 22.28 fossil bones per 0.1 m³ in the famed handaxe sites preserved in this same stratum. Thus,

the ~5x difference between artifact patches and scatters in upper M1 contrasted sharply with the ~293x difference in lower M7. The excavated archaeological signal in the LM7ds target interval also offered a strong contrast between patches and scatters, a pattern that was repeated in a paleolandscape target sand in upper M11' (Tryon and Potts in press; research in progress) and is apparent in all archaeological horizons above M1. Since this strong contrast between 'concentrations versus scatters' is maintained across diverse sedimentary and taphonomic contexts, we suggested that a notable shift in Acheulean hominin behavior occurred between 990,000 and 900,000 years ago in the Olorgesailie basin (Potts et al. 1999). The analysis of this shift from a relatively continuous to a highly discontinuous pattern of hominin tool-assisted use of the landscape is still in progress.

The paleolandscape approach has led us to recognize a wide variety of concentrations, including single-animal butchery sites (e.g., *Elephas recki* excavation Site 15 in the UM1p), sites with tool processing of skeletal parts from a diversity of animals, and overlapping butchery and plant-food processing sites (including butchered animal remains and anvils with plant phytoliths confined to surface pits). The effort to expand the archaeological window has also led us to investigate the highland lava source rocks. This work has resulted in 'by eye' and chemical definition and mapping of 14 volcanic rock types used by the toolmakers, the calculation of source-to-site distances, and the excavation of an Acheulean quarry active at the time of the 990,000-year-old upper M1 paleosol. At the quarry site, Acheulean toolmakers detached a series of huge flakes from a trachyte outcrop, tested each flake blank by thinning the striking platform, continued to strike flakes from the remaining periphery, and then discarded flawed blanks in the quarry and carried others away as basic handaxe/cleaver forms.

The second example, drawn from the Ph.D. dissertation research of Michael Noll (2000), extends upon Isaac's work on Acheulean biface morphological variation by applying Isaac's 'method of residuals' (Isaac 1986). This approach offers a logical chain of analysis that seeks to account for LCT morphological variation. The first step is to determine variations due to stone raw material properties, defined through mechanical engineering tests, and due to outcrop size and shape. Logistical constraints such as stone transport distance and tool curation, evaluated by source-to-site distances, are then considered. Aspects of tool function as reflected by edge maintenance and flaking intensity are the third step in the analysis. Finally, more complex cultural, stylistic aspects of stone technology are only considered at the end of this chain of analysis. Focusing on the assemblages that Isaac excavated, Noll (2000) makes a convincing case that a considerable portion of Acheulean biface variation at Olorgesailie arises from the different rock materials that were used, represented in varying proportions in the tool assemblages. In addition, the considerable difference between the small handaxes of M1 and the large handaxes of M7, which contributed to Isaac's idea of shifting cultural norms, are shown by Noll to reflect the use and continual rejuvenation/reduction of handaxe edges over a more prolonged time in M1 compared to the large handaxes that were rapidly buried in the ephemeral stream channels of M7. Noll thus rejects stylistic variability, including stochastic variation in cultural norms, as a viable explanation for LCT assemblage variations at Olorgesailie. Intriguing in this regard, however, is that Noll's study closely reflects the development of Isaac's own thinking. In the concluding 'muddle in the middle' chapter in his 1975 co-edited

volume, Isaac called for closer investigation of stone fracture mechanics in understanding tool typology, function, and explanations of artifact assemblage variation (Isaac 1975b:886). In one of the final papers he published, which laid out the 'method of residuals', Isaac (1986) favored mechanical constraints and least-effort flaking strategies in accounting for the variation in older Oldowan and Karari Industry assemblages at Koobi Fora. The same was discovered by Noll in his analysis of the 990,000 to 900,000-year-old Acheulean assemblages at Olorgesailie.

Taphonomy: Interpreting the Complexity of the Evidence

Isaac's experience in excavating Acheulean sites at Olorgesailie imparted a strong awareness of the complex manner in which various geological and biological processes can alter archaeological traces and thereby lead to misinterpretation. Conducted during his time at Olorgesailie, a study presented early in his move to the U.S. included what Isaac called 'a few simple experiments and systematic observations' focused on what we now call taphonomy and site formation. The study examined human versus fluvial influences on archaeological spatial patterns, the dispersal of bones from initial concentrations, and factors that can affect the proportions of skeletal elements (Isaac 1967).

Isaac noted that the majority of Acheulean artifact assemblages of the middle Pleistocene, including the archaeological remains at Olorgesailie, occur in alluvial sedimentary contexts, associated with fluvial sands and silts. This observation motivated field experiments in which replicas of handaxes and lava flakes were placed at points along an ephemeral stream channel to evaluate how water currents and burial by fluvial sediments affected the position and dispersal of these objects. The results indicated that archaeological assemblages are more complicated than implied by the standard division between 'primary' context (artifacts preserved where the toolmakers had left them) versus 'disturbed' (substantial redistribution of artifacts, of little value in archaeological interpretation). Later studies (e.g., Schick 1986; Petraglia and Potts 1994) provided further tests and analytical refinements consistent with the results of these initial experiments.

Isaac's second set of experiments, concerning the dispersal of bone debris, indicated that scavengers have a strong potential to remove and disperse bones from a concentration. Noting observations made by Merrick Posnansky, the former warden of the Olorgesailie Prehistoric Site, Isaac also emphasized that subaerial weathering can lead to substantial disintegration of bone; thus the presence of only a few bone splinters at a site does not preclude 'the consumption of fairly large quantities of meat' (Isaac 1967) - a statement indicating the importance he placed on faunal remains as evidence of hominin diet.

The third contribution of this taphonomically-oriented paper focused on bone accumulations made by humans and leopards in the caves of Mt. Suswa, Kenya. The main observation was that carcasses can be selectively transported and destroyed, leading to biased proportions of skeletal parts. While this result and the others presented in the paper may seem obvious to researchers today, these experiments and observations preceded the publication of foundational field experiments in skeletal part biases (Brain 1967, 1969) and fluvial winnowing (Schick 1986). Isaac's work gave an introductory hint at what are now basic principles in taphonomic research.

It may seem surprising, then, that the primary critique of Isaac's archaeological research came from a taphonomic standpoint, leveled as a broadside discharged by Binford (1977) in his review of

the Olorgesailie monograph (Isaac 1977). Based on an earlier comment by Isaac (1968b:254), it is likely that the two men had their first significant encounter at the 'Man the Hunter' symposium in April 1966. On that occasion, Binford heard Isaac's belief in the potentially positive contributions of hunter-gatherer ethnographic analogy to the interpretation of the Olorgesailie sites and his taxonomy of sites, including 'home bases' and 'campsite assemblages' (Isaac 1968b). By contrast, Isaac would have heard Binford's criticism of present-day analogy as a reliable source of archaeological interpretation (Binford 1968). Binford's later argument (1977) was that, given their fluvial context, the Olorgesailie sites were unlikely to be the original places of artifact accumulation; thus the artifact assemblages had no integrity as units of analysis relevant to hominin behavior.

The taphonomic experiments and Binford's critique foreshadowed one of Isaac's most prominent later research interests - the agents responsible for the accumulation of archaeological debris. Yet in his dissertation (Isaac 1968a) and later papers, he specifically favored hominin contributions. Isaac, for example, identified the accumulation of *Theropithecus gelada* bones from site DE/89B as remains that resulted from early human foraging, presumably derived from 'skillful hunting, which probably involved group coordination and cooperation' (Isaac 1977:199, 1971, 1975a). He did posit that fluvial processes probably acted as a secondary agent of artifact concentration. Yet the location of Acheulean remains indicated to Isaac a clear preference for camping places on sandy stream beds within abandoned or seasonal channels of the lowland basin (Isaac 1972b). Finally, the absence of hominin fossil remains at the Olorgesailie sites, in contrast with Bed I Olduvai archaeological sites, was tentatively taken to indicate that 'corpses were removed from occupation sites' (Isaac 1968b:260).

Recent Research

Isaac's initial work and later experiments by others regarding site formation have shaped our investigations at Olorgesailie over the past quarter century. One aim of our paleolandscape approach has been to sample a far greater diversity of sedimentary environments. The results show that clustered lithic assemblages, and Acheulean activities overall, were not always focused in sandy stream channels. Artifact concentrations also occur in paleosols, diatomaceous silts, and sand units broadly distributed over ancient landscapes rather than confined to stream channels (Potts et al. 1999). The richest concentrations of handaxes and other large cutting tools, nonetheless, are those excavated by Isaac in the sandy paleochannels at the base of M7. Our mapping and analysis of these paleochannels indicate, as Isaac posited, that they were shallow and likely seasonal. The inferred maximum flow depth of ~50 cm in these channels suggests that the fluvial regime was effective in dispersing and winnowing small artifacts, and in reorienting larger ones, but probably could not have caused the movement and dense aggregations of artifacts (Potts et al. 1999). This implies that Isaac's classic Acheulean excavations probably do sample places where hominins repeatedly stopped and discarded artifacts.

In each excavated paleolandscape layer, furthermore, we have found that the spatial distribution of animal bones consistently tracks that of stone tools. While this finding could imply that Acheulean toolmakers were the principal agent of bone accumulation in the Olorgesailie region over hundreds of thousands of years, Isaac's excavation at site DE/89B provides an important test case. At this site, the remains of several dozen *Theropithecus* individuals and more than 4,761 stone artifacts, including 514

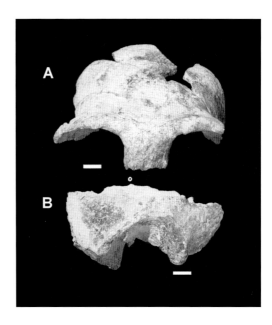

Figure 1.4 Hominin fossil KNM-OG-45500, consisting of a brow ridge (a) and left temporal bone (b) of a single individual, the first hominin fossil discovered at Olorgesailie since the Leakey's initiated systematic research in the basin in 1942. The discovery was made along the volcanic ridge (see Figure 1.3, in the distance behind the Hyena Hill excavations) where sediments of lower Member 7 abut the high ground that connected in early and middle Pleistocene times to Mt. Olorgesailie. Scale bar = 1 cm.

bifaces, were unearthed. Although our detailed reanalysis of the assemblages is in progress, no butchery or percussion marks have been found that match those in experimental tool-mark reference collections. At present, we see the attraction of the monkeys and hominin toolmakers to dry-season watering holes as potentially independent matters – i.e., the hominins left handaxes at these channel water sources, located near the transition zone between the lowlands and highlands, whereas the monkeys were preyed upon by carnivores at the same dry-season pools. Isaac commonly referred to this idea as a 'common amenity' interpretation.

This idea that hominins consistently discarded stone tools before they made the transition from lowland foraging zones to the highland outcrop sources is called the 'highland hypothesis' (Potts et al. 1999). The implication is that evening resting areas, and possibly the primary occupation sites, were located in the highlands rather than the lowland streams or lake margins. Isaac (1976, 1978a) correctly noted that predator avoidance was a vital aspect of early hominin survival strategies. The absence of hominin fossils from all the lowland areas where fossil animal bones are abundantly preserved suggested to us that hominins consistently avoided the evening 'predation arenas' focused at stream and lakeside watering places. The highland hypothesis thus led us to excavate where lowland sediments abut the volcanic ridges that lead to the high ground of Mt. Olorgesailie; in the first year of these excavations, we uncovered the first fossil hominin at Olorgesailie since the Leakeys initiated research in the 1940s (Figure1.4; Potts et al. 2004). According to the highland hypothesis, then, the dearth of hominin remains is attributed not to the removal of corpses from living sites but to an effective predator avoidance strategy – i.e., a pervasive tendency to stay away from the lowland water sources at night.

Paleoenvironments and Ecology: Understanding the Context of Early Hominins

Although the paleoecology of Pleistocene hominins was not a particularly important thrust of his Olorgesailie research, Isaac considered ecological context important to assess in the study of human evolution. Isaac's main foray into the biotic setting of Olorgesailie consisted of faunal lists for each excavated site (Isaac 1977). He also

recruited colleagues to try to recover fossil pollen, an effort that continues to be elusive to this day; and the lack of a pollen record, according to Isaac (1969), meant that climatic history was difficult to discern. Instead, he considered subtle ecological dynamics on a local scale to have played a greater role in evolution than climatic variation.

As in other basins in east Africa, tectonic activity and basin deformation was considered the most obvious factor in shaping the environmental history of Olorgesailie. Isaac (1978a) posited, nonetheless, that fresh water conditions persisted throughout the deposition of the Olorgesailie Formation, with a lake and a water outlet mechanism centered in the southwestern part of the basin, particularly in the Koora Graben (Figure 1.1, Locality G).

Recent Research

Over the past several decades, the number and quality of paleoenvironmental techniques for generating ecological reconstructions and climatic sequences have expanded enormously. The study of environmental dynamics and biotic context has, as a consequence, been the centerpiece of research at Olorgesailie since the early 1990s. Our analyses have focused on the spatial and stratigraphic distributions of lithofacies, diatoms, stable isotopes (primarily $\delta^{13}C$), and geochemical signals such as clay minerals and Rare Earth Elements (e.g., Behrensmeyer et al. 2002; Owen et al. 2008; Potts et al. 1999; Sikes et al. 1999; Trueman et al. 2006). Comparison of the paleoenvironmental sequence with broader east African and worldwide deep-sea records has also been an important aspect of our work (e.g., Potts 1996b, 2001; Behrensmeyer 2006; Trauth et al. 2007).

A synthesis of the aquatic and terrestrial evidence has led us to focus attention on environmental variability as a critical setting for human evolution in east Africa and elsewhere. Throughout the Olorgesailie Fm, the predominant picture is of large-scale revisions in landscapes, hydrology, and vegetation over time. In this context, there were associated shifts in fauna, extinction of the mid-Pleistocene large grazing community, and ultimately the demise of the Acheulean within the southern Kenya Rift. These observations, in turn, have led us to formulate a framework for hominin evolution in which environmental dynamics over short- to long time scales, involving minor to large-scale revamping of the conditions of natural selection, are hypothesized to have had a greater role in hominin and overall faunal evolution than any one ancestral setting or subtle shifts in biological balances (Potts 1996a, b, 1998a, b, 2007; Trauth et al. 2007; Owen et al. 2008). Revisions in landscapes and resources included, for example, the episodic expansion and apparent disappearance of the lake rather than the persistence of a fresh-water source. The deposition of even a modest thickness (~10 cm) of volcanic ash would have killed the grass in the lowland, leading to the periodic dispersal of the large mammal grazing community and its later reassembly in the area.

Isaac's demonstration of basin faulting during deposition implies that the impact of tectonics can mimic the influence of climate change (Behrensmeyer et al. 2002). Faulting and the rich record of volcanic tephra deposition, in fact, indicate that the overall tempo of environmental change was susceptible to many episodic influences combined with millennial- to precessional-scale climate variability. Thus, the environmental revisions that resulted from the intersection of climatic, tectonic, and volcanic factors very likely contributed to notable shifts in the combinations of grazing taxa recorded in the main fossiliferous beds (Figure 1.2; Potts 2007).

Ultimately, between ~490 ka and ~390 to 340 ka, a substantial faunal turnover occurred in which the large grazing taxa typical of the Olorgesailie Fm disappeared in the southern Kenya rift, and the extant sister taxa of these extinct mid-Pleistocene lineages began to dominate (Potts and Deino 1995). These observations helped instigate the idea that habitat instability can influence survival conditions, and the resulting pattern of natural selection may thereby promote the evolution of behavioral versatility (i.e., adaptability). This concept of variability selection (Potts 1996a, b, 2007) may offer the most viable explanation for the replacement of the specialized grazing fauna by species with flexible dietary and grouping strategies. An interesting finding in our latest research is that this taxonomic and ecological turnover occurred around the same time as the Acheulean was replaced by smaller, diverse lithic technologies and behaviors typical of the Middle Stone Age.

Evolution of Behavior: Reconstructions and Transitions in Extinct Ways of Life

In numerous publications, Isaac established his philosophy regarding archaeological inference: "The documents of archaeology are fossilized traces of human behavior in the form of artifacts and structures, food refuse and modifications to the environment" (Isaac 1972a:382). His research at Olorgesailie shaped his views on the connection between excavated data and behavioral reconstruction. The socio-economic traits that distinguish humans from other organisms formed the foundation of his explanations. Although his most far-reaching paper concerning the food sharing hypothesis of human evolution was published in 1978 (Isaac 1978b), his thinking on the topic was shaped by his 1960s research at Olorgesailie (e.g., Isaac 1968a, b). He referred to the sites in the Olorgesailie museum compound as 'excavated parts of stone age camps' where the bone-stone clusters resulted from the transport of food back to the area for social sharing (Isaac 1978a). He saw the Acheulean, furthermore, as a stable adaptive system involving tools, hunting, food sharing and division of labor, which created new selection pressures that promoted brain enlargement and profound evolutionary change (Isaac 1972a:402).

As a result of later research, though, Isaac thought that his interpretation of food sharing and home bases was not fully consistent with his parallel insight that early and middle Pleistocene behaviors were 'proto-human', i.e., an antecedent to, rather than a reflection of, modern human hunter-gatherer life. According to Isaac (1983): "My guess is that in various ways, the behavior system was less human than I originally envisioned". In this proto-human behavior system, nevertheless, a distinctly human type of division of labor, pair-bonding, and sharing were crucial components. Isaac saw the evolution of essentially a human hunter-gatherer socio-economic system as driven by these fundamental features from the Pliocene through the middle Pleistocene (Isaac 1982, 1983).

Recent Research

Isaac's emphasis on socio-economic drivers of hominin adaptation reflected his overarching concept of human evolutionary studies, to which archaeology could make unique and critical contributions. The tight integration that Isaac emphasized among tools, diet, social behavior, cognition, and life history largely signified what I have termed an intrinsic account of human evolution, in which the evolution of each aspect of behavior is largely generated by other behaviors (e.g., toolmaking in response to bipedality, cognitive advances in relation to toolmaking, language in

response to cognitive evolution, etc.) rather than by environmental drivers that were external to the adaptive system (Potts 1998b). The basic structure of Isaac's original food-sharing hypothesis thus brought together all at once an intertwined set of distinctively human characteristics, with food sharing (built upon the division of labor and pair bonding) as the glue on which the evolution of each characteristic depended. Later concepts of human evolution have considered some of these features of hominin adaptation (locomotion, pair bonding, tools, brain evolution, language, etc.) as essential to disentangle, to see as independently evolved over a prolonged period, and added to the adaptive profile of hominin species at different points in time.

Especially relevant to this point is the fact that Isaac's research at Olorgesailie dealt with what has appeared to be a stable plateau in human evolution, a lengthy era for which one could describe the principal mode of site location, diet, and ecology of the toolmakers. It seems paradoxical, then, that a person who generated enormous interest in the evolution of behavior began his career focused on what has typically been perceived as one of the most unchanging eras of early hominin behavior, the Acheulean of the mid-Pleistocene.

The latest phases of research at Olorgesailie have led toward a somewhat different perspective. Over hundreds of thousands of years, the handaxe makers exhibited an impressive ability to accommodate to diverse landscapes and rates of change in their surroundings. In this environmental context, as documented most completely so far at Olorgesailie, handaxes and other aspects of the Acheulean tool kit now appear to reflect a considerable degree of adaptability to change in the surroundings rather than a lengthy era of behavioral stagnation. As a result, this focus on hominin adaptability has suggested several important evolutionary questions, which are a focus for interdisciplinary investigation: How does adaptability evolve? How responsive is any given behavioral system to environmental dynamics? What limitations to adaptability lead to the demise of one technological and behavioral system and a transformation to a novel one? An investigation in terms of environmental dynamics can thus inspire new questions pertinent to the evolution of hominin behavior.

One of the most intriguing research developments in this regard is that, over the past decade, our team's work on the post-Olorgesailie Fm sediments has unearthed an informative series of Middle Stone Age (MSA) sites in the time frame between 490 and 220 ka. Detailed analysis of the excavated remains and repeated testing of the geochronological framework will be the focus of future publications. The vital point here is that the archaeological sequence at Olorgesailie now offers a surprising opportunity to test the factors underlying one of the most important transitions in middle Pleistocene hominin adaptation. This transition involved the dissolution of one behavioral system (the Acheulean) based on large cutting tools and the use of nearby rock sources, followed or paralleled by the emergence of a novel system of behavior (the early MSA) that relied on smaller, diverse tool kits, including projectile points, an interest in rare and distant lithic sources, greater mobility, and very likely a degree of symbolic expression through the use of coloring materials (research in progress by A. Brooks, J. Yellen, A. K. Behrensmeyer, A. Deino, and R. Potts).

Perhaps the most important implication of this latest research phase at Olorgesailie, then, is that Paleolithic archaeology may make its most important contributions by focusing on behavioral transitions rather than on specific industries like the Acheulean. The training of archaeologists

is overwhelmingly oriented toward gaining expertise in specific technological phases or relatively stable adaptive patterns, such as the Oldowan, the Acheulean, or the MSA, rather than toward understanding the most important intervals of technological and adaptive change. Isaac's impressive vision concerning the archaeology of human origins would appear to be better fulfilled as students and professional researchers focus their study on the key transitions in Pleistocene hominin adaptation.

Glynn Isaac's Influence: Past and Future

After his research was completed at Olorgesailie, Isaac inspired the field of paleoanthropology by his capacity to synthesize the diverse threads of the human evolutionary narrative and by his vision of the linkages between locomotor, foraging, social, cognitive, technological, and dietary adaptations (e.g., Isaac 1983). These elements of the narrative formed a unifying understanding of human evolution. His view that archaeological remains represent 'fossilized behavior' was an immense eye-opener to many of us drawn to his early writings (e.g., Isaac 1969, 1971) because it broadened archaeological research beyond the classification of artifact types and industries. This view played an important role in laying the foundation for paleoanthropology as a union of disciplines that, together, can address the adaptive aspects of human evolution over the past 2.5 million years.

Isaac considered natural selection as the core explanation of the behavioral aspects of human evolution, and he urged others to think in those terms. He saw the fundamental question of human origins as a 'puzzle' that called upon paleoanthropologists to identify the patterns of natural selection that transformed 'protohumans into humans' (Isaac 1978b). This point was a compelling one from the outset of my studies, and it has actively shaped the recent phases of research at Olorgesailie. Paleolandscape excavations, lengthy stratigraphic records that imply strong variation in adaptive conditions over time, and turnover in the southern Kenya rift faunal community, in fact, have inspired a concept of natural selection that focuses on environmental instability, in which variable contexts of selection can eventually favor adaptive versatility over habitat-specific ways of life (e.g., Potts 1996a, b, 1998a, b, 2007; Trauth et al. 2005, 2007; Owen et al. 2008). The habitat instability so apparent at Olorgesailie is now also recognized in many other long sedimentary sequences in east Africa (e.g., Ashley 2007; Lepre et al. 2007; Campisano and Feibel 2007). As a result, the east African record of environmental dynamics has raised reasonable doubt about the role of an ever-persistent ancestral habitat in shaping human evolution.

Ultimately, Isaac (1975a:882) offered the clearest statement on how archaeological studies can contribute to an understanding of hominin adaptation: 'our understanding of adaptation can only be as good as the best worked out case studies of archaeological traces of human activity in relation to a reconstructed local paleoenvironment'. His work in the Olorgesailie basin was the first case study in his career, and it created a rigorous foundation for novel hypotheses and methods of analysis in the pursuit of human origins.

Acknowledgments

During my last meeting with Glynn, in the summer of 1985, he offered his enthusiasm and thoughtful commentary regarding the new survey we had just finished at Olorgesailie and for the research my team was planning; it meant the world to me to have Glynn's support. I thank the National Museums of Kenya, particularly Dr. I. O. Farah, E. Mbua, F. K. Manthi, and M. Muungu, for our long-term collaboration, logistical support, and a terrific base of operations for the Olorgesailie research project. I am grateful to Jennifer Clark for assistance with the figures, and to Jeanne Sept and David Pilbeam for their invitation to contribute to this tribute to a man whose thinking enlarged archaeology's role in the study of human evolution.

References

Ashley, G. M.

2007 Orbital rhythms, monsoons, and playa lake response, Olduvai Basin, equatorial East Africa (ca. 1.85-1.74 Ma). *Geology* 35:1091-1094.

Behrensmeyer, A. K.

2006 Climate change and human evolution. *Science* 311:476-478.

Behrensmeyer, A.K., R. Potts, A. Deino, and P. Ditchfield

2002 Olorgesailie, Kenya: A million years in the life of a rift basin. In *Sedimentation in Continental Rifts*, edited by R. W. Renaut and G. M. Ashley, SEPM Special Publication 73:97-106.

Binford, L.

1981 *Bones: Ancient Men and Modern Myths.* Academic Press, New York.

Binford, L. R.

1968 Methodological considerations of the archeological use of ethnographic data. In *Man the Hunter*, edited by R. B. Lee and I. DeVore, pp. 268-273. Aldine, Chicago.

1977 Olorgesailie deserves more than the usual book review. *Journal of Anthropological Research* 33:493-502.

Brain, C. K.

1967 Hottentot food remains and their bearing on the interpretation of fossil bone assemblages. *Scientific Papers of the Namib Desert Research Station* 32:1-11.

1969 The contribution of Namib Desert Hottentots to an understanding of australopithecine bone accumulations. *Scientific Papers of the Namib Desert Research Station* 39:13-22.

Brooks, A., K. Behrensmeyer, J. Yellen, A. Deino, W. Sharp, and R. Potts

2007 After the Acheulean: Stratigraphy, dating, and archaeology of two new formations in the Olorgesailie basin, southern Kenya Rift. *Paleoanthropology, Internet Journal of the Paleoanthropology Society*, A5.

Campisano, C. J., and C. S. Feibel

2007 Connecting local environmental sequences to global climate patterns: Evidence from the hominin-bearing Hadar Formation, Ethiopia. *Journal of Human Evolution* 53:515-527.

Clark, J. D., and C. V. Haynes, Jr.

1970 An elephant butchery site at Mwanganda's Village, Karonga, Malawi, and its relevance for Palaeolithic archaeology. *World Archaeology* 1:390-411.

Deino, A., and R. Potts

1990 Single crystal 40Ar/39Ar dating of the Olorgesailie Formation, southern Kenya rift. *Journal of Geophysical Research* 95(no. B6):8453-8470.

1991 Age-probability spectra for examination of single-crystal 40Ar/39Ar dating results: Examples from Olorgesailie, southern Kenya rift valley. *Quaternary International* 7/8:81-89.

Isaac, G. L.

1967 Towards the interpretation of occupation debris: Some experiments and observations. *Kroeber Anthropological Society Papers* 37:31-57.

1968a The Acheulean site complex at Olorgesailie, Kenya: Contribution to the interpretation of Middle Pleistocene culture in East Africa. Ph.D. Dissertation, Cambridge University.

1968b Traces of Pleistocene hunters: An East African example. In *Man the Hunter*, edited by R. B. Lee and I. DeVore, pp. 253-261. Aldine, Chicago.

1969 Studies of early culture in East Africa. *World Archaeology* 1:1-28.

1971 The diet of early man: Aspects of archaeological evidence from Lower and Middle Pleistocene sites in Africa. *World Archaeology* 2:278-299.

1972a Chronology and the tempo of cultural change during the Pleistocene. In *Calibration in Hominoid Evolution*, edited by W. W. Bishop and J. Miller, pp. 381-430. Scottish Academic Press, Edinburgh.

1972b Comparative studies of Pleistocene site locations in East Africa. In *Man, Settlement and Urbanism*, edited by P. J. Ucko, R. Tringham, and G. W. Dimbleby, pp. 165-176. Duckworth, London.

1975a Stratigraphy and cultural patterns in East Africa during the middle ranges of Pleistocene time. In *After the Australopithecines*, edited by K. W. Butzer and G. L. Isaac, pp. 495-542. Mouton, The Hague.

1975b Sorting out the muddle in the middle: An anthropologist's post-conference appraisal. In *After the Australopithecines*, edited by K. W. Butzer and G. L. Isaac, pp. 875-887. Mouton, The Hague.

1976 The activities of early African hominids. In *Human Origins: Louis Leakey and the East African Evidence*, edited by G. L. Isaac and E. McCown, pp. 482-514. Benjamin, Menlo Park, California.

1977 *Olorgesailie: Archaeological Studies of a Middle Pleistocene Lake Basin in Kenya*. Chicago University Press, Chicago.

1978a The Olorgesailie Formation: Stratigraphy, tectonics and the palaeogeographic context of the middle Pleistocene archaeological sites. In *Geological Background to Fossil Man*, edited by W. W. Bishop, pp. 173-206. Scottish Academic Press, Edinburgh.

1978b The food-sharing behavior of protohuman hominids. *Scientific American* 238(4):90-108.

1981 Stone Age visiting cards: Approaches to the study of early land-use patterns. In *Pattern of the Past*, edited by I. Hodder, G. L. Isaac, and N. Hammond, pp. 131-155. Cambridge University Press, Cambridge.

1982 Models of human evolution. *Science* 217:295-296.

1983 Aspects of human evolution. In *Evolution from Molecules to Men*, edited by D. S. Bendall, pp. 509-543. Cambridge University Press, Cambridge.

1986 Foundation stones: Early artifacts as indicators of activities and abilities. In *Stone Age Prehistory*, edited by G. N. Bailey and P. Callow, pp. 221-241. Cambridge University Press, Cambridge.

Lepre, C. J., R. L. Quinn, J. C. A. Joordens, C. C. Swisher III, and C. S. Feibel
2007 Plio-Pleistocene facies environments from the KBS Member, Koobi Fora Formation: Implications for climate controls on the development of lake-margin habitats in the northeast Turkana Basin (northwestern Kenya). *Journal of Human Evolution* 53:504-514.

Noll, M. P.
2000 Components of Acheulean lithic assemblage variability at Olorgesailie, Kenya. Ph.D. Dissertation, University of Illinois at Urbana-Champaign, Urbana, Illinois.

Owen, R. B., R. Potts, A. K. Behrensmeyer, and P. Ditchfield
2008 Diatomaceous sediments and environmental change in the Pleistocene Olorgesailie Formation, southern Kenya Rift Valley. *Palaeogeography, Palaeoclimatology, Palaeoecology* 269:17-37.

Petraglia, M. and R. Potts
1994 Water flow and the formation of Lower Paleolithic sites in Olduvai Gorge, Tanzania. *Journal of Anthropological Archaeology* 13:228-254.

Potts, R.
1989 Olorgesailie: New excavations and findings in Early and Middle Pleistocene contexts, southern Kenya rift valley. *Journal of Human Evolution* 18:477-484.

1994 Variables vs. models of early Pleistocene hominid land use. *Journal of Human Evolution* 27:7-24.

1996a *Humanity's Descent: The Consequences of Ecological Instability*. William Morrow & Co., New York.

1996b Evolution and climate variability. *Science* 273:922-923.

1998a Variability selection in hominid evolution. *Evolutionary Anthropology* 7:81-96.

1998b Environmental hypotheses of hominin evolution. *Yearbook of Physical Anthropology* 41:93-136.

2001 Mid-Pleistocene environmental change and human evolution. In *Human Roots: Africa and Asia in the Middle Pleistocene*, edited by L. Barham and K. Robson-Brown, pp. 5-21. Western Academic Press, Bristol.

2007 Environmental context of Pliocene human evolution in Africa. In *Hominin Environments in the East African Pliocene: An Assessment of the Faunal Evidence*, edited by R. Bobe, Z. Alemseged, and A. K. Behrensmeyer, pp. 25-48. Springer, New York.

Potts, R., and A. Deino
1995 Mid-Pleistocene change in large mammal faunas of the southern Kenya rift. *Quaternary Research* 43:106-113.

Potts, R., A. K. Behrensmeyer, and P. Ditchfield

1999 Paleolandscape variation in early Pleistocene hominid activities: Members 1 and 7, Olorgesailie Formation, Kenya. *Journal of Human Evolution* 37:747-788.

Potts, R., A. K. Behrensmeyer, A. Deino, P. Ditchfield, and J. Clark

2004 Small mid-Pleistocene hominin associated with East African Acheulean technology. *Science* 305:75-78

Schick, K. D.

1986 *Stone Age Sites in the Making: Experiments in the Formation and Transformation of Archaeological Occurrences.* BAR International Series, 319, Oxford.

Shackleton, R. M.

1978 Geological map of the Olorgesailie area. In *Geological Background to Fossil Man*, edited by W. W. Bishop, pp. 171-172 and insert map. Scottish Academic Press, Edinburgh.

Sikes, N., R. Potts, and A. K. Behrensmeyer

1999 Early Pleistocene habitat in Member 1 Olorgesailie based on paleosol stable isotopes. *Journal of Human Evolution* 37:721-746.

Trauth, M. H., M. A. Maslin, A. Deino, and M. R. Strecker

2005 Late Cenozoic moisture history of East Africa. *Science* 309:2051-2053.

Trauth, M. H., M. A. Maslin, A. L. Deino, M. R. Strecker, A. G. N. Bergner, and M. Dühnforth

2007 High- and low-latitude forcing of Plio-Pleistocene East African climate and human evolution. *Journal of Human Evolution* 53:475-486.

Trueman, C. N., A. K. Behrensmeyer, R. Potts, and N. Tuross

2006 High-resolution records of location and stratigraphic provenance from the rare earth element composition of fossil bones. *Geochimica et Cosmochimica Acta* 70:4343-5355.

Tryon, C. A. and R. Potts

in press Approaches for understanding flake production in the African Acheulean. In *Reduction Sequence, Chaîne Opératoire, and Other Methods: The Epistemologies of Different Approaches to Lithic Analysis*, edited by G. Tostevin. Springer, New York.

2

Conversations with Glynn's Ghost: The Evolution of Paleolandscape Research at East Turkana

Anna K. Behrensmeyer

Introduction

Glynn Isaac was a great inspiration to young researchers involved in field work at East Turkana (then East Rudolf) during the 1970s and early 1980s. One of his seminal ideas was to explore patterns of artifact and bone preservation outside of the high-density patches that were the traditional focus of archeological interest and excavation. His early forays into documenting the "scatter between the patches" affected all of us and helped provide new meaning for my own dissertation research on patterns in unbiased samples of surface fossils. Glynn's ideas evolved into the landscape approach in African paleoanthropology, which has contributed a multitude of new insights and questions to the study of early human paleoecology and behavior. Many researchers who have worked on landscape archaeology and landscape paleontology over the past three decades can trace their interest back to Glynn's enthusiasm for understanding spatial patterns in early hominin behavior over a broad spectrum of scales, from individual sites to whole landscapes.

This paper is aimed at providing a fresh look at how the landscape approach championed by Glynn has evolved at East Turkana. It draws mainly upon archeological and paleontological landscape-oriented research in the Okote Member of the Koobi Fora Formation.

Background

It is not surprising that Glynn developed and implemented his novel landscape-scale research strategies in the context of East Turkana's unusually rich surface exposures of fossils and artifacts. On their own, these surface finds could have called into question the traditional assumption of dense patches with little of interest between them. Since the late 1960s, East Turkana has provided a wealth of fossils and archeological materials that have transformed understanding of human evolution, environmental change, and the development of the modern African fauna. Indeed, there is so much preserved in the channel and floodplain deposits of some stratigraphic levels in the Koobi Fora Formation that one can dig almost anywhere and find *in situ* material. The problem then becomes how to document, collect, and make sense of such an abundance of riches.

Landscape archaeology as a term appeared in publications as early as 1974 (Ashton and Rowley 1974) and has since become an accepted approach for historical and pre-historical archeologists in North America and Europe (Yamin and Metheny 1996; Chapman 2006). Glynn was responsible for its early development in East Africa, where during the 1970s he pioneered conceptual and methodological approaches for examining artifact and bone occurrences over large areas of ancient landscapes. He laid out the rationale for this agenda and his methodology for implementing it in two initial papers. The first of these, published with Jack Harris (1980; an earlier version appeared in 1975), reported the findings of his 'experiment' with surveys to determine artifact densities on the eroding outcrop surfaces of East Turkana and their relationship to both recent and paleogeographic features of the landscape. Glynn set the agenda for landscape archaeology by asserting that "...all traces of prehistoric activity are not necessarily concentrated in the patches of material that one can conveniently call sites and mark with a dot on the map" (Isaac and Harris 1980:19). The "Stone Age visiting cards..." paper (Isaac 1981) provides more substantive insights into Glynn's thinking about the importance of the "scatter between the patches" and established the theoretical underpinnings of his landscape approach.

From the outset, there was a necessity for new research strategies that included a presumed continuum of low- to high-density concentrations of artifact scatters and provided comprehensive assessments of the spatial patterns at scales of 10^2-10^7 m^2. Always one for clear definitions, Glynn characterized scatters as less than one piece per 10,000 m^2 (100 x 100 m) and patches as 1–100 pieces/m^2 (Isaac 1981), noting that these early guidelines were "very rough guesses". The "visiting cards" were any remains and traces of human activity (artifacts, hominin skeletal remains, post-holes, hearths, etc.). Whether these were clustered or widely dispersed, Glynn was convinced that their spatial patterns would contain important information about "early proto-human behavior". His initial model was an analogy to the physics of matter, with the smallest particle being a single artifact, bone, or other trace of human behavior, the "atom" a related cluster of such particles, such as a set of conjoined pieces representing a stone knapping event, or the fragments of a broken bone. At the level of a "molecule or compound," he imagined clusters of datum points representing different episodes or actions, "clusters of clusters," which constitute most archeological sites.

At the fourth hierarchical level are sets of sites that form landscape-scale patterns, or a 'regional system'. Glynn emphasized that the concept of this level must include scatters between sites, and that it "could in theory be made up entirely of unclustered fundamental datum points" (Isaac 1981:213). As pointed out by Rogers (1997), Glynn realized that the artifacts and bones themselves, rather than high-density concentrations at sites, should be the most basic unit of interest. To reveal this hierarchy in the archeological record, Glynn proposed collecting spatial information about all fundamental datum points across a landscape, so that the four levels of the "visiting card" hierarchy described above could emerge from the analysis (Isaac 1981).

Glynn's first attempt at implementing this approach was a rather elaborate "stratified random procedure" focused on sampling surface materials along the Karari Escarpment of East Turkana (Figure 2.1), where positions of 100 m wide linear transects were determined randomly within seven 2-km-long baseline segments. These transects were oriented perpendicular to the map trace of the escarpment in Area 131 and thus sampled across the stratigraphic levels exposed in the escarpment's eroded face (Isaac and Harris 1980; Isaac 1981). In spite of the limitations of this pilot study, Glynn concluded that less than 10% of the artifacts in these deposits were concentrated in sites, while over 90%, and presumably a wealth of associated information about ancient behavior and land use, lay between these sites. This finding was a major wake-up call for traditional, site-focused archaeology. Glynn and Jack Harris also identified particular stratigraphic levels where artifacts were consistently abundant and uncovered an intriguing pattern regarding the distribution of hand axes, which appeared to be more common outside of the main sites than within them.

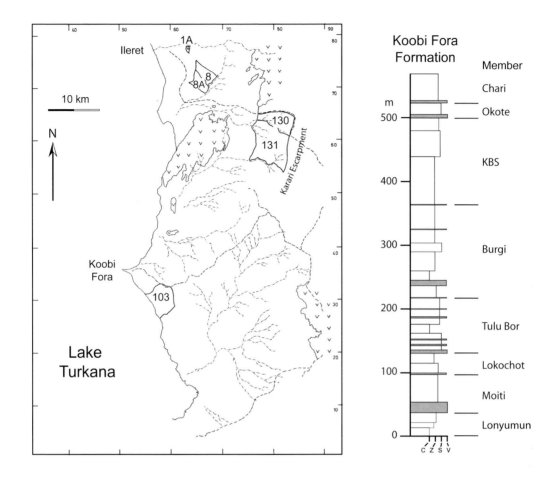

Figure 2.1 Map of East Turkana and stratigraphy of the Koobi Fora Formation. Map shows the fossil-collecting areas (numbers) where paleolandscape research has been done in the Okote Member. Marginal numbers refer to East Africa Grid, Belt H, roads are dashed lines, "V" pattern indicates volcanic basement (modified from Figure 1.2 in Leakey and Leakey 1978). Stratigraphic section is a generalized composite showing volcaniclastic units (tuffs, gray shading) used for chronostratigraphic correlation within and beyond the Turkana Basin (modified from Figure 2 in Feibel et al. 1989). Lower scale: c = clay, z = silt, s = sand, v = volcaniclastic.

As Glynn was well aware, the net result of accumulated "visiting cards" on any preserved land surface would always be a tangle of intersecting trajectories and material resulting from the movements of quadrupeds or bipeds living their lives in relation to the available resources – primarily food, water, and shelter (Isaac 1980, 1983). Nevertheless, he believed that even a tangled spatial patterning of artifacts and bones in time-averaged landscape samples could tell us new and important things about our ancestors and the ecosystems of which they were a part. He infused his students and colleagues with excitement at this prospect, and a number of them went on to dissertations and other research projects that took on the challenges of Turkana landscape archaeology and site formation processes, including Diane Gifford (1977), Nick Toth (1982), Zefe Kaufulu (1983), John Barthelme (1985), Kathy Schick (1986), Nikki Stern (Stern et al. 2002), and Henry Bunn (Bunn et al. 1980; Bunn 1981, 1994; for a thorough review, see Isaac and Isaac 1997, Chapter 1). Later the torch passed to the next generation through Jack Harris to Mike Rogers (1997; Rogers et al. 1994), David Braun, (Braun 2006; Braun et al. 2008), Pobiner et al. (2008), and Jack McCoy (2009). Of course, while encouraging and nurturing others, Glynn himself pursued evidence for landscape-scale human behavior up to 1985 with continuing field research at East Turkana (Isaac and Isaac 1997).

Simultaneously, paleontological and taphonomic investigations by Behrensmeyer and colleagues at East Turkana (Behrensmeyer 1975, 1985; Behrensmeyer and Laporte 1982; Isaac and Behrensmeyer 1997) targeted the Okote Member in three different areas, Ileret, Koobi Fora, and the Karari Escarpment (Figure 2.1). They documented lateral variability of vertebrate taxa in the fossil-bearing strata and related this to environmental parameters such as habitat structure (open versus closed vegetation, alluvial floodplains versus lake margins). Diane Gifford (now Gifford-Gonzalez) initiated research on recent site formation and landscape-scale vertebrate taphonomy at East Turkana during the 1970s (Gifford and Behrensmeyer 1977), which contributed to understanding of time-averaging and site preservation processes. Behrensmeyer and René Bobe led a landscape taphonomy and paleoecology project to sample fossil vertebrates in the earlier Lokochot and Tulu Bor Members in 2002–2003 (Behrensmeyer et al. 2004). Later

paleontological work by Francis Kirera (2007) focused on spatial distribution patterns of bovid fossils in the Koobi Fora Formation using GIS.

Rick Potts, who as a graduate student worked with Glynn's team in 1977 at East Turkana, went on to develop the landscape approach at Olorgesailie, where Glynn had worked before taking on the archeological research at East Turkana (Potts 1989, 1994; Potts et al. 1999). Rob Blumenschine, another of Glynn's Ph.D. students, investigated carcass availability in the modern Serengeti ecosystem (Blumenschine 1986) and subsequently applied a landscape approach to lower Bed II at Olduvai Gorge (Blumenschine and Masao 1991; Blumenschine et al. 2003). Many others have gone on to implement this approach in different parts of the African continent and beyond. A comprehensive review is beyond the scope of this volume, thus my goal is to focus on what came from the energy and purpose that Glynn unleashed at East Turkana through his extended team of students and colleagues. It is a tribute to this energy that some 40 years since its inception, landscape archaeology and paleontology at East Turkana are still going strong.

If he could rejoin us now, Glynn would be excited by the subsequent findings at East Turkana and their implications as well as the inevitable controversies. It is unfortunate indeed that we do not have his ever-inquisitive and incisive mind to ask the hard questions, to help us make sense of our findings, and to urge us on to new insights about early resource utilization by hominins and associated fauna. Because Glynn made a strong impression on so many of us during our formative years, it is not hard to imagine what he would have said and the questions he would have raised with regard to continuing paleolandscape research at East Turkana. I offer some fanciful (a word he enjoyed using) reactions from "Glynn's Ghost" about these changes as a way to reconnect with his thinking and to share with others the unique and lively spirit he brought to this field of study.

First of all, what is the question?

If there was any lesson drilled into the heads of all of Glynn's students, it was to start with a question. For Glynn and most other archeologists of the time, a fundamental question was, "How were hominins patterning their behavior over ancient landscapes?" The answers could tell us a lot about their ability to plan, socialize, acquire and retain resources, and control their environment – some of our most basic of human traits. Particularly tantalizing was the prospect of seeing a shift through time from hominins that responded passively to environmental parameters to those that made deliberate choices about landscape-use. Such a shift might be manifested, for example, in a change from more-or-less randomly distributed artifacts across a paleolandscape indicating a low level of organized resource use to more dense patches reflecting hominins' ability to control raw materials and food acquisition and to repeatedly return to particular places. Glynn characterized this as a transition between resource controlled and socially controlled "nodes of activity" on a landscape (Isaac 1981), predicting that evidence for this transition would be manifested in the quantity and morphology of artifacts and numbers of bones at these nodes (Stern et al. 2002; Rogers 1997; Braun et al. 2008) and also on the landscape between them (Isaac 1981, 1983).

It appeared to Glynn, and to all of us who were caught up in his enthusiasm at the time, that testing these ideas was feasible, especially at East Turkana. One only had to recognize paleolandscapes and accurately measure and analyze artifact and bone distributions along them.

A fundamental big question always generates subquestions, so in the following sections, I will focus on a few of these important to the study of the material record of an ancient landscape.

What is a paleolandscape?

This question is fundamental to the conceptualization of early hominin activities on that landscape. Ideally, we could strip away modern vegetation and overlying strata to reveal a buried soil horizon that represents the surface of a single, three-dimensional landscape extending, perhaps, for many square kilometers. Here we would find the discarded remains of hominin activities amidst well-preserved vertebrate skeletal debris and even plants. If we could hover over such a landscape and look down into the soil at will, we could document patterns of artifacts and bones in relation to geographic features such as river channels, ponds, elevated areas, vegetation types. We could then put all of this information into layers (as in an ArcGIS computer array) to provide evidence for the density and patchiness of artifacts and bones and their juxtaposition across the ancient land surface. Then the real work of figuring out how to interpret these patterns in terms of hominin land use and behavior would begin.

In reality, of course, what we have are disconnected patches of strata that represent successive land surfaces of variable duration sandwiched between rapidly accumulated sedimentary deposits that buried them (Figure 2.2). Some of these buried land surfaces can be traced laterally for hundreds of meters, kilometers, or in some cases much wider areas because they are marked by unique volcanic ash deposits (e.g., Stern et al. 2002; Potts et al. 1999; Rogers 1997; Behrensmeyer et al. 2004). These surfaces can sample floodplains that pass laterally into levees, channels, or lake margins, providing the opportunity to document variability of artifacts and fossils

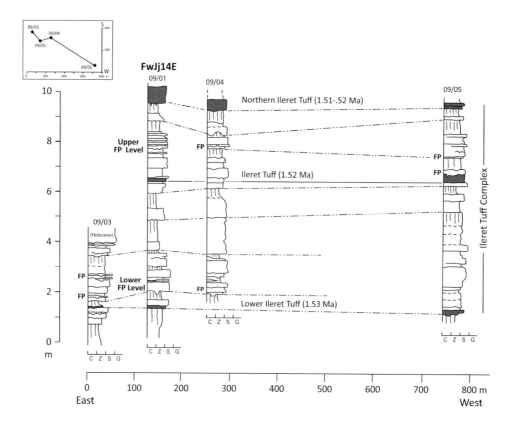

Figure 2.2 Example of a series of eight or more buried paleo-landsurfaces (lines with dashes and dots) represented on a two-dimensional panel diagram of the Ileret Tuff complex in Area 1A. This is based on preliminary microstratigraphy by the author in 2009. The "Upper and Lower FP levels" indicate the position of known hominin footprints (Bennett et al. 2009); "FP" shows that footprint levels can be traced laterally over hundreds of meters. Wavy vertical lines indicate paleosols that underly the paleo-land surfaces; these are interbedded with well-stratified silts and sands representing rapidly aggrading floodplain deposits that buried these surfaces. Sediment texture scale: C = clay, Z = silt, S = sand, G = gravel; tuffs in dark gray. Inset shows the actual plan positions of the sections.

in relation to physiographic features that likely were important to the hominins and other animals. Practically speaking, therefore, a paleolandscape is any laterally extensive, isochronous sedimentary interface or contact that bears evidence for a pause in deposition long enough to grow plants and develop pedogenic features in the underlying sediments. Plant growth can happen in days to weeks in modern environments, but

preservable soil features usually take longer, i.e., hundreds to thousands of years. The surface of the soil at any instant in time forms the actual landscape, but of course there is constant turnover of the soil through bioturbation, incremental additions of water- and wind-blown sediment, and larger-scale erosion and re-deposition before the whole package is finally buried and preserved. Therefore, in documenting a paleolandscape, it is important to pay attention to the sediments below the preserved surface and also to those immediately above it, both of which may contain artifacts and bones that accumulated during the "life" of the landscape.

What makes a paleolandscape suitable for study?

Glynn was fascinated with the potential for correlating artifact and bone distributions to features of the paleolandscape and always supported the need for careful microstratigraphy around his major "patch" excavations as well as the broader landscape samples (e.g., Kafulu 1983; Rogers 1997; Isaac and Isaac 1997; Stern et al. 2002). The ideal target level is a single sedimentary horizon or unit that can be traced across different depositional environments and topographic situations, preferably with a marker horizon as a reference above or below the paleolandscape surface. Clearly the highest resolution, least equivocal source of paleolandscape data comes from excavated sites along one well-defined paleolandscape surface. The problem is that such excavations take time and the samples outside of high-density patches are usually small.

Given a sequence of sedimentary rocks of the right age and a scatter of surface artifacts, careful consideration of stratigraphy and physiography of the modern outcrops are starting points for identifying a target paleolandsurface. General geological reference sections are helpful, but it is always better to measure new stratigraphic logs extending below and above the target interval, thereby providing evidence for the genesis of the deposit (e.g., a transgressive lake shoreline or a fluvial crevasse-splay that buried a land surface) as well as information on marker beds (e.g., datable tuffs, caliches) that can be correlated with an area's established stratigraphy. The optimal target is a single relatively thin, well-exposed fossil and

artifact-bearing stratum associated with a paleosol representing a buried land-surface (Figure 2.2). This can be documented with a series of detailed lateral sections from one end of the sampled paleolandscape to the other, correlated to make two-dimensional panel diagrams or three-dimensional fence diagrams that show the relationship of artifact and bone distributions in relation to paleoenvironmental gradients, such as channel to levee to floodplain transitions.

The physiography of the present, eroding landscape is crucial for maximizing the information derived from surface sampling. A gradually sloping bench or narrow platform just below the productive zone is ideal for catching naturally eroded fossils and for careful systematic searching (Figure 2.3). Too flat, and the search area may have a lag of surface fossils derived from many different eroding layers, but too steep and there will be little to find because it has all rolled downhill.

Glynn: That's all very well, and I certainly agree about the targeted lateral approach, but what if you want to trace an interval that is not so well-exposed?

Response: I would let the outcrops be my guide, opting for a controllable situation rather than trying to impose quantitative methods on uncooperative substrates.

Glynn: Ahh, you want to make things easy, when your question should dictate where to sample, regardless of thorns, skinned knees, and occasional vertigo. Why not clear the brush and clean off the slope, if you need to get down to the outcrops, rather than giving up on a potentially exciting level? Of course, when all else fails, you can DIG – put in test trenches to trace the landscape from outcrop to outcrop.

Geologists and archeologists have become much more adept over the past decades at

Figure 2.3 Schematic diagram showing the controlled lateral survey approach to sampling surface fossils and artifacts in order to minimize time averaging and limit the stratigraphic sources of surface fossils. Note that the fossils and artifacts eroded from approximately 4 m of strata would accumulate along the topographic bench. The discovery of in situ *materials along such eroding slopes can pinpoint source horizons for further investigation and excavation (e.g., see Figure 2.4).*

recognizing preserved paleolandscapes because of advances in the understanding of paleosols and features indicating a significant pause (hiatus) in terrestrial sedimentation, and also because of increased work at microstratigraphic levels of resolution. In her study of the surface of the "Blue Tuff" in the Okote Mb., Stern identified, surface sampled, and then excavated a ~30 m portion of a single preserved landscape surface, revealing patterns of artifact and bone distributions relating to a channel edge (Stern et al. 2002). In many cases, however, sampling artifacts and bones along a single unit over distances of tens to thousands of meters is a never-ending challenge because of outcrop discontinuities, changing lithofacies, and availability of geologists willing to work at this level of stratigraphic resolution.

When a number of excavated sites are at approximately the same stratigraphic level, as in the Okote Mb. site complex on the Karari Escarpment (Kafulu 1983; Rogers 1997;

Rogers et al. 1994; Braun et al. 2008), these often are characterized as representing approximately the same time, even though the artifact samples may span hundreds to thousands of years. In this example, patterns in the artifact assemblages emerged in spite of time-averaging (Braun et al. 2008), indicating that hominins made consistent choices about core size and level of reduction of tools over ~100 m stretch of landscape perpendicular to a paleochannel. In this study (Braun et al. 2008), broadening the definition of paleolandscape to accommodate the limitations of the record may not negatively affect interpretations of hominin activities relative to landscape features. In fact, one could even argue that this landscape sample provides stronger evidence for the observed patterns, since to be detectable, these patterns must have persisted through the period of time-averaging. Other questions might not fare as well from the blurring effects of time-averaging, however, and it is always important to understand (and if one is an author, discuss) what is really meant by the term, "a single paleolandscape".

How does one sample a paleolandscape?

After formulating the big questions and identifying target intervals, Glynn's next challenge to his students was to come up with a sound sampling strategy with testable hypotheses, and then go after answers with a lot of hard work (and minimal attention to physical hardship). Glynn began with surface sampling as a way to assess the frequency of artifacts eroding from the subsurface, and he initially assumed this would provide useful information about the scatter between the patches. Later he and others combined surface sampling with excavation to determine the actual source unit(s) for these surface materials.

Surface Sampling

When Glynn initiated his scatter between the patches surface sampling, he took pains to lay out random, 100 m-wide transects over a target interval along the Karari Escarpment (Isaac and Harris 1980). Randomization is statistically valuable, but in practice resulted in difficult sampling situations on steep, bush-covered slopes or ravines. In pre-GPS times, it often took considerable effort just to locate the randomized point on the ground based on the air photograph. With a number of stoic helpers, he systematically walked the transects and recorded every artifact or bone encountered, flagging and then surveying them in (with pre-laser transit). He also tried randomly placed 5 x 5 m squares along the target outcrops, recording and in some cases collecting everything in these squares. Because of variability in the steepness of the slopes, the juxtaposition of multiple stratigraphic sources, and the mix of artifact, fossil, and rock debris on these outcrops, the resulting

information was unsatisfying and hard to quantify. It was clear, however, that many artifacts and fossil bones were scattered at low density between the richer patches, which encouraged him to press on to refine the method.

Choosing places to sample along a target level is critical to understanding the preserved spatial patterning of artifacts and bones. Today, of course, one can select points for surface transects or excavations using GPS coordinates, at random or using a standardized grid system. On the ground, however, this may lead to logistical difficulties, just as Glynn's points on an air photo did. The best strategy is to assess benefits versus costs of the different landscape sampling methods for any given target level. These can be random, along a grid, continuous, or "strategic," in which targets areas with evidence of *in situ* artifacts or bones, or where outcrops are particularly well suited for surface sampling (Figure 2.3).

Mapping of the total area surveyed, as well as the positions of all bones and artifacts, is easily done now with the GPS as well, with + 5 m accuracy with hand-held models. The tedious process of determining area searched from air photos that Glynn had to use (e.g., cutting out and weighing bits of paper!) is long gone. The flipside is that one must be very careful about data recording, including downloading, back-up storage, labeling, and associating the GPS points with the objects collected or recorded. Although the time spent doing the survey is vastly shorter than before, the "curatorial" time back at camp is more labor-intensive. Having electronic rather than pencil-on-paper files leaves one at risk of computer crashes, dead battery issues, and communication break-downs regarding labeling. Previously each artifact was given a number, a brief description and recorded by hand in a notebook along with the transit survey information. Now, each artifact number is associated with a waypoint in a file that is downloaded from a GPS or hand-held computer unit. There are of course great advantages to the latter in terms of transmitting, sharing and analyzing the data.

Glynn: Good Lord! You mean that we can simply press a button and mark the position of anything we find? That is extraordinary! But... I worry about the resolution issue – a +5 m error is not acceptable for accurate spatial documentation. We could do better with the transit. On the other hand, I can see that this GPS vastly increases the scale of the paleolandscape we can sample.

Surface sampling has become accepted practice in paleolandscape research because it provides a relatively straightforward way to document large samples, assess dispersed artifact and bone distributions, and identify potential buried surfaces for more intensive study. The utility of such samples was hotly debated (Stern 1993) because surface artifacts and fossils typically represent such poorly constrained periods of time-averaging. We have learned quite a lot since Glynn's initial surface sampling (Isaac and Harris 1980) about what works and what doesn't from subsequent research at East Turkana (Isaac and Isaac 1997; Rogers 1997; Rogers et al. 1994; Stern 1993; Stern et al. 2002, Kirera, 2007), as well as Olorgesailie (Potts et al. 1999), Olduvai (Blumenschine and Masao 1991; Blumenschine et al. 2003; Domingo-Rodriquez et al. 2010 and references therein), and Aramis in Ethiopia (White et al. 2009) and even in the Miocene of Pakistan (Behrensmeyer and Barry 2005). Careful choice of a target level and topography can reduce problems with time-averaging in surface samples, though as Glynn might have said, there is no substitute for "truth from digging".

Excavations

Glynn wore a battered hat with the words, "I dig holes". He was an especially strong advocate for what all archeologists know - that nothing compares with well-excavated and recorded fossil and artifact assemblages for quality of data on spatial patterns. The advantages over surface samples include: 1) precise data on the original, in situ, buried x-y-z coordinates of each piece; 2) control on association with very small preserved remains and debitage through screening of sediment by square and spit; 3) information on original orientation and up versus down position of bones and artifacts; 4) assurance that taphonomic damage was pre- or post-burial and not acquired during erosion onto the outcrop surface; 5) information on fine-scale variation in sedimentary structures and grain sizes associated with the bones and artifacts. With such data, it is possible to reconstruct hominin behavior, activities of other organisms, and depositional processes that formed and preserved the site. When excavations are expanded outside of a high density patch to multiple sites along a target paleolandscape, they provide critical windows on the density and distribution of the material record left across that landscape.

Most of the next generation landscape research at East Turkana utilized a combination of both surface and excavation to establish scatter and patches patterns (Stern et al. 2002; Rogers 1997; Bunn 1994; Braun et al. 2008, 2010; Pobiner et al. 2008; McCoy 2009). Rogers (Rogers et al. 1994; Rogers 1997), for example, used this approach to broaden the scale of landscape study from the densely concentrated artifact assemblages of the Okote Member Site complex on the Karari Escarpment to surface and excavated samples from the same interval at Ileret, ~20 km to the northwest (Figure 2.1). He was able to analyze trends on the scale of hundreds of meters and tens of kilometers within a constrained time interval and showed that few artifacts on the surface between dense patches generally corresponded to few or none *in situ*. This led him to the conclusion that there was a consistently low level of inter-site artifact scatter in the Okote Member (Rogers et al. 1994; Rogers 1997; Braun et al. 2008). Glynn would have been very happy with this long distance comparison, as he emphasized from the beginning that it was important to investigate landscape patterns over a hierarchy of scales.

> *Glynn*: What Mike and Dave have discovered about the "lack" of scatter between the patches, both on the Karari and at Ileret, is fascinating. Given the number of artifacts on the outcrop surfaces of the Karari, I would not have predicted that randomly placed excavations between the patches would reveal so little *in situ* scatter. It is doubly fascinating that this pattern seems to hold true at Ileret, likely further accentuated by the greater distance to sources of raw materials.
>
> *Response*: What do you think this means about hominid behavior and "foresight" at the time?
>
> *Glynn*: Well, other explanations are possible, but I am intrigued by their suggestion that the hominids were clever about curating raw materials and struck flakes, leaving little debris across the landscape. One might even guess that they were aware of the consequences of not having a sharp edge to use when they needed one. Perhaps they had invented carrying bags, or pockets!

The discovery of numerous hominin fossils at Ileret (Leakey and Leakey 1978) and Glynn's excavation of archeological site FwJj1 (Isaac and Isaac 1997) in the Okote Member (Ileret Tuff interval), focused attention to the geological and paleontological context of this well-delimited limited slice of time. The Okote Tuff interval at Ileret is about 8 m thick, and we now know it represents <200 kyr based on radiometric dating of the bounding tuffs (5.3–5.1 Ma; Bennett et al. 2009; Figure 2.2). Although artifact density is extremely low in this interval at Ileret (Rogers et al. 1994; Rogers 1997) except at FwJj1, hominin body fossils, several hominin footprint layers (Bennett et al. 2009), and cutmarked bone just above the Ileret Tuff (Pobiner et al. 2008) provide ample evidence that hominins occurred in this area. This adds support to the hypothesis that hominins were controlling the acquisition, retention, and discard of artifacts on the landscape in interesting and deliberate ways.

Landscape Paleontology and Taphonomy

"Landscape" taphonomy at East Turkana began in the early 1970s with controlled sampling of surface bone assemblages (Behrensmeyer 1975). From the beginning of Glynn's research at East Turkana, an important part of both site and inter-site archaeology was the parallel need to establish baseline patterning that was non-human, in other words, the "natural" background of bone distributions and taphonomic modification on paleolandscapes. This was the focus of a project in the late 1970s that targeted one stratigraphic interval – the Koobi Fora – Okote – Ileret Tuff complex, using multiple sampling strategies to document lateral variation in lithofacies, bones and artifacts, including 10 x 10 m squares, continuous lateral surface sampling, and excavations (Behrensmeyer and Laporte 1982; Behrensmeyer 1985). Glynn was predictably supportive and enthusiastic about this work, because his corollary questions included how patches of bones could form independent of humans and how taphonomic features of such patches would compare with nodes of human activity. He was particularly pleased when a number of our paleontological excavations turned up scattered artifacts as well as fossil bones, and one of them, PE/18, became FxJj64 (Isaac and Isaac 1997), subsequently yielding an informative array of archeological materials that he classified as a "mini-site" (Isaac et al. 1981). Another of these, PE/11 at Ileret (Figure 2.4) was included in Rogers' comparison of Okote level landscape patterns (Rogers et al. 1994; Rogers 1997). Taphonomic surface surveys and excavations by Bunn (1994) turned up cut-marked bones in Area 103 in the KBS and Okote Members, where no artifacts had been found *in situ*, and later work by Pobiner (Pobiner et al. 2008) in Area 103 and Area 1A at Ileret added to the evidence for widely distributed meat-eating by hominins even in the absence of associated artifacts. Subsequent surface sampling and excavations in the older Upper Burgi Member (McCoy 2009; Braun et al. 2010) indicate that a similar pattern of widely distributed cut-marked bone extends back at least to ~2.0 Ma. A new phase of landscape taphonomy and paleoecology 2002–2003 targeted the older Tulu Bor and Lokochot Members of the Koobi Fora Formation (Behrensmeyer et al. 2004).

What have we learned?

Glynn: It's clear that the evolution of technology has made it possible to collect vast amounts of high quality data on paleolandscapes. I'm very glad to hear that the approach has caught on at Plio-Pleistocene sites up and down the Rift Valley. But tell me, what has all this research done in 25 years to illuminate the scatter between the patches?

We return to the question, "How were hominins patterning their behavior over ancient landscapes?". Research at East Turkana has shown that there is a distinctly bimodal pattern to the artifact patches and the scatter in the Okote Member, i.e., dense concentrations of artifacts

and very low density inter-site scatters, with little gradient between (Rogers et al. 1994; Rogers 1997; Braun et al. 2008). There is also good evidence that hominins were doing different things with stone tools depending on the paleoenvironmental context (channel versus floodplain) and the proximity of raw materials (Braun and Harris 2003; Braun et al. 2008). Surface surveys and excavations indicate that Okote Member hominins were careful curators of their tools, ranging into areas where they processed carcasses, but left only cutmarks behind (Bunn 1994; Pobiner et al. 2008). These findings represent important advances in establishing hominin landscape-use patterns and are a direct consequence of the research that Glynn set in motion.

Paleoanthropologists and archaeologists working in East Africa today take seriously the entire spectrum of artifact and bone densities as informative and important, also a tribute to Glynn's pioneering work. Although the surface artifact scatters between the patches that he so enthusiastically sampled in the early days turned out to be time-averaged mixtures of multiple levels offering little reliable information on hominin behavior, we have learned to control surface sampling to reduce time-averaging and increase the information quality of surface surveys. We now understand more about the complexity of the palimpsests of "visiting cards" that accumulated on any given paleolandscape but still are working on how to characterize this complexity using multiple excavations as well as surface sampling, accompanied by ever-more sophisticated quantitative methodologies.

Glynn probably would not have been surprised to learn that, despite his students' early dedication to examining the scatter, subsequent research has tended to gravitate back to the patches, where there are large and interesting samples of artifacts and bones amenable to statistical

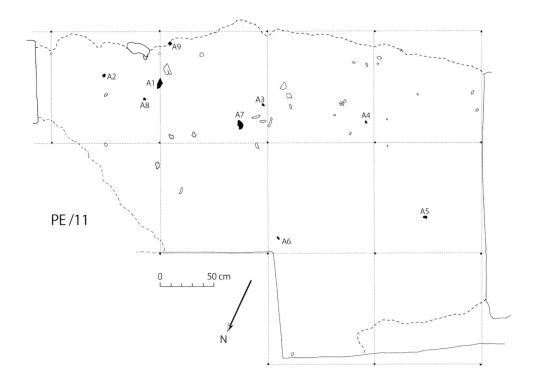

PE /11

0   50 cm

N

Figure 2.4 PE /11 in Ileret Area 8A, a 1979 paleontological excavation by the author's team, showing 31 in situ fossil bones (hollow shapes) and nine artifacts (black, with "A#") in a ~9 m² area of a paleolandsurface. Fossil bones and teeth were in highly variable condition, and some of the bones and artifacts were conjoining pieces. At least four mammalian taxa (pigmy hippo, suid, large bovid, very large mammal) occur in this scatter along with fish remains and crocodile teeth (see also Isaac 1997:55). This is an example of the kind of low-density artifact and bone scatter that Glynn would have called a "mini-site" (Isaac et al. 1980) representing a limited set of activities and miminal time-averaging. Rogers (1997:156) revisited this site and found four additional surface artifacts.

analysis. As a consequence of this and decades of intensive experimental and taphonomic research, we now know a lot more about spatial variation of artifact morphology and functions from patch to patch and how high-density hominin-generated bone concentrations differ from other causes of bone patchiness. There is no question that there is still much to learn from studying low density distributions between sites, but this is labor intensive, potentially high risk, and potentially hard to justify as a dissertation project with an advisor

who is not a reincarnation of Glynn Isaac. What might energize Glynn the most, if he were here, is the challenge of synthesizing the accumulated knowledge of the past 25 years - a task he would have welcomed and at which he excelled.

It is easy to imagine how Glynn would have enjoyed querying and parrying with us about the evolving landscape approach at East Turkana and beyond. He would have approached such discussions with characteristic humility and humor, "I will try to be delighted when knowledge is advanced by the elimination or modification of even my most cherished hypotheses!" (Isaac 1980:227). He would have evolved himself, pushing forward to test his ideas against the accumulating data, so it is hard to know where we would be now if he had been with us through this time. Hardest to imagine is how his conceptual framework might have changed and what new questions he would have asked. Perhaps this is the greatest tribute - that we cannot know what he would have been thinking - the mark of an irreplaceable mind that inspired so many of us to continue along pathways he laid out while he was with us.

Acknowledgments

I am very grateful to Jeanne Sept and David Pilbeam for the patience and good faith that helped bring to completion an overly ambitious idea for my contribution to this volume. I also might not have pushed through to the end without Barbara Isaac's unfailing confidence in my ability to do so. It is probably presumptuous of me to attempt to recreate Glynn's reactions and put these into "his" words, but I decided to try this in order to bring him back, in some small measure, for those who knew him and for readers who never had that privilege. It was a satisfying if time-consuming task to reconnect with the memories, theses, and journal articles, and from these to distill a summary of paleolandscape research at East Turkana since 1985. I don't doubt that this history could have been more comprehensive and take full responsibility for any oversights or errors. I sincerely thank Jack Harris for including me in the continuing work at East Turkana and Brian Richmond for his support and collaboration at FwJj14E. Mike Rogers, Dave Braun, Jack McCoy, Briana Pobiner, and Francis Kirera provided essential input and references. I am indebted to the Kenya National Museum for its long-term support of research at East Turkana and to the 1978-1979 field team - Léo Laporte, Hilde Schwartz, Mahmood Raza, Andy Cohen, Mutete Nume, Peter Hetz, and the many skilled Kenyan excavators and logistical staff who made our research possible. Original research included in this paper was supported by NSF EAR 77-23149 and NSF BCS-0924476.

References

Aston, M., and T. Rowley
1974 *Landscape Archaeology: An Introduction to Fieldwork Techniques on Post-Roman Landscapes.* Newton Abbot.

Barthelme, J. W.
1985 *Fisher-Hunters and Neolithic Pastoralists in East Turkana, Kenya.* Cambridge Monographs in African Archaeology 13, BAR International Series 254.

Behrensmeyer, A. K.
1975 The taphonomy and paleoecology of plio-pleistocene vertebrate assemblages east of Lake Rudolf, Kenya. *Bulletin of the Museum of Comparative Zoology* 146(10):473-574.

1985 Taphonomy and the paleoecologic reconstruction of hominid habitats in the Koobi Fora Formation. In *L'environment des hominides au Plio-Pleistocene*, edited by Y. Coppens, pp. 309-324. Foundation Singer-Polignac, Paris.

Behrensmeyer, A. K., and J. C. Barry
2005 Biostratigraphic surveys in the Siwaliks of Pakistan: A method for standardized surface sampling of the vertebrate fossil record. *Palaeontologica Electronica* (Special issue in honor of W. R. Downs): Vol. 8, Issue 1:15A:24p, 839kb. http://palaeo-electronica.org.

Behrensmeyer, A. K., and L. F. Laporte
1982 Lateral facies analysis of Koobi Fora Formation, Northern Kenya: Summary of research objectives and goals. In *Paleoecology of Africa*, vol. 14, edited by J. Coetzee and E. M. van Zinderen Bakker, pp. 147-150. A. A. Balkema, Rotterdam.

Behrensmeyer, A. K., R. Bobe, C. J. Campisano, and N. Levin
2004 High resolution taphonomy and paleoecology of the Plio-Pleistocene Koobi Fora Formation, northern Kenya, with comparisons to the Hadar Formation, Ethiopia. *Journal of Vertebrate Paleontology* 24 (Supplement to #3):38A.

Bennett, M. R., J. W. K. Harris, B. G. Richmond, D. R. Braun, E. Mbua, P. Kiura, D. Olago, M. Kibunjia, C. Omuombo, A. K. Behrensmeyer, D. Huddart, and S. Gonzalez
2009 Early hominin foot morphology based on 1.5 million year-old footprints from Ileret, Kenya. *Science* 323, 1197-1201.

Blumenschine, R. J.
1986 *Early Hominid Scavenging Opportunities: Implications of Carcass Availability in the Serengeti and Ngorongoro Ecosystems*. BAR International Series 283.

Blumenschine, R. J., and F. Masao
1991 Living sites at Olduvai Gorge, Tanzania, Preliminary landscape archaeology results in the basal Bed II lake-margin zone. *Journal of Human Evolution* 21:451-462.

Blumenschine, R., C. R. Peters, F. T. Masao, R. J. Clarke, A. L. Deino, R. L. Hay, C. C. Swisher, I. G. Stanistreet, G. M. Ashley, L. J. McHenry, N. E. Sikes, N. J. van der Meerwe, J. C. Tactikos, A. E. Cushing, D. M. Deocampo, J. K. Njau, and J. I. Ebert
2003 Late Pliocene *Homo* and hominid land use from western Olduvail Gorge, Tanzania. *Science* 299:1217-1221.

Braun, D. R.
2006 The Ecology of Oldowan Technology: Perspectives from Koobi Fora and Kanjera South (Kenya). Unpublished Ph.D. Dissertation, Department of Anthropology, Rutgers University, New Jersey.

Braun, D. R., and J. W. K. Harris
2003 Technological developments in the Oldowan of Koobi Fora: Innovative techniques of artifact analysis. In *Oldowan: Rather More than Smashing Stones, First Hominid Technology Workshop (2001), Bellaterra, Spain*, edited by J. M. Moreno, R. M. Torcal, I. de la Torre Sainz, pp. 117-144. Centre d'Estudis del Patrimoni Arqueològic de la Prehistòria, Treballs d'Arquelogia, 9. Barcelona, Spain.

Braun, D. R., M. J. Rogers, J. W. K. Harris, and S. J. Walker
2008 Landscape-scale variation in hominin tool use: Evidence from the Developed Olduwan. *Journal of Human Evolution* 55:1053-1063.

Braun, D. R., J. W. K. Harris, N. E. Levin, J. T. McCoy, A. I. R. Harries, M. K. Bamford, L. D. Bishop, B. G. Richmond, and M. Kibunjia
2010 Early hominin diet included diverse terrestrial and aquatic animals 1.95 Ma in East Turkana, Kenya. *Proceedings of the National Academy of Sciences* 107:10002-10007.

Bunn, H. T.
1981 Archaeological evidence for meat-eating by Plio-Pleistocene hominids from Koobi Fora and Olduvai Gorge. *Nature* 291:574-577.

1994 Early Pleistocene hominid foraging strategies along the ancestral Omo River at Koobi Fora, Kenya. *Journal of Human Evolution* 27:247-266.

Bunn, H. T., J. W. K. Harris, G. Ll. Isaac, Z. Kaufulu, E. Kroll, K. Schick, N. Toth, and A. K. Behrensmeyer
1980 FxJj50: An early Pleistocene site in northern Kenya. *World Archaeology* 12(2):109-136.

Chapman, H.
2006 *Landscape Archaeology and GIS*. Stroud, New York.

Domínguez-Rodrigo, M.
2010 Distinguishing between apples and oranges again: A response to Pobiner (2008) on the importance of differentiating between epistemically trivial and non-trivial analogies. *Journal of Taphonomy* (in press).

Feibel, C. S., F. H. Brown, and I. McDougall
1989 Stratigraphic context of fossil hominids from the Omo Group deposits: Northern Turkana Basin, Kenya and Ethiopia. *American Journal of Physical Anthropology* 78:595-622.

Gifford, D. P.
1977 Observations of Modern Human Settlements as an Aid to Archaeological Interpretation. Unpublished Ph.D. Dissertation, Department of Anthropology, University of California, Berkeley, CA.

Gifford, D. P., and A. K. Behrensmeyer
1977 Observed formation and burial of a recent human occupation site in Kenya. *Quaternary Research* 8:245–266.

Isaac, G. Ll.
1980 Casting the net wide: A review of archaeological evidence for early hoominid land-use and ecological relations. In *Current Argument on Early Man*, edited by L.-K. Konigsson, pp. 226–251. Pergamon, Oxford.

1981 Stone Age visiting cards: Approaches to the study of early land-use patterns. In *Pattern of the Past*, edited by I. Hodder, G. Isaac, and N. Hammond, pp. 131–55. Cambridge University Press, Cambridge.

1983 Some archaeological contributions towards understanding human evolution. *Canadian Journal of Anthropology* 3(2):233–243.

Isaac, G. Ll., and A. K. Behrensmeyer
1997 Chapter 2: Geological context and paleoenvironments. In *Koobi Fora Research Project Volume 5: Plio-Pleistocene Archaeology*, edited by G. Ll. Isaac and B. Isaac, pp. 12–53. Clarendon Press, Oxford.

Isaac, G. Ll., and J. W. K. Harris
1975 The scatters between the patches. Paper presented at the Kroeber Anthropological Society Annual Meetings, Berkeley, CA.

1980 A method for determining the characteristics of artefacts between sites in the Upper Member of the Koobi Fora Formation, East Lake Turkana. In *Proceedings of the VIIIth Pan African Congress on Prehistory, Nairobi, 1977*, edited by R. E. Leakey, and B. A. Ogot, pp. 19–22. International Louis Leakey Memorial Institute for African Prehistory, Nairobi.

Isaac, G. Ll., and B. Isaac (Editors)
1997 *Koobi Fora Research Project Volume 5: Plio-Pleistocene Archaeology*. Clarendon Press, Oxford.

Isaac, G. Ll., J. W. K. Harris, and F. Marshall
1981 Small is Informative: The application of the study of mini-sites and least effort criteria in the interpretation of the early Pleistocene archaeological record at Koobi Fora, Kenya. *Las industrias mas antiguas, Proceedings of the X Congreso*, edited by J. Clark, and G. Ll. Isaac, pp. 102–119. Union Internacional de Ciencieas Prehistoricas y Protohistoicas, Mexico.

Kaufulu, Z. M.
1983 The Geological Context of Some Early Archaeological Sites in Kenya, Malawi and Tanzania: Microstratigraphy, Site Formation and Interpretation. Unpublished Ph.D. Dissertation, Department of Anthropology, University of California, Berkeley, CA.

Kirera, F.

2007 GIS Model of Koobi Fora Plio-Pleistocene Paleoenvironment: Northern Kenya. Unpublished Ph.D. Dissertation, Environmental Dynamics Program, University of Arkansas, Fayetteville.

Leakey, M. G. and R. E. L. Leakey (Editors)

1978 *Koobi Fora Research Project, Vol. l: The Fossil Hominids and an Introduction to their Context l968-75.* Clarendon Press, Oxford.

McCoy, J. T.

2009 Ecological and Behavioral Implications of New Archaeological Occurrences from Upper Burgi Exposures at Koobi Fora, Kenya. Unpublished Ph.D. Dissertation, Department of Anthropology, Rutgers University, New Jersey.

Pobiner, B. L., M. J. Rogers, C. M. Monahan, and J. W. K. Harris

2008 New evidence for hominin carcass processing strategies at 1.5 Ma, Koobi Fora, Kenya. *Journal of Human Evolution* 55:103-130.

Potts, R.

1989 Olorgesailie: New excavations and findings in Early and Middle Pleistocene contexts, southern Kenya rift valley. *Journal of Human Evolution* 18:477-484.

1994 Variables vs. models of early Pleistocene hominid land use. *Journal of Human Evolution* 27:7-24.

Potts, R., A. K. Behrensmeyer, and P. Ditchfield

1999 Paleolandscape variation and early Pleistocene hominid activities: Members 1 and 7, Olorgesailie Formation. *Journal of Human Evolution* 37:747-788.

Rogers, M. J.

1997 A Landscape Archaeological Study at East Turkana, Kenya. Ph.D. Dissertation, Department of Anthropology, Rutgers University, New Brunswick, NJ.

Rogers, M. J., J. W. K. Harris, and C. S. Feibel

1994 Changing patterns of land-use by Plio-Pleistocene hominids in the Lake Turkana Basin. *Journal of Human Evolution* 27:139-158.

Schick, K. D.

1986 *Stone Age Sites in the Making: Experiments in the Formation and Transformation of Archaeological Occurrences.* BAR International Series, 319, Oxford.

Stern, N.

1993 The structure of the Lower Pleistocene archaeological record: A case study from the Koobi Fora Formation. *Current Anthropology* 34:201-225.

Stern, N., N. Porch, and I. McDougall

2002 FxJj43: A window into a 1.5-million-year-old palaeolandscape in the Okote Member of the Koobi For a Formation, northern Kenya. *Geoarchaeology* 17(4):349-392.

Toth, N.

1982 The Stone Technologies of Early Hominids at Koobi For a, Kenya: An Experimental Approach. Unpublished Ph.D. Dissertation, Department of Anthropology, University of California, Berkeley, CA.

White, T. D., B. Asfaw, Y. Beyene, Y. Haile-Selassie, C. O. Lovejoy, G. Suwa, and G. WoldeGabriel

2009 *Ardipithecus ramidus* and the paleobiology of early hominids, *Science* 326:64, 75-86.

Yamin, R., and K. B. Metheny (Editors)

1996 *Landscape Archaeology: Reading and Interpreting the American Historical Landscape.* University of Tennessee Press, Knoxville.

3

GLYNN LLYWELYN ISAAC: A RECOLLECTION AND RETROSPECTIVE ASSESSMENT

Diane Gifford-Gonzalez

A retrospective assessment of Glynn Isaac's influence on archaeology could readily take a straightforward topical emphasis, listing his research foci and directions, methodological approaches, and so forth. However, reflecting on Glynn's legacy, I have concluded that it is impossible to discuss these without first talking about the person and the personality. If themes that Glynn articulated resonate over time, this owes much to the way he championed archaeology, taught systematic research methods, and engaged with other scholars, from undergraduates to graduate students to his colleagues and peers on several continents.

As a "re-entry" graduate student with a background principally in physical anthropology, what attracted me to working with Glynn was his intellectual openness, his stimulating teaching, and, despite the famous twinkling eye, his seriousness of purpose in learning more about the past. Although he had tremendous fun doing it, archaeology was never just a "fun pastime" for him - it was an endeavor about which he was passionate. As he described it in a piece on the needs of paleoanthropological researchers and research in developing nations, "Paleoanthropology, with its arduous fieldwork and its elusive fragmental evidences, is a pursuit to which, regardless of ability, not all temperaments are equally well suited. How are countries to find the youngster that have the particular flair and drive?" (Isaac 1985:348). That "flair and drive" in his own personality was

what inspired others to join in the arduous fieldwork and nearly equally arduous analytic phases of archaeological research.

Glynn never deviated from the conviction, formed early in his studies, that archaeological evidence can tell a story about the human past well beyond typology and chronology. He was blessed with an extraordinarily openness of mind and gift for encouraging networks of communication. He also was willing to admit that some of his initial inferences about the behavior of ancient human ancestors were probably quite wrong (e.g., Isaac 1983), while at the same time resourcefully seeking other ways to use the evidence to arrive at better-founded inferences. In this, Glynn possessed the audacity typical of the extraordinary achievers, be they scholars, athletes, or entrepreneurs: the eye for opportunity, the willingness to seize every promising one, and the courage to risk very public failure.

As I educated myself more in the history of archaeological theory and its national variants, I could of course trace Glynn's intellectual heritage in the environmentally inclined, theoretically eclectic Cambridge school of the age of Graham Clark. However, recollecting course work and conversations with him, I understood that during the Cambridge years Glynn and his age-mate David Clarke departed from that canon, thinking innovatively about archaeology in a broader register, human evolution, as well as about how science works. His interest in models, both abstract and visual, willingness to go back to basics in developing a new perspective on Plio-Pleistocene sites (e.g., Isaac 1983; Isaac and Crader 1981; Isaac et al. 1981), and to ruthlessly subject his own assumptions to the test of empirical evidence, doubtless stemmed from those experiences.

General Influences and Trajectories

In preparing this piece, I re-read some of Glynn's last publications, to look for his next anticipated directions. I was already aware that, at the time Lewis Binford published his stinging critiques of paleoanthropological archaeology (e.g., 1977, 1981), in part specific attacks on Glynn's and Mary Leakey's inferences, Glynn and his students were already engaged in rigorously challenging his earlier ideas about the hominin behaviors that produced east African Plio-Pleistocene sites. Nonetheless, with nearly 30 years' distance, I was struck by how much his 1983 article, "Bones in contention: Competing explanations for the juxtaposition of early Pleistocene artifacts and faunal remains", anticipated the emergence of a major and, to use David Clarke's (1973) phrase, critically self-conscious research strategy in archaeology, with its attendant challenges. The best way to convey this is with a brief exegesis of direct quote from that piece:

> A key element in the new research movement is the concern to cope with multiple, rival hypotheses at levels of interpretation that range from the specifics of mode of deposition and details of damage and breakage patterns to more indirect constructs such as 'hominid meat-eating', 'food transport', 'sharing', and 'home-bases'. For each set of competing hypotheses a series of test implications must be defined. Appropriate evidence must then be collected and scrutinized for goodness-of-fit with

the test implications. However, very often the rival hypotheses are not mutually exclusive and simple Popperian falsification is not possible. Research then winds up as the acquisition of data that can be used in assessing the relative importance of a series of interacting factors.

The inspiration for the recognition of items in the set of rival hypotheses must in general come from observations of configurations and processes in the contemporary world, including those induced by human, animal and geological agencies. The working-out of test implications also involves experiments and ethnographic, taphonomic and geological observations (Isaac 1983:4).

The research program described would today be classed as actualistic ("from observations of configurations and processes in the contemporary world"). It acknowledges that qualitative differences exist between various levels of observations of processes and their products ("is the concern to cope with multiple, rival hypotheses at levels of interpretation that range from... deposition and... damage and breakage patterns to more indirect constructs"). It also recognizes that some outcomes of complex interactions might be equifinal ("rival hypotheses are not mutually exclusive").

Although Glynn once declared in the midst of a seminar with Jim O'Connell and myself around 1977, "Basically, I am an empiricist", he firmly grasped that actualistic research could supply more resilient hypotheses for, and challenge implicit assumptions in, archaeological research design. Certainly, he was not the only archaeologist to realize the fundamental necessity and efficacy of observing present-day causal relationships as a foundation for understanding the past. Binford, of course, had turned to ethnoarchaeology in the 1970s, after realization that his "nomothetic" methods were wholly inadequate for grappling with the French Mousterian (Binford 1983). Longacre, after his initial optimistic "inference" of matrilocality at Grasshopper Pueblo (1976), realized he had no real-world basis for linking ceramic styles with sociological categories and began a multi-decade ethnoarchaeological inquiry with Kalinga potters (e.g., Longacre and Skibo 1994). Although prefigured by an insightful article in the 1950s (Kleindienst and Watson 1956), the shift toward a view of contemporary observations as a "source side" (Wylie 1985) for analogues occurred over the late 1960s through late 1970s, among "New Archaeologists" and scholars such as Glynn, whose intellectual background differed profoundly from that of most processualists. Resorting to studies of human behavior and non-human site formation processes in the contemporary world was an idea whose time had arrived, independently, at a number of scholarly addresses, as is often the case in science, though sorting out how it was best applied took another few years (e.g., Binford 1981; Gifford 1981; Gould and Watson 1982; Wylie 1982).

Personal Influences and Trajectories

I was one of the UC Berkeley Old World Prehistory program's first guinea pigs for actualistic research, a role pioneered by Diana Crader in her ethnoarchaeological research in the Luangwa Valley, Zambia (1974, 1983). Although never officially funded by the Koobi Fora Research Project, I was encouraged by Glynn in 1972-1974 to develop an ethnoarchaeological dissertation project with the Dassanetch people of northeastern Lake Turkana. He and Richard Leakey kindly permitted me use of KFRP gear, facilities, and vehicles during the project's off-season. This fieldwork opportunity during the monsoonal part of the year led to unexpected,

firsthand, actualistic observations of fluvial processes, site formation, and vehicular rescue and resuscitation, beyond my wildest expectations. My study also established a pattern of cooperative interaction with local Dassanetch people, who later worked with Nick Toth (1982) and Kathy Schick (1986) on their experimental lithic technology and site formation research, as part of the KFRP's program by the late 1970s.

Later, once clean and dry, I found myself intellectually confronted by a lack-of-fit between my empirical evidence for site formation and a myriad of mutually contradictory discussions by cultural anthropologists, archaeologists, and paleontologists about the use and abuse modern analogues. I have recently recounted my intellectual journey toward a personal vision of this matter, the rather irreverent tone of which I believe Glynn would have approved, in the *SAA Archaeological Record* (Gifford-Gonzalez 2010). While writing an overview of taphonomy and paleoecology for Michael Schiffer (Gifford 1981) and simultaneously reviewing the draft of Binford's (1981) *Bones: Ancient Men and Modern Myths*, I went through a major paradigm shift concerning why we "study the present to understand the past". Binford and I virtually simultaneously (Binford 1981; Gifford 1981) introduced the term "actualism" to encompass that great range of contemporary observational studies which Glynn had already been prescribing as aids to archaeological exegesis (for the real scoop on this simultaneity, see Gifford-Gonzalez 2010).

My own methodological specialization is zooarchaeology, a specialization whose time had come, emerging from multiple sources in disparate places over the 1960s to the 1970s. Both Glynn and Desmond Clark strongly encouraged several of us 1970s graduate students to study vertebrate remains, methodologically hybridizing archaeological problem definition and zoological and paleontological osteology. Glynn and Desmond's motivation obviously stemmed from the already well-established tradition of archaeobotanical and archaeofaunal analysis at Cambridge. We hybrids now breed true, producing our own little zooarchaeologists, but in those days, like counterparts at Harvard, and universities and museums in Florida, Michigan, Oregon, and Illinois, we shuttled between various departments to gain the requisite knowledge. Glynn, Desmond, and Clark Howell vastly facilitated this by developing a series of interdisciplinary studies courses with counterparts from paleontology and geology, a gesture that only someone entangled in a university bureaucracy for 35 years can fully appreciate.

From early in his career (Isaac 1967, 1971), Glynn had recognized the information available from vertebrate remains in very old archaeological sites. His own interest in vertebrate taphonomy was strong, and encouraged us students to combine insights from paleontological taphonomy with archaeological analysis. His 1983 piece ruefully recounts his own experiments, in which wild hyenas repeatedly "destroyed the evidence". Despite these taphonomic mishaps, this natural landscape experimental approach influenced research careers of such Berkeley-trained zooarchaeologists as Rob Blumenschine (1985) and Curtis Marean (1990).

Unlike most of Glynn's students in the 1970s and 1980s, my temporal focus has always been on the Late Pleistocene and Holocene, especially on the emergence of pastoralism in east Africa. Glynn engaged in active research on the Holocene for a time, when he focused on the Central Rift of Kenya, on which a cohort of his students, including Harry Merrick (1975) and Charles Nelson (1975), wrote their dissertations on sites from the Middle and Later Stone Age and what has come to be called the Pastoral Neolithic.

I began investigating early east African pastoralism with the analysis of the fauna from Prolonged Drift (GrJi1), excavated by Charles Nelson, with funding Glynn secured from the Wenner-Gren Foundation. Glynn coined the site's rather odd name in the aftermath of Louis Leakey's visit there. Glynn was convinced that they had relocated Leakey's Long's Drift locality on the floodplain of the Nderit River, near Lake Nakuru (Leakey 1931). After Louis vehemently denied that Nelson's site had any connection with Long's Drift, Glynn rather puckishly assigned the colloquial name of Prolonged Drift to the site, with emphasis on the second syllable.

A team of graduate and undergraduate students actually began preliminary analysis of the site's huge faunal assemblage, comprising over 150,000 specimens, in Berkeley in 1971 as part of the interdisciplinary studies series. I then completed definitive identifications of about 5,000 specimens at the National Museum of Kenya in Nairobi, during intervals between my ethnoarchaeological investigations with the Dassanetch people of Ileret. Glynn had originally thought I would be able to meld the two projects into a dissertation. However, I found the points of contact between the two few and challenging, given our rudimentary state of knowledge of early pastoralist sites in east Africa, mine being only the second zooarchaeological analysis of such a site (Gramly 1972). I opted to complete the dissertation on site formation (Gifford 1977) before turning to writing up the Prolonged Drift fauna (Gifford et al. 1980), the name of which project, by the end of the 1970s, seemed to be turning from a noun to a verb. Notwithstanding the fact that these two projects did not come together in a single publication, my experience with Inkoria section's agropastoral way of life offered a memorable living example of the range of possible pastoralist adaptations that still informs my thinking about early pastoralism.

Although my own work diverged from the research foci of many of Glynn's students and colleagues, the bones themselves always formed a bridge between of Plio-Pleistocene archaeologists and those of us working on Holocene sites. By the end of the 1980s, I began to wonder why, in structural terms, zooarchaeologists working on the early sites just could not agree on the meaning of patterning in their faunal data, despite great advances in recognizing the traces of various taphonomic agents. To sort this out in my own mind, I returned to the question of analogical inference in archaeology, this time drawing much inspiration from the perspective on biological science articulated by Ernst Mayr (1982).

In the resulting article, "Bones are not enough: Analogues, knowledge, and interpretive strategies in zooarchaeology" (Gifford-Gonzalez 1991), I argued that some of the problem involved a conflation of interpretive levels, further obscured by underspecified words such as "agent". The distinctions I drew among *causal process*, *actor*, and *effector* were, I believe, useful in sorting out the multiple levels of causation involved in "agent". Based largely upon my reading of Mayr's discussion of key differences in inference in the physicochemical versus the biological sciences, I further argued that archaeologists must understand the distinction between analogical inferences based simple cause-effect relations, e.g., cut mark identification, and those involving the probabilistically operating processes typical of ecosystems, primate social systems, and so forth.

On re-reading the previously quoted passage from Glynn's 1983 article, I realize that he was anticipating the problems of equifinality and probabilistic inference in what I termed "behavioral context" and "ecological context". To cope

with such less determinative analogical relationships, researchers are obliged to use multiple, independent lines of evidence, as also advocated by Lyman (1987), Behrensmeyer et al. (1986), and Wylie (1989).

In sum, despite the divergence of my own temporal research focus from Glynn's, I can trace how our intellectual paths crossed and crossed again. I owe him not only my start as an archaeologist, but also the example he set for standards of documentation and curation of materials, for rigorous analysis, for treating students as able junior colleagues, and for remaining resilient and open-minded in the face of challenges. Through my career, which now has encompassed working in western North America as well as Africa, I have always moved recursively between hands-on analysis of archaeological materials and actualistic "fact-checking", seeing these as part of one investigative process.

Lasting Legacies

Looking beyond my own career and asking about Glynn's lasting contribution to archaeology, his legacy must be in the careers and productivity of his students, if only because of the untimely termination of his own creativity. From the late 1970s, Glynn's students in paleoanthropological archaeology pursued a diversity of actualistic research, including projects by Bunn, Kroll, Schick, Sept, Stern, and Toth – whom paleoanthropologist Andrew Hill once collectively described as the "Berkeley Monosyllabs" (A. Hill, personal communication 1978) – but he should talk. Glynn's Harvard students, Jack and Helen Fisher and Martha Tappen, undertook actualistic research on foragers and landscape-scale bone deposition in Africa.

Diana Crader in her own career and in her encouragement of her undergraduate, Rob Blumenschine, combined contemporary observations and zooarchaeological analysis. Jack Harris, though sticking to archaeological excavation and analysis, encouraged his own students, including David Braun, Purity Kiura, and Brian Pobiner, to combine actualistic and archaeological analyses, as did Henry Bunn with his students, Larry Bartram, Chris Monahan, Jim Oliver, and Travis Pickering. The same can be said for one of Glynn's later students, Fiona Marshall and her students, Lisa Hildebrand, Ruth Shahack-Gross, and Lior Weisbrod, and Nick Toth, Kathy Schick, and Travis Pickering's student, Charles Egeland. Stanley Ambrose, his students, and colleagues have also engaged in actualistic research to clarify aspect of Later Stone Age and Pastoral Neolithic assemblages in east Africa (Ambrose and Sikes 1991; Balasse and Ambrose 2005). Thus, two and even three generations of intellectual progeny have developed and expanded Glynn's vision of the value of actualistic research for building a casebook of reference materials for understanding more of early hominin lifeways.

No retrospective could be complete without noting the central role Glynn played, along with Desmond Clark, in developing the first cadre of African researchers. Well in advance of his 1985 chapter noted earlier in this piece, Glynn offered consistent and supportive outreach to African students, who were able to complete doctoral degrees in archaeology at UC Berkeley. Glynn saw himself as a son of Africa, despite, or perhaps because of, his youth in an increasingly segregated South Africa, he respectfully reached out to black Africans seeking advanced training. These doctoral students came to include heads of archaeology departments in museums and universities, as well as of antiquity services in a number of Anglophone countries, including, Bassey Andah, Yusef Juwayeyi, Jonathan Karoma, Zefe Kaufulu, Gadi Mgomezulu, Francis Musonda, and John Onyango-Abuje. This tradition has been continued by a number of his students in their own doctoral and postdoctoral programs.

It is a poignant exercise to write a retrospective on the life of a mentor that ended some twenty years earlier than one's own present length of days. Among the most indelible impressions Glynn made upon others was of his quickness of mind and his intensity and gusto in living. I do wonder whether, had he the time to slow down, as we all do with age, and gather in the wisdom of his years, what linkages and connections he might have made. I know he would be ecstatic that everything from bone isotopes (e.g., Lee-Thorp et al. 2003) to tapeworms (Hoberg et al. 2001) suggests early carnivory in genus *Homo*. Glynn would have been fascinated by the latest on Underground Storage Organs (Dominy et al. 2008), the study of which he had suggested to Anne Vincent (1985) who was doing fieldwork with the Hadza in the 1980s, and keen to know there's a logical functional reason for early hominin selectivity for coarse-grained lithic raw materials – and one that probably correlates with heightened carnivory (Braun et al. 2009).

I will close by offering a poem I wrote in October 1985 and read at his Berkeley memorial, which seeks to capture in a different register some aspects of that person whose life influenced the lives and career paths of so many.

Stone of Memory

This stone of memory
is not smooth.
It is finely wrought along the edges
with intersecting scars.
Here is fellowship, here amusement,
here exasperation,
here, anger,
and here, along this edge, gratitude
and love.
This stone of memory
is not smooth.
It is finely touched
with small facets of experience.
Here we sit side by side on a foggy Berkeley morning,
sorting bones.
    Here is your joke, next to my cartoon,
    in the margin of an exam.
Here you turn to me as the car crests the edge of the Rift, saying,
"Here's the *real* Africa!"
    Here I walk four fast miles behind you over blazing outcrops,
    cursing, a blister on my heel.
Here we sit in the blessed evening cool of Koobi Fora,
silent, as waves lap the shore.
    Here you turn, clear eyes smiling, knapsack on your back,
    stride down a gully,
    disappear from sight.

This stone of memory
is not smooth.
Its edges trace the outlines of my pain.
Yet I hold it tight,
As the only relict of those times,
now that you, so strongly present,
become the past.

References

Ambrose, S. H., and N. E. Sikes
1991 Soil carbon isotope evidence for Holocene habitat change in the Kenya Rift Valley. In *Science* 253(1402-1405).

Balasse, M., and S. H. Ambrose
2005 Mobilité altitudinale des pasteurs néolithiques dans la vallée du Rift (Kenya): Premièrs indices de l'analyse du [delta] δ^{13}C de l'émail dentaire du cheptel domestique. *Anthropozoologica* 40(1):147-166.

Behrensmeyer, A. K., K. D. Gordon, and G. T. Yanagi
1986 Trampling as a cause of bone surface damage and pseudo-cutmarks. *Nature* 319:768-771.

Binford, L. R.
1977 Olorgesailie deserves more than the usual book review. *Journal of Anthropological Research* 33(4):493-502.

1981 *Bones: Ancient Men and Modern Myths.* Academic Press, New York.

1983 The challenge of the Mousterian. In *In Pursuit of the Past*, edited by L. R. Binford, pp. 79-94. Thames and Hudson, New York.

Blumenschine, R. J.
1985 Early Hominid Scavenging Opportunities: Insights from the Ecology of Carcass Availability in the Serengeti and Ngorongoro Crater. Tanzania. Ph.D. Dissertation, Anthropology, University of California, Berkeley.

Braun, D. R., T. Plummer, J. V. Ferraro, P. Ditchfield, and L. C. Bishop
2009 Raw material quality and Oldowan hominin toolstone preferences: Evidence from Kanjera South, Kenya. *Journal of Archaeological Science* 36(7):1605-1614.

Clarke, D. L.
1973 Archaeology: Loss of innocence. *Antiquity* 47:6-18.

Crader, D. C.
1974 The effects of scavengers on bone material from a large mammal: An experiment conducted among the Bisa of the Luangwa Valley, Zambia. In *Ethnoarchaeology*, edited by C. B. Donnan and C. W. Clewlow, pp. 161-173, UCLA Institute of Archaeology, Monograph 4.

1983 Recent single-carcass bone scatters and the problem of "butchery" sites in the archaeological record. In *Animals and Archaeology I. Hunters and their Prey*, edited by J. Clutton-Brock and C. Grigson, pp. 10-141. BAR International Series 163, Oxford.

Dominy, N. J., E. R. Vogel, J. D. Yeakel, P. Constantino, and P. W. Lucas

2008 Mechanical properties of plant underground storage organs and implications for dietary models of early hominins. *Evolutionary Biology* 35:159-175.

Gifford, D. P.

1977 Observations of Modern Human Settlements as an Aid to Archaeological Interpretation. Ph.D. Dissertation, Anthropology, University of California, Berkeley.

1981 Taphonomy and paleoecology: A critical review of archaeology's sister disciplines. In *Advances in Archaeological Method and Theory*, edited by M. B. Schiffer, pp. 365-438. vol. 4. Academic Press, San Francisco.

Gifford, D. P., G. L. Isaac, and C. M. Nelson

1980 Evidence for predation and pastoralism at Prolonged Drift, a Pastoral Neolithic site in Kenya. *Azania* 15:57-108.

Gifford-Gonzalez, D.

1991 Bones are not enough: Analogues, knowledge, and interpretive strategies in zooarchaeology. *Journal of Anthropological Archaeology* 10:215-254.

2010 Ethnoarchaeology - looking back, looking forward. *The SAA Archaeological Record* 10(1):22-25.

Gould, R. A., and P. J. Watson

1982 A dialogue on the meaning and use of analogy in ethnoarchaeological reasoning. *Journal of Anthropological Archaeology* 1:355-381.

Gramly, M. R.

1972 Report on the teeth from Narosura. *Azania* 7:87-91.

Hoberg, E. P., N. L. Alkire, A. d. Queiroz, and A. Jones

2001 Out of Africa: Origins of the Taenia tapeworms in humans. *Proceedings of the Royal Society of London: Biological Sciences* 268(1469):781-787.

Isaac, G. L.

1967 Towards the interpretation of occupation debris: Some experiments and observations. *Kroeber Anthropological Society Papers* 37:31-57.

1971 The diet of early man: Aspects of archaeological evidence from Lower and Middle Pleistocene sites in Africa. *World Archaeology* 2(3):278-299.

1983 Bones in contention: Competing explanations for the juxtaposition of early Pleistocene artifacts and faunal remains. In *Animals and Archaeology: 1. Hunters and their Prey*, edited by Clutton-Brock, J. and C. Grigson, pp. 3-19. vol. 163. BAR International, Oxford.

1985 Ancestors for us all towards broadening international participation in paleoanthropological research. In *Ancestors: The Hard Evidence*, pp. 346-351. Alan R. Liss, Inc., New York.

Isaac, G. L., and D. C. Crader
1981 To what extent were early hominids carnivorous? An archaeological perspective. In *Omnivorous Primates: Gathering and Hunting in Human Evolution*, edited by G. Teleki, pp. 37-103. Columbia University Press, New York.

Isaac, G. L., J. W. K. Harris, and F. Marshall
1981 Small is informative: The application of the study of mini-sites and least effort criteria in the interpretation of the Early Pleistocene archaeological record at Koobi Fora, Kenya. *Procedimientos de la Union Internacional de Ciencias Prehistóricas y Protohistóricas, X Congreso*, pp. 101-119. Mexico City.

Kleindienst, M. R., and P. J. Watson
1956 "Action archaeology": The archeological inventory of a living community. *Anthropology Tomorrow* V(1):7-78.

Leakey, L. S. B.
1931 *The Stone Age Cultures of Kenya Colony*. Cambridge University Press, Cambridge.

Lee-Thorp, J. A., M. Sponheimer, and N. J. van der Merwe
2003 What do stable isotopes tell us about hominid dietary and ecological niches in the Pliocene? *International Journal of Osteoarchaeology* 13:104-113.

Longacre, W. A.
1976 Population dynamics at the Grasshopper Pueblo, Arizona. In *Demographic Anthropology: Quantitative Approaches*, edited by E. B. W. Zubrow, pp. 169-183. School of American Research Advanced Seminar. University of New Mexico, Albuquerque.

Longacre, W. A., and J. M. Skibo
1994 *Kalinga Ethnoarchaeology: Expanding Archaeological Method and Theory*. Smithsonian Institution Press, Washington, D. C.

Lyman, R. L.
1987 Archaeofaunas and butchery studies: A taphonomic perspective. *Advances in Archaeological Method and Theory* 10:249-337.

Marean, C. W.

1990 Late Quaternary Paleoenvironments and Faunal Exploitation in East Africa. Ph.D. Dissertation, Anthropology, University of California, Berkeley.

Mayr, E.

1982 *The Growth of Biological Thought: Diversity, Evolution, and Inheritance.* Belknap Press, Cambridge.

Merrick, H. V.

1975 Change in Later Pleistocene Lithic Industries in Eastern Africa. Ph.D. Dissertation, Anthropology, University of California, Berkeley.

Nelson, C. M.

1975 A Comparative Analysis of Later Stone Age Occurrences in East Africa. Ph.D. Dissertation, Anthropology, University of California, Berkeley.

Schick, K.

1986 Processes of Paleolithic Site Formation: An Experimental Study. Ph.D. Dissertation, Anthropology, University of California, Berkeley.

Toth, N.

1982 The Stone Technologies of Early Hominids at Koobi Fora, Kenya: An Experimental Approach. Ph.D. Dissertation, Anthropology, University of California, Berkeley.

Vincent, A. S.

1985 Plant foods in savanna environments: A preliminary report of tubers eaten by the Hadza of northern Tanzania. *World Archaeology* 17(2):131-148.

Wylie, A.

1982 An analogy by any other name is just as analogical: A commentary on the Gould-Watson dialogue. *Journal of Anthropological Archaeology* 1(14):382-401.

1985 The reaction against analogy. *Advances in Archaeological Method and Theory* 8:63-111.

1989 Archaeological cables and tacking: The implications for practice for Bernstein's "Options beyond objectivism and relativism". *Philosophy of Social Science* 19:1-18.

4

FACTORS AFFECTING VARIABILITY IN EARLY STONE AGE LITHIC ASSEMBLAGES: PERSONAL OBSERVATIONS FROM ACTUALISTIC STUDIES

Nicholas Toth and Kathy Schick

Introduction

Glynn Isaac (Figure 4.1) was one of the brightest people we have ever had the pleasure to know and work with. We have very fond memories of Glynn's good-natured demeanor, his incisive intellect, his rigorous work ethic, his toughness in the field, and his infectious sense of humor. It was an honor to know him, to study under him, and to work closely with him as members of the Koobi Fora research program in northern Kenya. His contributions to paleoanthropology are broad and varied, perhaps best seen in *Olorgesailie: Archaeological Studies of a Middle Pleistocene Lake Basin in Kenya* (1977a), "The Earliest Archaeological Traces" in *The Cambridge History of Africa: Volume 1, From the Earliest Times to c. 500 BC* (1982), the edited volume *The Archaeology of Human Origin: Papers by Glynn Isaac* (Isaac 1989) which contains many of his seminal publications (Isaac 1967, 1971a, 1971b, 1972a, 1972b, 1976a, 1976b, 1977b, 1978, 1981a, 1981b, 1983a, 1983b, 1984, 1985, 1986, Isaac et al. 1981; and, *Koobi Fora Research Project, Volume 5: Plio-Pleistocene Archaeology* (Isaac and Isaac 1997).

A strong advocate of actualistic studies in archeology, Glynn invited us to join the Koobi Fora project to conduct experimental archeological research towards our doctorates, one of us to conduct experiments into the manufacture and use of early stone artifacts (Toth 1982) and the other to conduct research into site formation

Figure 4.1 Our favorite photo of Glynn Isaac, taken in 1979 at Koobi Fora site FxJj50 during the filming of the BBC television documentary series "The Making of Mankind". All of the excavated stone artifacts and animal bones have been placed in their original spatial positions, and Glynn is discussing the evolutionary significance of this archeological site.

processes (Schick 1984). Over the past three-and-a-half decades we have carried out a wide range of actualistic studies in order to help understand the nature and variability of Early Stone Age occurrences, especially focusing on the Oldowan and Acheulean Industrial Complexes, as well as conducting archeological field work in Africa, Europe, and East Asia.

The potential factors influencing lithic variability discussed in this chapter are primarily based on the result of our actualistic work focusing on Early Stone Age archeology, a career trajectory profoundly influenced by Glynn Isaac, and we summarize our collective efforts in what follows. This research has focused especially on sites at Koobi Fora in Kenya (Bunn et al. 1980; Schick 1984; Toth 1982, 1985a, 1985b, 1987a, 1987b, 1997; Toth and Schick 2009c) dating to between 1.9 and 1.4 million years ago at the early Gona sites (EG 10 and EG 12) in Ethiopia (Toth et al. 2006) dating to between 2.6 and 2.5 million years ago in the Nihewan Basin of northern China (Clark and Schick 1987; Schick et al. 1991; Toth and Schick 2009c; Xie et al. 1994) dating to perhaps 1.2 million years ago; at sites in the Middle Awash of Ethiopia dating to 1.0 million years ago and approximately 300,000 years ago (Clark et al. 1994; de Heinzelin et al. 2000; Schick and Clark 2003; Schick and Toth in press a), and has included studies of archeological site formation processes (Isaac et al. 1997; Schick 1984, 1986, 1987a, 1987b, 1991, 1992, 1997, 2001; Schick and Toth 1993; Toth and Schick 2009c).

Our research has also included experimental investigations into the production of quartz spheroids (Schick and Toth 1994; Toth and Schick 2009c), limestone spheroids (Sahnouni et al. 1997; Toth and Schick 2009c), basalt lava spheroids (Toth and Schick 2009c), pitted anvils and hammers (Toth and Schick 2009c), Acheulean artifacts (Schick and Toth 1993, in press a; Toth 2001; Toth and Schick 2009c, in press), kinesiological studies (Dapena et al. 2006; Toth and Schick 2009c), electromyography of hand and arm muscle activity (Marzke et al. 1998; Toth

and Schick 2009c), brain imaging studies during stone tool activities (Stout et al. 2000, 2006, 2009; Toth and Schick 2009c), and meat-cutting efficiency using different types of stone artifacts (Toth and Schick 2009c, in press).

We have also conducted actualistic studies in teaching modern apes (bonobos) to make and use stone tools (Savage-Rumbaugh et al. 2006; Schick et al. 1999; Toth and Schick 2009c, 2009d; Toth et al. 1993, 2006; Whiten et al. 2009), ethnoarcheological research with modern stone tool-makers in Irian Jaya, New Guinea (Toth and Schick 2009c; Toth et al. 1992), ethnoarcheological and experimental research into the use of bamboo (Jahren et al. 1997; Toth and Schick 2009c), bone modification studies (Pickering et al. 2000, Schick et al. 1989, 2007; Toth and Schick 2009c; Toth and Woods 1989; White and Toth 1989, 1991, 2007), and the relationships between brain reorganization/encephalization and the archeological record (Toth and Schick 2010). We have also discussed the Early Stone Age archeological record and its evolutionary implications in Schick and Toth 1993, 1994, 2006, in press b; Toth 1990; Toth and Schick 1986, 1993, 2007, 2009a, 2009b, 2009d, 2010).

Functional Considerations

Of prime importance in determining the nature of early lithic assemblages is consideration of what resources may have been exploited by early hominins and how a simple material culture could be employed to more efficiently exploit those resources. It is a hard, cold fact that prehistorians do not have a very good understanding of the precise behaviors and daily activities of early hominins and the full range of tool use in the course of those activities. We can model the types and distribution of edible resources and water in an African environment (e.g., Blumenschine 1989; Blumenschine and Peters 1998; Blumenschine et al. 2003, 2008; Peters and Maguire 1981; Peters and Blumenschine 1995; Peters and O'Brien 1981; Peters et al. 1984; Vincent 1985a, 1985b; Sept 1984, 1986, 1992, 1994, 2001). We can also look at contemporary hunter-gatherers as well as field studies of modern apes for models. However, in general, with the exception of modified faunal remains, we usually have little direct evidence of the resources that were exploited by early hominin populations, and are often left with concentrations (or scatters) of flaked and battered lithic artifacts and their associated geological and environmental context (sedimentology, fauna, pollen, phytoliths, carbonate, and bone isotopes, etc.).

Interestingly, for example, two very different foraging patterns might produce similar lithic signatures. If hominins were carrying around flaked lithic materials in anticipation of cutting off flesh and breaking bones of animal carcasses through hunting or scavenging, then concentrations of lithic material (percussors, cores, flakes) would occur where such carcasses were encountered or where they were transported to (for reasons of safety or shade, or if the precursors of home bases existed). If hominins were foraging with digging sticks and re-sharpening them with acute-edged choppers or flakes at the site of underground food resources, then one might get a similar lithic signature of percussors, cores, and flakes. If fossil bone survived, the carcass processing site should show modified bones (cut-marks, percussion striae, and notching), but if bone did not survive these two lithic assemblages could show appreciable equifinality.

Another important observation that has been made is that many Early Stone Age lithic assemblages associated with fossil bones show little if any hominin-induced modification

(Domínguez-Rodrigo et al. 2007). At Olduvai Gorge Beds I and II only site FLK Zinj shows appreciable butchery evidence. The other major sites, such as DK (with some of the best fossil bone preservation we have ever seen) show little or no such modification. This suggests that these lithic assemblages were not necessarily butchery-related, or, if they were, that the hominins were not producing butchery modification to the bones characteristic of modern butchery experiments.

Over time the frequency of retouched flakes increases at many Early Stone Age sites. We interpret this as an indication of more intensive cutting activities and resharpening of flake edges. Experiments have shown that once a natural flake edge dulls or becomes coated with dried animal tissues, retouching this form into an Oldowan denticulated "scraper" makes a very effective butchery knife, not quite as sharp as an unmodified flake edge, but with a longer use-life. We have postulated that with more habitual and intensive animal butchery through time, hominin populations would do one of two things: either stress such denticulated retouched flakes (Developed Oldowan, Tayacian, Nihewan Basin industries, etc.) or else develop large Acheulean cutting tools such as handaxes or cleavers, which have been demonstrated experimentally to be very efficient heavy-duty butchery tools.

Woodworking (e.g., making spears or digging sticks from branches or saplings) usually requires an acute-edged core for chopping and shaping (for example, making a pointed end) and flakes or retouched flakes for final shaping. Flakes can also be used as chisel-like implements in conjunction with a hammerstone for shaping wood or splitting bamboo. Such flake chisels are transformed into *outils écaillés* with light to moderate chipping on their sides or ends (Toth and Schick 2009a).

Nut-cracking with stone hammers and anvils is an activity that can also affect the nature of a lithic assemblage. Nut-cracking, as practiced by modern chimpanzees in the wild, historical hunter-gatherers, and, presumably, prehistoric populations, is essentially a bipolar technique in which a hard-shelled nut (rather than a lithic core) is placed on an anvil and struck from above with a hammer. Over time, this technique produces pitted anvils and sometimes pitted hammerstones as well, but unlike the rough type of pitting that forms as a by-product of bipolar stone flaking, the pitting produced by nut-cracking tends to be smooth. Such smooth-pitted stones (associated with the broken remains of five species of nuts) were found at the Acheulean site of Gesher Benot Ya'aqov in Israel (Goren-Inbar et al. 2002). Smooth-pitted stones, according to Goren-Inbar et al. are also found at the site of Ubeidiya in Israel, with no evidence of *outils écaillés* bipolar cores.

Hominin Taxa: Cognitive and Biomechanical Abilities

During the time period discussed here, (between 2.6 to 1.4 million years ago), there are perhaps eight hominin taxa in Africa: *Australopithecus garhi* (East Africa), *Australopithecus (Paranthropus) aethiopicus* (East Africa), *Australopithecus africanus* (South Africa), *Homo habilis* (East Africa and possibly South Africa), *Homo rudolfensis* (East and Central Africa), *Australopithecus (Paranthropus) boisei* (East and Central Africa), *Australopithecus (Paranthropus) robustus* (South Africa), and *Homo ergaster/erectus* (much of Africa and into Eurasia). These taxa had different cranial capacities, dental and postcranial morphologies, and probably differed significantly in their diets and foraging patterns and in their cognitive complexities and biomechanical skills as well. If the

different taxa used lithic material culture in different ways, they might leave different archaeological signatures. Experiments with modern African apes show that they can learn to make and use simple flaked stone tools (Savage-Rumbaugh et al. 2006; Schick et al. 1999; Toth and Schick 2009c; Toth et al. 1993, 2006; Whiten et al. 2009), so that is very likely that all early hominins had the cognitive and motor capabilities to manufacture and use stone tools if they needed them and wanted them.

We still have no clear idea of the extent and nature of tool use among these taxa, but many anthropologists reason that the genus *Homo*, with larger brains and reduced dentition over time, suggests a more technology-reliant lineage. By one million years ago, all australopithecine taxa disappear in the fossil record, with *Homo erectus* and subsequent later forms of *Homo* as the principal tool-using hominins. In time, we may have enough sites with the direct co-occurrence of stone tools with hominin fossils to see clearer patterns of association in the Pliocene and early Pleistocene.

Technological Considerations
Shared Cultural Traits and Geographical Distribution

Even in the Early Stone Age it is possible that hominin populations in closer geographical proximity may have shared more cultural traits. This type of patterning can be seen in wild chimpanzees subspecies today (Toth and Schick 2009c, 2009d; Whiten et al. 2009). Such shared cultural norms could be manifested in the Early Stone Age in particular types of recurrent artifacts localized in time and space, for example in the recurrent pattern of unifacial cores at the early Gona sites (Semaw et al. 2009), the Karari core-scrapers from the Okote Member at Koobi Fora (Harris and Isaac 1976), the use of bipolar flaking (evidenced by rough pitted anvils and bipolar *outils écaillés* cores) at some sites at Olduvai Gorge (Leakey 1971, 1994), or perhaps the polished and striated pointed bone tools from South Africa (Brain 1981; Brain and Shipman 1993; Pat Shipman pers. comm.). Experiments have shown that the vast range of Oldowan core forms can be produced by a least-effort strategy of flake production, often incorporating alternate bifacial flaking of cores, but not necessarily planning ahead much beyond each individual flake removal (Toth 1982, 1985b).

Once Acheulean forms such as handaxes and cleavers were invented between 1.7 and 1.5 million years ago, a clear cultural marker that differentiates it from Oldowan/Mode 1 industries emerged (if the larger raw materials were available). Such artifacts, in our opinion, show a clearer element of style of a preconceived form. The spread of these technologies can be traced from Africa through the Near East and into Europe and Central Asia, while simpler Mode 1 industries characterized much of Eastern Europe and the East Asia. Some specialized methods were developed with a restricted geographical and temporal distribution, such as the Tabalbalat-Tachenghit method of producing special cleaver-flakes in North Africa or the Kombewa "Janus" flake method of detaching a large flake from an even larger flake blank in Central and East Africa and the Near East.

Raw Material Types

Even with simple stone technologies, the different types of rock types used by early hominins could greatly affect the types of artifacts produced. Easily-flaked raw materials such as chert, chalcedony, obsidian, quartz, and fine-grained ignimbrite and quartzite are normally much easier to reduce, producing higher debitage to core ratios, proportionately more non-cortical flakes,

and often smaller cores such as discoids and polyhedrons. Some of these cores may be so small that they may be hard to identify as cores and be placed in a debitage category. Raw materials such as medium-grained lavas and quartzites can be much more difficult to flake (especially for knappers with less skill), often resulting in larger exhausted cores with fewer flake scars and more chopper-cores.

Quartz cores tends to quickly "snowball" into polyhedron and subspheroid forms, with appreciable battering from hammerstone percussion that might be mistaken for use-wear (Schick and Toth 1994; Toth and Schick 2009c). Angular quartz chunks used as percussors will become battered spheroids in a few hours of knapping. Heavily weathered basalt often produces spheroidal weathering clasts with a softer exterior and harder interior, and using these globular clasts as percussors for knapping will, over several hours of use, produce battered spheroids. Harder-cortex basalt clasts from fresher lava flows and found in stream as water-worn cobbles will tend to produce classic hammerstones with selective areas of battering (Toth and Schick 2009c).

Fine-grained limestone, as at Ain Hanech, Algeria, has unusual flaking qualities (Sahnouni et al. 1997). Unlike most other raw materials, flakes tend to have obtuse external platform angles (the angle between the striking platform and the dorsal surface) and acute internal platform angles (the angle between the striking platform and the ventral surface). Because of this unusual fracture pattern, cores have a tendency to become "faceted spheroids" with intensive reduction.

Higher-quality raw materials often produce more identifiable whole and broken flakes, while lower-quality raw materials (coarser-grained, less homogeneous, or having incipient weathering fracture) often produce higher proportions of angular fragments and chunks. Early tool-making hominins often selectively retouched flakes of higher-quality raw materials, suggesting that they were targeting these rock types for their superior cutting edges before and after resharpening.

Size of Raw Material

The size of raw material available could greatly affect the nature of a lithic assemblage. Access to only small cobbles will tend to produce chopper-cores and debitage. Access to larger cobbles will tend to produce a wider range of cobble-core forms (choppers, discoids, polyhedrons) as well as second-generation cores made on larger flakes that have been detached (discoids and heavy-duty scrapers). Larger cores that are heavily reduced will tend to produce larger quantities of non-cortical flakes than smaller cores as well.

With the advent of the Acheulean, it was critical to find boulders of lava, quartz, or quartzite, obsidian, or other raw materials from which to detach large flakes to serve as blanks for handaxes, cleavers, and picks, or to flake large cobbles, nodules, or tabular chunks of stone to make such typical Acheulean forms. In areas where accessible raw materials only came in smaller sizes such technologies would be impossible, and Acheulean hominins would, of necessity, produce assemblages that would be classified as Developed Oldowan.

Mode of Flaking

Experiments have shown that there are predictable patterns of flake production from different modes of flaking cobble cores as well as cores made on large flake blanks. The proportion of different flake types in a lithic assemblage can yield important clues to patterns of knapping and stages of flaking represented. Six major types of flakes are:

Flake Type	Description
1	Cortical platform, all cortex dorsal surface
2	Cortical platform, partial cortex dorsal surface
3	Cortical platform, non-cortical dorsal surface
4	Non-cortical platform, all cortex dorsal surface
5	Non-cortical platform, partial cortex dorsal surface
6	Non-cortical platform, non-cortical dorsal surface

A predominance of unifacial flaking of cobbles at some sites (e.g., Pliocene sites at Gona and Hadar, Ethiopia) will produce flakes with cortical striking platforms (Types 1-3). A predominance of alternate bifacial flaking of cobbles can produce flakes almost exclusively with non-cortical striking platforms (Types 4-6), with the exception of the very first flake removed from each cobble (Type 1). Normally there is some combination of both unifacial and bifacial flaking at most early sites.

Stages of Flaking Reduction

One important line of evidence that has been greatly elucidated by actualistic studies is an appreciation that at many Early Stone Age occurrences only partial, often later stages in the reduction of cores are represented (Toth 1982, 1985b; Toth et al. 2006). Later stages of flaking can be seen by one or more of lines of evidence.

In later stages of flaking cores tend to be more reduced (well over half the mass of a cobble or large flake blank removed) with more flake scars, more of the circumference flaked, and less surface cortex. Side choppers tend to be more numerous than end choppers, and polyhedrons and discoids are more common. There are more cores of indeterminate original blank form since evidence of a cobble or large flake blank has been obliterated. The heavily reduced cores tend to have a lower step and hinge scar/total scar ratio than earlier stages (Toth et al. 2006).

Flakes from later stages of reduction tend to exhibit less cortex on their dorsal surfaces, with higher proportions of Type 3 flakes (cortical platforms, non-cortical dorsal surfaces, from intensive unifacial flaking of cobbles) and type 6 flakes (non-cortical platforms and dorsal surfaces, from intensive bifacial and polyfacial flaking). Dorsal surfaces of flakes from later stages also have a lower step and hinge scar/total scar ratio relative to early stages, and less mean cortex (often means of less than 20% cortex).

Table 4.1 contrasts flake type proportions from earlier and later stages of flaking from two experimental assemblages. The first is predominantly bifacial flaking of cobbles and large flakes with Koobi Fora basalt, and the second is unifacial and bifacial reduction of Gona lava cobbles (predominantly trachyte).

Transport and Curation

When early hominins decided to move from one activity area to the next, they often had to decide what stones or artifacts to take with them, especially if moving to a foraging area with no nearby raw materials. This would have affected the nature of the lithic assemblages both at the first activity area and the next. Based on personal experience, stones that early hominins would have chosen to transport would have included modified cobbles for hammerstones, larger cores for later flake production, and larger, sharper flakes (assuming they are not blunted from use). Removal of such items from a lithic assemblage could affect assemblage composition and this residual assemblage would have less complete sequences of refitting artifacts. Curation of quartz, soft basalt, and limestone percussors could also produce more subspheroids and spheroids.

Skill Levels

The level of knapping skill by early tool-makers, within sites and between sites, could have a profound effect on the nature of lithic assemblages. Presumably, in any social group, there was for each tool-maker a learning curve in the acquisition of skill, with novices or less experienced individuals also contributing to the lithic assemblages. It is possible to assess various levels of skill in modern stone knappers by comparing novice knappers to experienced ones in experimental and ethnographic settings, and by comparing the

Table 4.1 Early vs. late stages of flaking from two experimental lithic assemblages

Stages of Flaking	Bifacial: Koobi Fora Basalts Flake Types						Unifacial + Bifacial: Gona Lavas Flake Types					
	1	2	3	4	5	6	1	2	3	4	5	6
Early	.06	14	.10	.11	.36	.22	.10	.39	.19	.03	.23	.06
Late	.00	07	.04	.03	.43	.44	.01	.11	.28	.01	.31	.28
Change	-.06	-.07	-.06	-.08	+.07	+.22	-.09	-.28	+.09	-.02	+.08	+.22

Left: flake types from predominantly bifacial flaking of cobbles and large flakes of basalt from Koobi Fora, Kenya (from Toth 1982, 1985b). Note especially the great rise in Type 6 (totally non-cortical) flakes in the later stages. Even further reduction of cores would produce much higher proportions of Type 6 flakes, as seen at many Plio Pleistocene Koobi Fora sites. Right: flake types from unifacial and bifacial reduction of lava cobbles (predominantly trachyte) at Gona, Ethiopia (from Toth et al. 2006). Note especially the increase in Type 3 (cortical butt, non-cortical dorsal surface) and Type 6 (totally non-cortical) flakes in the later stages. The dorsal cortex index went from 40% in the earlier stages to 16% in the later stages. If large, sharp flakes were then transported away from the later stage assemblage, the dorsal cortex index dropped to 12% (identical with the Gona EG sites). If only the reduction of unifacial cores is considered (all flakes having cortical platforms), more extensive reduction produced a dramatic rise in Type 3 flake proportions from .28 (earlier stages) to .70 (later stages, comparable to the Gona proportion of .61).

skill level of experienced African apes in an experimental setting with those of modern humans.

For simple Mode 1 technologies, skilled stone knappers, relative to less skilled knappers, are able to reduce cobbles much more efficiently, generating large quantities of flakes and fragments and producing a lower core to debitage ratio. End choppers tend to be transformed into side choppers as reduction proceeds. Skilled knappers can often select more suitable sizes and shapes of cobbles to be flaked, and have a better eye for selecting higher-quality raw materials. They tend to generate higher hammerstone percussion velocities, recognize overhangs and acute angles better, and can place hammerstone blows with more accuracy in the most appropriate place. Edges of cores have less battering from unsuccessful hammerstone blows. Such battering can be mistaken for utilization. They are better able to maintain acute edges on cores as reduction proceeds, and can often rejuvenate a core by driving off a large flake and creating new angles to be exploited. Flake scars tend to be more invasive as well. Surprisingly, in experiments skilled knappers (modern humans) and less skilled knappers (bonobos) produced very similar step and hinge scars/total scar ratios in early stages of flaking. Later stages of flaking produced much lower step and hinge scar/total scar ratios. As the frequency of steps and hinges is often (and perhaps erroneously) interpreted as an index of skill, it seems that stage of reduction and flaking qualities of different raw materials needs to be taken into consideration.

Skilled knappers, in our experience, tend to do more bifacial and polyfacial flaking of cobbles than less skilled tool-makers. This is due to their ability to recognize and exploit the scars of previous flake removals as striking platforms to remove flakes from the other face of a cobble.

Because of more intensive reduction of cobbles, skilled knappers can create a wider range of core forms than less skilled knappers. Besides chopper-cores, there is a tendency to reduce cobbles into polyhedrons and sometimes discoids. And, because larger, thicker flakes are often produced by skilled knappers, these can also serve as blanks for the production of core forms such as discoids and heavy-duty scrapers.

Skilled knappers produce, on average, larger flakes, more flakes that are end-struck versus side-struck, and more parallel-sided (as opposed to divergent) flakes in planform shape. These flakes tend to produce more usable cutting edge relative to mass. External platform angles on flakes (the angle formed between the striking platform and the dorsal surface) tend to be more acute. Skilled knappers can produce (surprisingly) lower numbers of whole flakes to other debitage (snaps, splits, and angular fragments). These lower numbers are likely due to the fact that skilled knappers produce higher impact forces, driving off larger flakes but shattering others in the process. Experiments have shown that knapping stone with low-velocity impact (the minimum force required for stone fracture) produced significantly more whole flakes relative to other debitage as compared to high-velocity impact. High velocity impact also produced more split and snapped flakes.

Techniques of Flaking

The type of flaking technique used by early hominins could affect the nature of a lithic assemblage. Hard-hammer freehand percussion, in which a stone hammer is held in one hand (normally the dominant hand) and the core in the other, was probably the most common knapping technique employed in the Oldowan Industrial Complex. Experimentally, this technique tends to produce the typical range of Oldowan core forms (choppers, polyhedrons, discoids, heavy-duty scrapers) and flakes characterized by thick striking platforms and prominent bulbs of percussion. Battered hammerstones or subspheroids/spheroids are the typical percussors.

Bipolar flaking, in which the core is set on a stone anvil and struck from above with a hammerstone can be done in two major ways. Probably the more common was vertical bipolar flaking, in which the long axis of a core, pebble or cobble, or thick flake is placed vertically on an anvil and struck from the top. This technique, seen at some sites in Bed II and subsequent beds at Olduvai Gorge, Tanzania and at Zhoukoudian Cave in China, normally produces flat, barrel-shaped cores, *outils écaillés*, with signs of flaking and crushing on both ends of the core, and flat flakes with thin or shattered striking platforms and flat release surfaces with very diffuse bulbs of percussion. Anvils exhibit central rough pits, while hammerstones tend to have one or, for elongated hammers, two off-set rough pits. With horizontal bipolar flaking a flat core is rested horizontally on an anvil and then flaked with a hammerstone. This technique does not tend to create the pitted hammers and anvils, and produces cores with steep edges and sometimes abrasion and micro-flaking on the side in contact with the anvil. This is a technique that was apparently used at the sites at Orce in Spain (de Lumley et al. 2009; Fajardo 2009) estimated to be about 1.2 million years old.

Anvil technique, in which a hand-held core is struck on a stationary anvil stone, can produce fracture patterns similar to that of freehand hard-hammer percussion. It is probably harder to identify this technique in the archaeological record, but cobbles or chunks with appreciable battering, crushing, and micro-flaking around its perimeter could one indication of this technique being employed.

Throwing one cobble against another can fracture even the most intractable spherical cobble. Such fracture often splits a cobble neatly in two with flat release surfaces and pronounced areas of crushing at the point of percussion. Hammerstones can also fracture in this way, but will usually exhibit more cortical battering near the point of fracture. Such split cobbles can then be further reduced by exploiting the newly-created acute angles as striking platforms.

Other Considerations

Many other factors could influence the variability seen in lithic assemblages at prehistoric sites. Larger group sizes occupying a locality, for example, would probably produce more numerous and denser concentrations of lithic materials. Re-occupation of favored localities (due to nearby food or water resources, safety from predators or other dangers, shade provided by trees or bushes) would also produce denser concentrations of artifacts, and possibly a thicker occupation layer if frequented over a number of seasons or years with occasional light flooding and burial, or occasional deposition of air-borne volcanic ash. Proximity to trees would provide shade during the mid-day, protection from predators such as large cats, hyenas, or hunting dogs by climbing into high branches (Hunt, pers. comm.), and probable sleeping quarters at night. Site FLK Zinj at Olduvai Bed I may in fact have sedimentary evidence of a large uprooted tree (Leakey 1971). Caves, such as the Transvaal localities in South Africa, may have been hominin sleeping areas as is seen in modern baboons in the area today (Brain 1981).

Proximity to Raw Material
Proximity to raw material could greatly affect both quantity and the size of cores and flakes in a lithic assemblage. One might predict that, all

things being equal, archeological sites situated in close proximity to raw materials tend to often have larger, less reduced cores and earlier stages of reduction, as it is easy to replenish stone. Flakes tend to be larger, with a higher dorsal surface cortical index (mean percentage of cortex area on flake dorsal surfaces). Retouch tends to be lighter and less frequent, as re-sharpening of tool edges is less necessary.

In the case of sites situated more distantly (e.g., several kilometers) from raw material sources, all stone for percussors, cores, and debitage must have been transported there. Lithic assemblages tend to have cores that are smaller and more heavily reduced, with less cortex and often more flake scars. Retouch on flake edges tends to be more intensive as pieces are re-sharpened more and more. Percussors may be less frequent as they are rare and highly valued, probably carried more habitually by hominin tool-makers. It is likely that hominins moving from such sites to other activity areas without nearby stone would especially take away percussors, larger cores, and larger flakes (either as potential cores or for their longer cutting edges). Such depletion of these larger artifact forms would lower the mean size of cores and flakes and further reduce the amount of cortex on dorsal surfaces of flakes (Toth et al. 2006).

Proximity to Water

Hominins, like most primates, are very water-dependent animals. Prior to sophisticated water storage and transport technologies, early hominins would have been strongly tethered to water sources (streams, lakes, deltas, water holes, springs, etc.) during their foraging patterns, affecting the nature and density of lithic assemblages. Activity areas requiring stone tools in proximity to water would often have denser concentrations of stone artifacts, while activity areas more distant from water sources would have more ephemeral, possibly heavily reduced, lithic assemblages. As most Early Stone Age sites are, by their nature, in depositional situations, most were buried by stream floods and sediments or by transgressive lake deposits requiring water as the major agent of sedimentation (the possible exception being air-born volcanic ash falls). Sites more distant from water sources (e.g., in highlands above an alluvial valley or in drier montane environments) tend to be in more erosional and non-depositional situations, so their archeological visibility would be much lower.

Predator Danger

Proximity to dangerous animals (especially crocodiles, mammalian predators such as large cats and hyenas) or even other hominin groups could greatly influence where and when tool-using hominins chose to forage, process, and consume foodstuffs as well as influence where they would rest at mid-day or sleep at night (Blumenschine et al. 2007). Presumably large concentrations of artifacts and animals bones (e.g., site FLK Zinj at Olduvai Bed I) were in areas far enough away so that an immediate threat of predation was not an issue.

It is unlikely that early hominins slept in close proximity to areas where they processed carcasses or parts of animals and left stone artifacts, as the smell of fresh organics such as greasy bones would almost certainly attract carnivores and other scavenging animals in a relatively short period of time. Until the controlled use of fire, sleeping in trees was probably the norm for many early hominin groups [although in karstic areas like the Transvaal of South Africa sleeping in caves, as suggested by Brain (1981) may have also been common].

Bone Taphonomy Considerations

While stone artifacts tend to be relatively indestructible over millions of years if buried reasonably quickly in a sedimentary context, animal bones are much more susceptible to a range of taphonomic agents of modification and destruction. These agents include ravaging by carnivores, rodents, or crocodiles; chemical weathering (in either highly acidic or highly alkaline environments); sediment compaction and breakage; trampling by animals; root etching; and, bacterial erosion. Archeological sites with quantities of stone artifacts but with little or no fossil animal bones do not necessarily mean that there was no association originally; the conditions for bone preservation and fossilization may not have been present. The absence of evidence is not necessarily evidence of absence. We suspect, for example, that animal bones were probably associated with the early sites at Gona, Ethiopia, but that conditions for bone preservation were generally very poor.

Hydraulic Action and Postdepositional Considerations

Most Early Stone Age sites are buried in sediments deposited by floodwaters. Water action can radically change the nature and spatial distribution of lithic assemblages, from almost pristine sites, to sites highly disturbed and winnowed of the smaller and lighter artifacts, to sites totally swept downstream from hominin activity areas. Archeological sites that have been hit hard by water action before burial tend to have, relative to undisturbed original conditions, a higher core to debitage ratio, a higher whole flake to fragment ratio, a higher mean debitage size, a reduction in Type 3 and Type 6 flakes, and a somewhat higher dorsal cortex flake index. The sedimentology of a hydrologically disturbed site may suggest higher flow energies in the forms of coarser sedimentary matrix (sands or gravels),

Table 4.2 A computer-generated simulation of water action on a lithic assemblage, based on experimental replicative experiments in basalt at Koobi Fora, Kenya (Toth 1982, 1985b).

Degree of Disturbance	Flake Types					
	1	2	3	4	5	6
None	.03	.09	.09	.07	.34	.38
Moderately Disturbed	.04	.21	.05	.08	.46	.16
Change-	+.01	+.12	-.04	+.01	+.12	-.22

The top row shows the flake type distribution for a non-disturbed site with all stages of flaking represented. The bottom row shows the effect of water action on this assemblage, preferentially removing the smaller, lighter debitage. Note that Type 3 (cortical platform, non-cortical dorsal surface) and especially Type 6 (totally non-cortical) flakes tend to be winnowed away with water action, dramatically increasing the proportions of Type 2 (cortical platform, partial cortex dorsal surface) and Type 5 (non-cortical platform, partial cortex dorsal surface) flakes.

cross-bedding, clustering of larger artifacts, preferential artifact/bone long axis orientation (parallel or perpendicular to stream flow), and preferential artifact/bone dipping upstream.

Table 4.2 shows the effect of increased water action on a large experimental Oldowan assemblage, based upon a computer-generated simulation of the winnowing away the smaller, lighter artifacts and changing flake type frequencies (Toth 1982, 1985b). Table 4.3 shows the effect of high water action on an experimental "mega-site", winnowing away over 90% of the artifacts and increasing core proportion and mean debitage size. Table 4.4 shows three Koobi Fora sites in the Okote Member (ca. 1.5 million years ago) that appear to represent three degrees of water disturbance. FxJj50 (low degree of disturbance), FxJj18GL (moderate degree of disturbance), and FxJj33 (very high degree of disturbance). With increased disturbance, core proportions, mean debitage size, and flake/fragment ratio all increase.

Table 4.3 Results of water action on a large experimental floodplain "mega-site" (100 m², almost 5,000 artifacts) that was hit hard by water action before burial and subsequent excavation

Degree of Disturbance	Number of Artifacts	Survivorship of Artifacts	Core Proportion	Mean Debitage Size
None	4,655	1.00	.04	22.1 mm
High	396	.09	.10	34.5 mm

From Schick 1984, 1986, 1987a, 1987b. Note that only nine percent of the original artifacts were recovered in the excavation, the proportion of cores in the assemblage more than doubled after water action, and mean debitage size increased by over 14 mm after water action.

Table 4.4 Three Early Pleistocene sites at Koobi Fora, Kenya interpreted as showing very different degrees of water action and disturbance

Site	Inferred Disturbance	Core Proportion	Mean Debitage Size	Flake:Fragment Ratio
FxJj 50	Low	.04	23.8 mm	.57
FxJj 18GL	Moderate	.10	47.1 mm	.88
FxJj 33	Very High	.88	86.9 mm	1.75

FxJj 50 = low disturbance; FxJj 18GL = higher disturbance; FxJj 33 = very high disturbance. Note that greater disturbance by water increases the core proportion in the assemblage, increases mean debitage size, and increases the whole flake to fragment ratio (from Schick 1984, 1986, 1987a, 1987b; Toth, 1882, 1985b).

After deposition, bioturbation can move artifacts vertically or horizontally, possible increasing the overall thickness of an archeological horizon. Hovers (2003) has also demonstrated that Early Stone Age artifacts such as thinner flakes have a tendency to fragment into several pieces *in situ*, presumably from sediment compaction. This might decrease the whole flake to fragment ratio at archaeological sites over time.

Conclusions

Glynn Isaac was a major proponent of the importance of actualistic studies in our understanding of the archeology of human origins. In this chapter we have tried show how actualistic studies can help identify major factors that could influence variability in Early Stone Age lithic artifact assemblages, and provide criteria useful in assessing the effect of these factors. We are now profoundly aware that a myriad of variables can affect patterns seen at early sites, including functions of stone tools, hominin cognitive and motor skills, cultural norms, raw material size and flaking characteristics, distance to raw materials sources, stages of reduction represented, flaking techniques, artifact transport and curation, bone taphonomy, hydrological disturbance before burial, and post-depositional disturbance. Careful analysis of prehistoric archeological lithic assemblages, combined with a robust experimental program of stone artifact manufacture and use and other actualistic studies, can give us a much better appreciation of the relative impact of these diverse factors and put us in a much better position to make informative inferences from the traces left behind by early hominins.

Acknowledgements

First and foremost, we would like to acknowledge our intellectual and personal debt to Glynn Isaac. We thank the editors of this volume for inviting us to contribute this chapter. Aspects of this research was funded through a number of granting agencies, including the National Science Foundation, The Harry Frank Guggenheim Foundation, The Fulbright Fellowship Program, The Social Science Research Council, The Luce Fund, The National Institute of Health, The Stone Age Institute, the Center for Research into the Anthropological Foundations of Technology (CRAFT) at Indiana University, The L. S. B. Leakey Foundation, the Wenner-Gren Foundation For Anthropological Research, The National Geographic Society, the Ligabue Research Center (Venice), The Kroeber Society (UC Berkeley), and the Boise Fund (Oxford), as well as through our own personal funds. Finally, we would like to thank the governing board members and advisory board members of the Stone Age Institute and CRAFT for their long-term encouragement, friendship, and support.

References

Blumenschine, R. J.

1989 A landscape taphonomic model of the scale of prehistoric scavenging opportunities. *Journal of Human Evolution* 18:345-371.

Blumenschine, R. J. and C. R. Peters

1998 Archaeological predictions for hominid land use in the paleo-Olduvai Basin, Tanzania during lowermost Bed II times. *Journal of Human Evolution* 34:565-607.

Blumenschine, R. J., C. R. Peters, S. D. Capaldo, P. Andrews, J. K. Njau, and B. L. Pobiner

2007 Vertebrate taphonomic perspectives on Oldowwant hominin land use in the Plio-Pleistocen Olduvai Basin, Tanzania. In *Breathing Life into Fossils: Tahonomic Studies in Honor of C. K. (Bob) Brain*, edited by T. Pickering, K. Schick, and N. Toth, pp. 161-179. Stone Age Institute Press, Gosport, Indiana.

Blumenschine, R. J., F. T. Masao, J. C. Tacktikos, and J. I. Ebert

2008 Effects of distance from stone source on landscape-scale variation in Oldowan artifact assemblages in the Paleo-Olduvai Basin, Tanzania. *Journal of Archaeological Science* 35:76-86.

Blumenschine, R. J. , C. R. Peters, F. T. Masao, R. L. Clarke, A. L. Deino, R. L. Hay, C. C. Swisher, I. G. Stanistreet, G. M. Ashley, L. J. McHenry, N. E. Sikes, N. J. van der Merwe, J. C. Tactikos, A. E. Cushing, D. M. Deocampo, J. K Njau, and J. I Ebert

2003 Late Pliocene *Homo* and hominid land use from western Olduvai Gorge, Tanzania. *Science* 299:1217-1221.

Blumenschine, R. J., F. T. Masao, J. C. Tacktikos, and J. I. Ebert

2008 Effects of distance from stone source on landscape-scale variation in Oldowan artifact assemblages in the Paleo-Olduvai Basin, Tanzania. *Journal of Archaeological Science* 35:76-86.

Brain, C. K.
1981 *The Hunters or the Hunted: An Introduction to African Cave Taphonomy.* University of Chicago Press, Chicago.

Brain, C. K., and P. Shipman
1993 Bone tools from Swartkrans. In *Swartkrans: A Cave's Chronicle of Early Man*, edited by C. K. Brain, pp. 195-215. Transvaal Museum Monograph No. 8.

Bunn, H., J. W. K. Harris, G. Isaac, Z. Kaufulu, E. Kroll, K. Schick, N. Toth, and A. K. Behrensmeyer
1980. FxJj 50: an early Pleistocene site in northern Kenya. *World Archaeology* 12(2):109-136.

Clark, J. D., and K. Schick
1987 Context and content: Impressions of Palaeolithic sites and assemblages in the People's Republic of China. *Journal of Human Evolution* 17:439-448.

Clark, J. D., J. de Heinzelin, K. D. Schick, W. K. Hart, T. D. White, G. Woldegabriel, R. C. Walter, G. Suwa, B. Asfaw, E. Vrba, and Y. H.-Selassie
1994 African *Homo erectus*: Old radiometric ages and young Oldowan assemblages in the Middle Awash Valley, Ethiopia. *Science* 264:1904-1910.

Dapena, J., W. Anderst, and N. Toth
2006 The biomechanics of the arm swing in Oldowan stone flaking. In *The Oldowan: Case Studies into the Earliest Stone Age*, edited by N. Toth and K. Schick, pp. 333-338. Stone Age Institute Press, Gosport, Indiana.

Domínguez-Rodrigo, M., R. Barba, and C. P. Egeland
2007 *Deconstructing Olduvai: A Taphonomic Study of the Bed I Sites*, Springer, Dordrecht, The Netherlands.

2000 *The Acheulean and the Plio-Pleistocene Deposits of the Middle Awash Valley, Ethiopia.* Musée Royale de l'Afrique Centrale, Tervuren, Belgium. de Lumley, H., D. Barsky, and D. Cauche

2009 Archaic stone industries from East Africa and southern Europe: Pre-Oldowan and Oldowan. In *The Cutting Edge: New Approaches to the Archaeology of Human Origins*, edited by K. Schick and N. Toth, pp. 55-91. Stone Age Institute Press, Gosport, Indiana.

Fajardo, B.
2009 The oldest occupation of Europe: Evidence from southern Spain. In *The Cutting Edge: New Approaches to the Archaeology of Human Origins*, edited by K. Schick and N. Toth, pp. 115-136. Stone Age Institute Press, Gosport, Indiana.

Goren-Inbar, N., G. Sharon, Y. Melamed, and M. Kislev
2002 Nuts, nut cracking, and pitted stones at Gesher Benot Ya'aqov, Israel. *Proceedings of the National Academy of Sciences* 99:2455-2460.

Harris, J. W. K., and G. Isaac

1976 The Karari Industry: Early Pleistocene archaeological evidence from the terrain east of Lake Turkana, Kenya. *Nature* 262:102-107.

Hovers, E.

2003 Treading carefully: Site formation processes and Pliocene lithic technology. In *Oldowan: Rather More Than Smashing Stones*, edited by J. Martinez, R. Mora, and I. de la Torre, pp. 145-164. Universitat Autonoma de Barcelona, Barcelona.

Isaac, B. (Editor)

1989 *The Archaeology of Human Origins: Papers by Glynn Isaac.* Cambridge University Press, Cambridge.

Isaac, G.

1967 Towards the interpretation of occupation debris: Some experiments and observations. *Kroeber Anthropological Society Papers* 37:31-57.

1971a The diet of early man: aspects of archaeological evidence from Lower and Middle Pleistocene sites in Africa. *World Archaeology* 2:278-99.

1971b Whither archaeology? *Antiquity* 45:123-129.

1972a Chronology and tempo of cultural change during the Pleistocene. In *Calibration of Hominid Evolution*, edited by W. Bishop and J. Miller, pp. 381-430. Scottish Academic Press, Edinburgh.

1972b Early phases in human behavior: Models in Lower Palaeolithic archaeology. In *Models in Archaeology*, edited by D. Clark, pp. 167-199. Methuen, London.

1976a Early hominids in action: A commentary on the contribution of archaeology to understanding the fossil record in East Africa. *Yearbook of Physical Anthropology* 1975:19-35.

1976b Stages of cultural elaboration in the Pleistocene: Possible archaeological indicators of the development of language capabilities. In *Origins and Evolution of Language and Speech*, edited by S. Harnad, H. Stekelis and J. Lancaster, pp. 275-288. New York Academy of Sciences, New York.

1977a *Olorgesailie: Archaeological Studies of a Middle Pleistocene Lake Basin in Kenya.* University of Chicago Press, Chicago.

1977b Squeezing blood from stones. In *Stone Tools as Cultural Markers*, edited by R. Wright, pp. 5-12. Australian Institute of Aboriginal Studies, Canberra.

1978 The food-sharing behavior of proto-human hominids. *Scientific American* 238:90-108.

1981a Archaeological tests of alternative models of early hominid behavior: excavation and experiments. *Philosophical Transactions of the Royal Society of London*, Series B, 292:177-188.

1981b Stone Age visiting cards: Approaches to the study of early land-use patterns. In *Patterns of the Past*, edited by I. Hodder, G. Isaac, and N. Hammond, pp. 131-155. Cambridge University Press, Cambridge.

1982 The earliest archaeological traces. In *The Cambridge History of Africa, Volume 1: From the Earliest Times to 500 BC*, edited by J. D. Clark, pp. 157-247. Cambridge University Press, Cambridge.

1983a Aspects of human evolution. In *Evolution from Molecules to Men*, edited by D. Bendall, pp. 503-543. Cambridge University Press, Cambridge.

1983b Bones in contention: Competing explanations for the juxtaposition of Early Pleistocene artifacts and faunal remains. In *Animals and Archaeology: Hunters and their Prey*, edited by J. Clutton-Brock and G. Grigson, pp. 3-19. British Archaeological Reports International Series 163, Oxford.

1984 The archaeology of human origins: Studies of the Lower Pleistocene in East Africa 1971-1981. In *Advances in World Archaeology*, Vol. 3, edited by F. Wendorf and A. Close. Academic Press, New York.

1985 Ancestors for us all: Towards broadening international participation in palaeoanthropological research. In *Ancestors: The Hard Evidence*, edited by E. Delson, pp. 346-351. Alan R. Liss, New York.

1986 Foundation stones: Early artefacts as indicators of activities and abilities. In *Stone Age Prehistory: Studies in Memory of Charles McBurney*, edited by G. Bailey and P. Callow, pp. 221-241. Cambridge University Press, Cambridge.

Isaac, G. Ll., and B. Isaac (Editors)
1997 *Koobi Fora Research Project, Volume 5: Plio-Pleistocene Archaeology*. Clarendon Press, Oxford.

Isaac, G. Ll., J. W. K. Harris, and F. Marshall
1981 Small is informative: The application of the study of mini-sites and least-effort criteria in the interpretation of the Early Pleistocene archaeological record at Koobi Fora, Kenya. In *Las Industrias mas Antiguas*, edited by J. D. Clark and G. Isaac, pp. 101-119. X Congresso Union International de Ciencias Prehistoricas y Protohistoricas, Mexico.

Isaac, G. Ll., J. W. K Harris, Z. M. Kaufulu, and K. Schick
1997 Applications of the observations and experiments to the Koobi Fora Cases. In *Koobi Fora Research Project, Vol. 5: Plio-Pleistocene Archaeology*, edited by G. Ll. Isaac, assisted by Barbara Isaac, pp. 256-261. Clarendon Press, Oxford.

Jahren, A.H., N. Toth, K. Schick, J. D. Clark, and R. G. Amundson

1997 Determining stone tool use: Chemical and morphological analyses of residues on experimentally manufactured stone tools. *Journal of Archaeological Science* 24:245-250.

Leakey, M.

1971 *Olduvai Gorge Volume 3: Excavations in Beds I and II, 1960-1963.* Cambridge University Press, Cambridge.

1994 *Olduvai Gorge: Excavations in Beds III, IV, and the Masek Beds, 1968-1971.* Cambridge University Press, Cambridge.

Marzke, M., N. Toth, K. Schick, S. Reece, B. Steinburg, K. Hunt, R. Linscheid, and K'N An

1998 EMG Study of hand muscle recruitment during hard-hammer percussion manufacture of Oldowan tools. *American Journal of Physical Anthropology* 105:315-332.

Peters, C. R., and R. J. Blumenschine

1995 Landscape perspectives on possible land use patterns for early hominids in the Olduvai Basin. *Journal of Human Evolution* 29:321-362.

Peters, C. R., and B. Maguire

1981 Wild plant foods of the Makapansgat area: A modern ecosystem analogue for *Australopithecus africanus* adaptations. *Journal of Human Evolution* 10:565-583.

Peters, C. R., and E. M. O'Brien

1981 The early hominid plant-food niche: Insights from an analysis of plant exploitation by *Homo, Pan,* and *Papio* in eastern and southern Africa. *Current Anthropology* 22:127-140.

Peters, C. R., E. M. O'Brien, and E. O. Box

1984 Plant types and seasonality of wild-plant foods, Tanzania to Southwestern Africa: Resources for models of the natural environment. *Journal of Human Evolution* 13:397-414.

Pickering, T., T. White, and N. Toth

2000 Cutmarks on a Plio-Pleistocene hominid from Sterkfontein, South Africa. *American Journal of Physical Anthropology* 111:579-584.

Sahnouni, M., K. Schick, and N. Toth

1997 An experimental investigation into the nature of faceted limestone 'spheroids' in the Early Palaeolithic. *Journal of Archaeological Science* 24:01-713.

Savage-Rumbaugh, S., N. Toth, and K. Schick

2006 Kanzi learns to knap stone tools. In *Primate Perspectives on Behavior and Cognition,* edited by David Washburn, pp. 279-291. American Psychological Association, Washington, DC.

Schick, K.

1984 Processes of Palaeolithic Site Formation: An Experimental Study. Ph.D. Dissertation, Anthropology Department, University of California at Berkeley.

1986 *Stone Age Sites in the Making: Experiments in the Formation and Transformation of Archaeological Occurrences.* BAR International Series 319, Oxford.

1987a Experimentally derived criteria for assessing hydrologic disturbance of archaeological sites. In *Natural Formation Processes and the Archaeological Record*, edited by D. T. Nash and M. D. Petraglia, pp. 86-107. BAR International Series 352, Oxford.

1987b Modeling the formation of Early Stone Age artifact concentrations. *Journal of Human Evolution* 16:789-807.

1991 On making behavioral inferences from early archaeological sites. In *Cultural Beginnings: Approaches to Understanding Early Hominid Life-Ways in the African Savanna*, edited by J. D. Clark, pp. 79-107. Dr. Rudolf Habelt GMBH, Monographien, Bandd 19, Bonn.

1992 Geoarchaeological analysis of an Acheulean Site at Kalambo Falls, Zambia. *Geoarchaeology* 7:1-26.

1997 Experimental studies of site formation processes. In *Koobi Fora Research Project, Vol. 5: Plio-Pleistocene Archaeology*, edited by G. Isaac and B. Isaac, pp. 244-256. Clarendon Press, Oxford.

2001 Examination of Kalambo Falls Acheulean Site B5 from a geoarchaeological perspective. In *Kalambo Falls Prehistoric Site, Volume 3: The Earlier Cultures: Middle and Earlier Stone Age*, edited by J. D. Clark, pp. 463-480. Cambridge University Press, Cambridge.

Schick, K., and J. D. Clark

2003 Biface technological development and variability in the Acheulean Industrial Complex in the Middle Awash region of the Afar Rift, Ethiopia. In *Multiple Approaches to the Study of Bifacial Technologies*, edited by M. Soressi and H. L. Dibble, pp. 1-30. University of Pennsylvania Museum of Archaeology and Anthropology, Philadelphia.

Schick, K., and N. Toth

1993 *Making Silent Stones Speak: Human Evolution and the Dawn of Technology.* Simon and Schuster, New York.

1994 Early Stone Age technology in Africa: A review and a case study into the nature and function of spheroids and subspheroids. In *Integrative Paths to the Past: Palaeoanthropological Advances in Honor of F. Clark Howell*, edited by R. Corruccini and R. Ciochon, pp. 429-449. Prentice-Hall, Englewood Cliffs, New Jersey.

2003 The origins of the genie: Human technology in an evolutionary context. In *Living with the Genie: Essays on Technology and the Quest for Human Mastery*, edited by A. Lightman, D. Sarewitz, and C. Desser, pp. 23-34. Island Press, Washington, D.C.

2006 An overview of the Oldowan Industrial Complex: The sites and the nature of their evidence. In *The Oldowan: Case Studies into the Earliest Stone Age*, edited by N. Toth and K. Schick, pp. 3-42. Stone Age Institute Press, Gosport, Indiana.

In press a *Acheulean Industries of the Lower and Middle Pleistocene, Middle Awash, Ethiopia. Proceedings, Colloque international: Les Culture à Bifaces du Pléistocène Inférieur et Moyen Dans le Monde. Émergence du Sens de l'Harmonie*, edited by H. de Lumley. Centre Européen de Recherches Préhistoriques de Tautavel, France.

In press b The evolution of technology. In *Companion to Paleoanthropology*, edited by D. Begun. Wiley-Blackwell, Hoboken, New Jersey.

Schick, K., N. Toth, and T. Daeschler
1989 An early paleontological assemblage as an archaeological test case. In *Bone Modification*, edited by R. Bonnichsen and M. Sorg, pp. 121-137. Center for the Study of the First Americans, Orono, Maine.

Schick, K., N. Toth, T. Gehling, and T. Pickering
2007 A taphonomic analysis of an excavated striped hyena den from the Eastern Desert of Jordan. In *Breathing Life into Fossils: Taphonomic Studies in Honor of C. K. Bob Brain*, edited by T. Pickering, K. Schick, and N. Toth, pp. 75-106. Stone Age Institute Press, Gosport, Indiana.

Schick, K., N. Toth, Q. Wei, J. D. Clark, and D. Etler
1991 Archaeological perspectives in the Nihewan Basin, China. *Journal of Human Evolution* 21:13-26.

Schick, K., N. Toth, G. Garufi, S. Savage-Rumbaugh, R. Sevcik, and D. Rumbaugh
1999 Continuing investigations into the stone tool-making capabilities of a bonobo (*Pan paniscus*). *Journal of Archaeological Science* 26:821-832.

Semaw, S., M. J. Rogers, and D. Stout
2009 Insights into late Pliocene lithic assemblage variability: The East Gona and Ounda Gona South Oldowan archaeology (2.6 million years ago), Afar, Ethiopia. In *The Cutting Edge: New Approaches to the Archaeology of Human Origins*, edited by Kathy Schick and Nicholas Toth, pp. 211-246. Stone Age Institute Press, Gosport, Indiana.

Sept, J. M.
1984 Plants and Early Hominids in East Africa: A Study of Vegetation in Situations Comparable to Early Archaeological Site Locations. Ph.D. Dissertation, University of California, Berkeley.

1986 Plant foods and early hominids at site FxJj 50, Koobi Fora, Kenya. *Journal of Human Evolution* 15:751-770.

1992 Archaeological evidence and ecological perspectives for reconstructing early hominid subsistence strategies. In *Archaeological Method and Theory, Vol. 4*, edited by M. B. Schiffer, pp. 1-56. University of Arizona Press, Tucson.

1994 Beyond bones: Archaeological sites, early hominid subsistence, and the costs and benefits of exploiting wild plant foods in east African riverine landscapes. *Journal of Human Evolution* 27:295-320.

2001 Modeling the edible landscape. In *Meat Eating and Human Evolution*, edited by C. B. Stanford and H. T. Bunn, pp. 73-98. Oxford University Press, Oxford.

Stout, D., N. Toth, K. Schick, J. Stout, and G. Hutchins
2000 Stone tool-making and brain activation: Positron emission tomography (PET) studies. *Journal of Archaeological Science* 27:1215-1223.

Stout, D., N. Toth, and K. Schick
2006 Comparing the neural foundations of Oldowan and Acheulean toolmaking: A pilot study using positron emission tomography (PET). In *The Oldowan: Case Studies into the Earliest Stone Age*, edited by N. Toth and K. Schick, pp. 321-331. Stone Age Institute Press, Gosport, Indiana.

Stout, D., N. Toth, K. Schick, and T. Chaminade
2009 Neural correlates of Early Stone Age toolmaking: Technology, language and cognition in human evolution. In *The Sapient Mind: Archaeology Meets Neuroscience*, edited by C. Renfrew, C. Frith, and L. Malafouris, pp. 1-19. Oxford University Press, Oxford.

Toth, N.
1982 The Stone Technologies of Early Hominids at Koobi Fora, Kenya: An Experimental Approach. Ph.D. Dissertation, University of California, Berkeley. Ann Arbor: University Microfilms.

1985a Archaeological evidence for preferential right-handedness in the Lower and Middle Pleistocene, and its possible implications. *Journal of Human Evolution* 14:607-614.

1985b The Oldowan reassessed: A close look at early stone artifacts. *Journal of Archaeological Science* 12:101-120.

1987a Behavioral inferences from Early Stone Age assemblages: An experimental model. *Journal of Human Evolution* 16:763-787.

1987b The first technology. *Scientific American* 255:112-121.

1990 The prehistoric roots of a human concept of symmetry. *Symmetry: Culture and Science* 1(3):257-281.

1991 The importance of experimental replicative and functional studies in palaeolithic archaeology. In *Cultural Beginnings: Approaches to Understanding Early Hominid Life-Ways in the African Savanna*, edited by J. D. Clark, pp. 109-124. Dr. Rudolf Habelt GMBH, Monographien, Bandd 19, Bonn.

1997 Chapter Seven: The artefact assemblages in the light of experimental studies. In *Koobi Fora Research Project, Volume 5: Plio-Pleistocene Archaeology*, edited by G. Isaac and B. Isaac, pp. 363-401. Oxford University Press, Oxford.

2001 Experiments in quarrying large flake blanks at Kalambo Falls. In *Kalambo Falls Prehistoric Site, Volume 3: The Earlier Cultures: Middle and Earlier Stone Age*, edited by J. D. Clark, pp. 600-604. Cambridge University Press, Cambridge.

Toth, N., and K. Schick
1986 The first million years: The archaeology of proto-human culture. In *Advances in Archaeological Method and Theory, Vol. 9*, edited by M. Schiffer, pp. 1-96. Academic Press, New York.

1993 Early stone industries and inferences regarding language and cognition. In *Tools, Language and Cognition*, edited by K. Gibson and T. Ingold, pp. 346-362. Cambridge University Press, Cambridge.

2009a African Origins. In *The Human Past: World Prehistory and the Development of Human Societies, Second Edition*, edited by C. Scarre, pp. 46-83. Thames & Hudson, London.

2009b Early hominids. In *The Oxford Handbook of Archaeology*, edited by B. Cunliffe, C. Gosden, and R. A. Joyce, pp. 254-289. Oxford University Press, Oxford.

2009c The importance of actualistic studies in Early Stone Age research: Some personal reflections. In *The Cutting Edge: New Approaches to the Archaeology of Human Origins*, edited by Kathy Schick and Nicholas Toth, pp. 267-344. Stone Age Institute Press, Gosport, Indiana.

2009d The Oldowan: The tool making of early hominins and chimpanzees compared. *Annual Review of Anthropology* 38:289-305.

2010 Hominin brain reorganization, technological change, and cognitive complexity. In *The Human Brain Evolving: Paleoneurological Studies in Honor of Ralph L. Holloway*, edited by Doug Broadfield, Michael Yuan, Kathy Schick, and Nicholas Toth, pp. 293-312. Stone Age Institute Press, Gosport, Indiana.

In press Why the Acheulean? Experimental studies of the manufacture and function of Acheulean tools. *Proceedings, Colloque International: Les Culture à Bifaces du Pléistocène Inférieur et Moyen Dans le Monde. Émergence du Sens de l'Harmonie*, edited by H. de Lumley. Centre Européen de Recherches Préhistoriques de Tautavel, France.

Toth, N., and M. Woods
1989 Molluscan shell knives and experimental cut-marks on bones. *Journal of Field Archaeology* 16:250-255.

Toth, N., J. D. Clark, and G. Ligabue
1992 The last stone axe-makers. *Scientific American* 267(1):88-93.

Toth, N., K. Schick, and S. Semaw

2006 A comparative study of the stone tool-making skills of *Pan, Australopithecus*, and *Homo sapiens*. In *The Oldowan: Case Studies into the Earliest Stone Age*, edited by Nicholas Toth and Kathy Schick, pp. 155-222. Stone Age Institute Press, Gosport, Indiana.

Toth, N., K. Schick, S. Savage-Rumbaugh, R.A. Sevcik, and D. M. Rumbaugh

1993 *Pan* the tool-maker: Investigations into the stone tool-making and tool-using capabilities of a bonobo (*Pan paniscus*). *Journal of Archaeological Science* 20:81-91.

Vincent, A.

1985a Plant foods in savanna environments: A preliminary report of tubers eaten by the Hadza of northern Tanzania. *World Archaeology* 17:131-148.

1985b Underground Plant Foods and Subsistence in Human Evolution. Ph.D. Dissertation, University of California, Berkeley.

Whiten, A., K. Schick, and N. Toth

2009 The evolution and cultural transmission of percussive technology: Integrating evidence from palaeoanthropology and primatology. *Journal of Human Evolution* 57:420-435.

White, T., and N. Toth

1989 Engis: Preparation damage, not ancient cutmarks. *American Journal of Physical Anthropology* 78:361-367.

1991 The question of ritual cannibalism at Grotta Guattari. *Current Anthropology* 32:118-138.

2007 Carnivora and carnivory: Assessing hominid toothmarks in zooarchaeology. In *Breathing Life into Fossils: Taphonomic Studies in Honor of C. K. Bob Brain*, edited by T. Pickering, K. Schick, and N. Toth, pp. 281-296. Stone Age Institute Press, Gosport, Indiana.

Xie, F., K. Schick, N. Toth, and J. D. Clark

1994 Study of refitting of stone artifacts at Cenjiawan site in 1986. *Journal of Chinese Antiquity* 3:86-102.

5

ARCHAEOLOGY OF HUMAN ORIGINS:
THE CONTRIBUTION OF WEST TURKANA (KENYA)

Hélène Roche

In September 1986, Richard Leakey took me for a short trip to West Turkana where, at that time, he was excavating the Nariokotome Boy skeleton with Kamoya Kimeu. But for the demise of Glynn Isaac the previous year this journey would certainly never have occurred - not at least as it was proposed to me. Having surveyed the Nachukui Formation's sedimentary outcrops in search of fossils from south to north their team had found lithic evidence suggesting the existence of archaeological sites - some possibly dating to very early periods. At the time Richard Leakey was Director of the National Museums of Kenya and he wanted to set up a program of archaeological exploration on behalf of the Museum, as he had with Glynn for Koobi Fora, while training young Kenyan archaeologists at the same time, which is what he suggested I should do in West Turkana. In a few hours I was shown about half a dozen potential sites - later to become the "no. 1" sites of the main complexes we should go to great lengths to explore. I have little doubt I was hardly the first to have made this trip, only to be offered the same propositions. I still wonder 25 years later how I could have been the only one to accept the deal. My first impressions - however positive they may have been - were still well below the true archaeological potential the West Turkana Plio Pleistocene sequence had in store...

Introduction

Two years before this first trip in West Turkana an article by Glynn Isaac (Isaac 1984) had come

out, which I have boldly borrowed as part of the title for this chapter. This article was later also included in a collected volume of his articles published by Barbara Isaac (Isaac 1989). This article was certainly not a recapitulation of all those he had written. Nor was it a complete sampling of the great variety of his centers of interest and the riches of his thought. Rather, it took as its base and starting point the most important publication on Olduvai (Leakey 1971); it synthesised research he and his Koobi Fora Research Project team had effected throughout the 1970s - which has been the subject of an impressive number of papers, Ph.D. research, and lastly a monograph (Isaac and Isaac 1997). To a great extent this paper also put in perspective the totality of Plio-Pleistocene archaeological research which at that point covered the period from 2 Ma to 0.7 Ma. (A more recent synthesis of current research can be found in an article by Plummer (2004), which looks at all the lines of research considered by G. Isaac in his time.) My goal here is briefly to present the work done by the West Turkana Archaeological Project in the light of Isaac's 1984 article, which I have always considered a "road map" for the archaeology of human origins. A few words, however, will put this presentation in context.

Nearly half of Isaac's long 1984 article had as its subject what was the hottest topic of the time: the formation of the sites (the "dense patches of artifacts plus bones"), their location in the landscape and the distribution of scatters of artifacts between these patches; it discussed the methodologies most suitable for dealing with these issues, and of course their related models at that time which were aggressively challenged by Binford (1981). To a lesser degree it discussed the lithic assemblages and their significance, as well as the diet of early man, the importance meat may have had in this diet, and problems of seasonality. At the end of the article it advocated the need for going "towards realistic ecologic models." All the same, since for me one of Glynn's remarkable qualities was he never imprisoned himself in a theoretical dogma anymore than he forced the facts to fit a hypothesis, I do not think he would ever have given up any line of research capable of nourishing his reflection and advancing his research. If in West Turkana great efforts have been made to further understanding of the processes of technical behaviour evolution, it was also obvious for us that it was absolutely necessary to define contexts and environments as closely as possible and measure the interactions of hominins with their environment - without necessarily adhering to an ecological paradigm. We have followed a sites-based (or sites-complex-based) exploration approach for several reasons. First, using a landscape approach, which some would prefer instead, requires thorough knowledge of the terrain one seeks to come to grips with. This was far from the case at the start of our research; whereas now it can not only be envisaged, but it will be perfectly complementary to all the work hitherto carried out. Secondly, when we began our work, West Turkana was an empty space on the archaeological map of the Turkana basin: filling it in was a priority. Since documentation of the Plio-Pleistocene was still poor, especially for the earliest sites, we felt it was necessary to augment the already existing stock of data. The perception we quickly acquired - that we were dealing

with assemblages different from what was already known in the "classic" Oldowan - reinforced this sentiment.

Painted with a very broad brush this then was the methodological landscape in which work began in West Turkana - fairly sporadically at first, for I was also occupied with the excavation of the Acheulean site of Isenya (Southern Kenya). On the institutional level the 1987/88 and 1991 missions were under the authority of the National Museums of Kenya (NMK) alone. Only in 1994 was the West Turkana Archaeological Project (WTAP) created as a joint program of the National Museum of Kenya (Nairobi) and the Mission Préhistorique Française au Kenya - as is still case.

The Archaeology in West Turkana
Although at the beginning of the 1980s the general chronostratigraphy of the Nachukui Formation had already been built up, a biostratigraphy established (Harris et al. 1988) and several major hominin fossils brought to light, on the archaeological level, as I have said, everything was still to do. This sedimentary formation is part of the Omo Group, as are the Formations of Shungura and Kibish to the north and Koobi Fora to the east. It should be remembered it is 712 m thick, and its eight upper members cover nearly 4 million years (4.05/07 Ma) without any interruption - a definite advantage for observing unbroken phenomena. The tephras, albeit less well defined and more altered than in the Shungura Formation, where they serve as a reference for the whole basin, are nevertheless present and invaluable for the chronostratigraphic framework. The formation appears as a long (circa 80 km NS) and narrow (circa 10 km WE) band of fluvio-lacustrine sediments dissected by a network of temporary water-courses flowing WE to reach the lake, the longest having their

sources in or even on the other side of the ranges of the Murua Rith and the Lapur - the natural geographic frontiers that border the basin to the West.

On the archaeological level our first task (1987/1988) was to confirm the presence of material *in situ* on the sites discovered by the National Museum of Kenya fossils hunters. Accordingly, test excavations were undertaken at Lokalalei 1, Kokiselei 1, Naiyena Engol 1, and Nadung'a 1. Thanks to Frank Brown, we quickly learned that Lokalalei was in the Kalochoro member (2.35/1.90 Ma), Kokiselei and Naiyena Engol in Kaitio (1.90/1.65 Ma), and Nadung'a in Nariokotome (1.30/0.7 Ma; Kibunjia et al. 1992). So it was evident we had the unique perspective of being able to observe hominins' presence in this part of the basin for a period of nearly 2 million years. The exploratory period continued - with surveys and test excavations - to evaluate the potential better. From 1997 it was followed by extensive excavation campaigns at a certain number of key-sites with a clearly identified goal: to scrutinize the evolution of hominin behaviour from the end of the Pliocene (now early Pleistocene) until the middle Pleistocene within a geographically well circumscribed territory and in environmental contexts reconstructed as precisely as possible. The aim was to obtain representative samples for the four chrono-cultural periods represented - early Oldowan and late Oldowan (early Pleistocene), early Acheulean (early Pleistocene), and middle Acheulean (early middle Pleistocene). Our sampling program was intended to allow comparisons first within and, then between these various periods within the Nachukui Formation, and later more broadly on the scale of the Turkana basin and the east-African region. Of course, many chance factors have interfered with this research design, which had been intended to progress quite logically - especially

the discovery of other exceptional sites, which at times made us sacrifice analysis, and consequently publication, for intensive fieldwork. However this may be, the work progressed during the nearly fifteen years that followed, enabling us to collect in total a quite considerable mass of information, of which a large part has been analysed but only a small part published except in the annual reports. A monograph is under way due to come out in 2012.

Oldowan Sites (Figure 5.1)

For many years work in what was then the final Pliocene was concentrated on the Lokalalei complex (2.4–2.3 Ma). Not only was Lokalalei 1 the first early Oldowan site to be explored, it was also the very first of all the West Turkana sites to be tested in 1987 (Kibunjia et al. 1992). It was then excavated extensively in 1991 by Mzalendo Kibunjia for his Ph.D., which he completed in 1998 (Kibunjia 1994, 1998; Kibunjia et al. 1992). It was during this 1991 campaign that M. Kibunjia showed me the site of Lokalalei 2C, which we would only excavate in 1996/1997 (Roche et al. 1999). Prospecting in the vicinity in subsequent years has led to discovering other sites (still not excavated) and remains of hominin teeth, including the first instance of an Early Homo in the Turkana basin (Prat et al. 2005), and to locating possible fossil sources of lithic raw material in the surrounding region (Harmand 2005, 2009a). An interpretation slightly divergent (Brown and Gothogo 2002) from the chronostratigraphy proposed in the announcement paper of the site (Roche et al. 1999) has resulted in a complementary chronostratigraphic study and paleoenvironmental interpretation (Tiercelin et al. 2010).

In 2002 Craig Feibel discovered Nasura 1, a new early Pleistocene site very precisely dated to 2.34 Ma thanks to its decimetric proximity to the

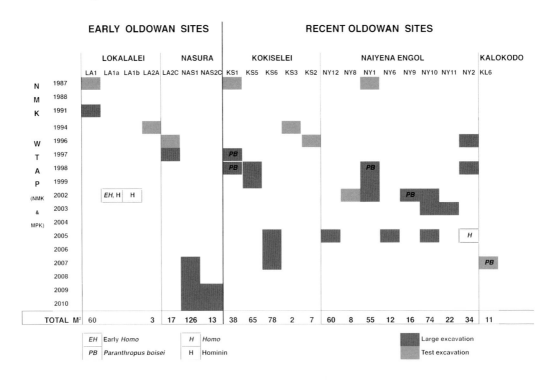

Figure 5.1 Excavations at the West Turkana Oldowan sites.

Kalochoro tuff (Harmand et al. in prep). As sites of this age are few and far between this was one of the exceptional discoveries referred to above which we were able to take advantage of. The excavation of Nasura 1 was started in 2007 and finished during the 2010 campaign. Contextual research in the surroundings very quickly revealed a very promising Nasura 2 now under exploration.

The sites of Lokalalei and Nasura are located in paleosols formed in the flood plain of a fluvial system (Paleo-Omo) that followed the paleo-lake Lonyumun (Feibel et al. 1989, 1991), and see below) - either in the far part of the flood plain where the marginal water-courses running WE join the NS axial system, or on the edge of a meandering paleo-river or near one of its broad arms.

I shall not go into the details of the chronology of the work on all the "classic" Oldowan sites, i.e. the same age as those of Olduvai Bed 1, of which the equivalent in the Nachukui Formation is the Kaitio member (1.9/1.65 Ma). This period seems to correspond to an important paleoanthropic presence in this part of the Turkana basin only partially reflected in the number of sites excavated. Two major complexes have been explored in particular (Kokiselei and Naiyena Engol) and a third has been started (Kalokodo). A recently terminated paleomagnetism program will allow a more precise chronostratigraphic positioning of most of these sites (Lepre et al. in prep.) distributed on either side of the Olduvai/Matuyama inversion (1.78 Ma). In fact, the Kaitio member is characterized by a succession of important shore deposits, and most of the sites are located

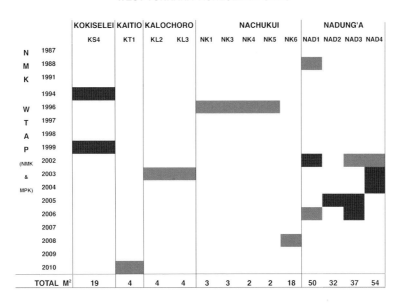

Figure 5.2 Excavations at the West Turkana Acheulean sites.

along transversal water-courses that drained the margins of the paleo-lake Lorenyang (Roche et al. 2003) and see below). All these sites line up in the Oldowan gradient – as much for their age as for their technological features – to which we shall return below. Lastly, it should be noted that until now no site has been discovered in the Natoo member (1.65/1.30 Ma), which is poorly exposed in the outcrops.

Acheulean Sites (Figure 5.2)

In this respect the Acheulean sites as a whole are less homogeneous. The beginnings of the Acheulean in West Turkana are illustrated quite spectacularly by the presence in the Kaitio member of a unique site, Kokiselei 4, stratified between the nearby Oldowan sites of Kokiselei 3 and Kokiselei 2. Kokiselei 4 is the earliest Acheulean site known in East Africa at present (Lepre et al. in prep.), and it is not disputed that

the technological characteristics of its lithic material place it in this techno-cultural complex. But it is the only one of this type and age in the Nachukui Formation and we have no satisfactory explanation for its uniqueness. All the other "Acheulean" sites in the formation are included in the Nariokotome member (1.30/0.7 Ma): either at the base with the sites of Nachukui 6 and Kaitio 1, both under exploration, or at the top, with all the other sites of Nachukui and those of Nadung'a, among them the already published site of Nadung'a 4 (Delagnes et al. 2006). The quotation marks above are justified by the fact that, with the exception of a broken handaxe tip in a test-pit in one of the late group sites (Nachukui 4), none of these "Acheulean" sites has yielded bifacial pieces *in situ*. The presence of bifacial pieces on the surface - never in large numbers in any case - does not satisfy our requirement for clarity, the less so given that the relation between surface and *in situ*

Table 5.1 West Turkana archaeological sites according to the sites complexes and the geological members

Sites Complex / Members	Age Limits	Nasura	Lokalalei	Kokiselei	Naiyena Engol	Kalokodo	Kaitio	Kalochoro	Nachukui	Nadung'a	T
NARIOKOTOME	0.70										
	1.30	-	-	-	-	-	1	3	6	8	18
NATOO	1.30										
	1.65	-	-	-	-	-	-	-	-	-	-
KAITIO	1.65										
	1.90	-	-	10	13	7	-	-	-	-	30
KALOCHORO	1.90										
	2.35	1	7	-	-	-	-	-	-	-	8
LOKALALEI	2.35										
	2.50	1	1?	-	-	-	-	-	-	-	2

context is not evident. No comparable bifacial pieces are known from surface assemblages associated with the earlier sites of Kaitio 1 and Nachukui 6. We have here then an important collection of sites Acheulean in age but for which the usual diagnostic features seem to be lacking.

Continuing Research

All periods considered, nearly 35 sites have been excavated, following various extensions. Most also contain faunal remains, but not all, and the nature of the fauna/lithic relationship was not necessarily behavioral. Eight of these sites also yielded hominin remains (Prat et al. 2003) – essentially teeth found on surface – except at Naiyena Engol 9 (Prat et al. in prep). It is notable that but for one exception (Naiyena Engol 2) the remains attributed to the genus *Homo*, including the first earliest occurrence at Lokalalei 1 (Roche et al. 2003) have been found in the early Oldowan sites, whereas the other teeth found in the late Oldowan sites should be assigned to the species *Paranthropus boisei*.

Fine-tuning the chronostratigraphic framework and interpreting the paleoenvironments –

tasks which Craig Feibel began in 1994 – continues collaboratively on three geographic scales: the region, complexes of sites, and sites. In addition site formation processes are studied by our team members using recording methods (Lenoble and Bertran 2004) which extend and improve upon those pioneered in the basin by G. Isaac and the Koobi Fora team from the mid 1970s onwards: measuring fabric of elongated objects *in situ*,[1] evaluation of their surface condition, dimensional sorting – after wet sieving in test zones – by vestige type (sediments, bone splinters, stone splinters), studying horizontal and vertical spatial distributions and, when the assemblage allows it, conjoining.

Parallel to our excavation work and as a necessary preliminary to a spatial approach we have also continued to carry out surveys to establish an exhaustive cartography of the anthropic occupation of this part of the basin between 2.4 and 0.7 Ma. The database of these inter-sites and inter-complexes surveys, in constant evolution, currently contains more than 300 occurrences. More recently, we have created a GIS for which all the GPS recordings and elevation digital models are accompanied by aerial photos of the sites, or survey zones, taken by kite or helium inflated balloon; the photographs are geo-referenced and can be draped on the elevation digital models. The excavation databases, which contain the spatial coordinates of each object, as well as the bases of analysis of the lithic and faunal remains, will also be connected to this GIS. Lastly, projects for evaluating the influence of the deposit types on sites distribution and of high resolution analysis of fine lacustrine sediments are also on the way.

Putting Some Results in Perspective

One of the major characteristics of the West Turkana sites is that they cover a very long time, during which the basin changed its hydrographic regime, as has been the case throughout its geological history (Brown and Feibel 1991). Thus during the early Oldowan time period (2.4/2.3 Ma) the basin was traversed by a paleo-river (paleo-Omo), while during the late Oldowan it was occupied by the paleo-lake Lorenyang, the member Kaitio consisting of deposits from this long lake phase. With the exception of Kokiselei 4, which is also in Kaitio Member, the Acheulean sites were located in a fluvial system before and after Lake Silbo, which is the last fossil lacustrine episode before Lake Turkana.

In both these configurations water-courses drained the alluvial plain from west to east, as is still the case intermittently today on temporary mode; when their flow was sufficient these rivers could carry abundant detritus from the ranges of highlands they originated in, which made them very good reserves for raw material. At the sites where the raw material supply has been studied, neighbouring paleo-channels have almost always been identified as possible sources (Harmand 2005, 2009a, b). A similar configuration has been found at other early

Oldowan sites (D. Braun et al. 2008; D. R. Braun et al. 2008; Goldman-Neuman and Hovers 2009; Stout et al. 2005), but in West Turkana it has been proven for every chrono-cultural period – from early Oldowan to the Acheulean. This makes us cautious in proposing models that hypothesize the transport and/or the caching of lithic raw materials.

For the same reasons – long time periods and quite wide geographic spaces – the sites are also located in varied environments: alluvial fan near the sources of rivers; various positions in the alluvial plain – on the edges of transversal channels, on a natural levee with sometimes a residual pool, or along an axial paleo-river; in the periphery of a lake, on a shore or up a beach, with or without residual pool. They are also characterized by a very large diversity of depositional processes. All sites with sufficient taphonomic study have shown evidence of perturbations – whether at the moment of deposit or subsequently (Lenoble in Roche et al. 2008, 2009). None, however, can be characterized as an "hydraulic jumble" (cf. Isaac 1984). But overlain on this great variability of contexts (induced by situations in the landscape, deposit conditions, and post-depositional processes) is evidence of highly probable differing site "functions" or "natures." This results in site contents contrasting in quantity and quality (Table 5.2) that have to be interpreted once all the natural phenomena involved in their formation have been untangled.

Without forgetting that the absence of fauna is perhaps only due to taphonomic processes, the sites can be classified as follows: sites with high lithic density and significant knapping activities on the spot with or without animal remains; sites with high lithic density, significant knapping activities on the spot, and remains of a single big animal; sites with low lithic density and limited knapping activities, with or without animal

Table 5.2 Examples of artifact density at some of the West Turkana archaeological sites

Sites	Abbrev.	Techno Cultural Complex	Total Exc. m²	Total Lithic Coord. in situ	Total Fauna Coord in situ	Dens/ m²
Lokalalei 2C	LA2C	Early Old.	17	2,122	390	147.5
Nasura 1	NAS1	Early Old	126	920	22	7.4
Nasura 2C*	NAS2C	Early Old	10	85	38	12.3
Kokiselei 6	KS6	Oldowan	78	5,584	420	76.9
Kokiselei 1	KS1	Oldowan	38	526	247	20.3
Kokiselei 5	KS5	Oldowan	65	1,765	298	31.7
Naiyena Engol 1	NY1	Oldowan	55	2,424	358	50.5
Naiyena Engol 2	NY2	Oldowan	31	276	101	12.1
Naiyena Engol 12	NY12	Oldowan	61	433	150	9.5
Nachukui 6*	NK6	Acheulean	18	447	28	26.3
Nadung'a 1	NAD1	Acheulean	50	1,887	0	37.7
Nadung'a 2	NAD2	Acheulean	32	402	0	12.5
Nadung'a 3	NAD3	Acheulean	37	641	4	17.4
Nadung'a 4	NAD4	Acheulean	53	6,797	145	130.0

*excavation in process

remains. So the data we have to compare across sites fall, as expected, under multiple parameters: chrono-cultural (four periods are represented); paleogeographic (fluvial or lake system, location of sites in the landscape); topographic (site position); taphonomic (process of site formation, post-depositional alterations); paleoanthropic (nature/function of site). This work is in progress.

As research advanced in West Turkana it quickly became apparent that technical behaviour which overlay a similar technical substrate could vary significantly from one site to another. To a certain extent it could be said this simple technical substrate (named "Mode 1 industry" by many colleagues) comes close to what Glynn Isaac tried to express by simplifying, from the terminological point of view, Plio-Pleistocene technical productions into three large categories: the flaked pieces, the detached pieces, and the pounded pieces (Isaac 1981, 1984). By means of a neutral and simplified terminology he quite rightly wanted to do away with descriptive labels loaded with functional subjectivity. At the same time he expressed the idea that "awareness that stones can be fractured at will is a threshold, and

as soon as one starts to do it, even in a simple-minded way, a wide range of possibilities is opened" (Isaac 1984). So the door was open to observe this "wide range of possibilities," and this is what we try to do through a technological approach founded on the concept of *chaîne opératoire*. Often poorly understood this conceptual framework can also be used as a tool for analysis; it is only a method for observations and reconstructing mental and motor processes that, in the case of knapping, go from raw material procurement to discarding the manufactured objects (the core reduction sequence being only a part of this "chain"). This dynamic overview of a technical system, however simple it may be, was perfectly expressed by Glynn Isaac, but certainly runs counter to the repeatedly updated postulate (Whiten et al. 2009) that in archaeology it is results, and not processes, that are observed.

In the Oldowan these differences of a technological nature - observable and observed - reside especially in the ways the flakes were produced and, most likely, in a repertoire of uses for the objects produced that one can but hypothesize. In West Turkana these differences appear synchronically as much as diachronically. Between the small flat beach pebbles with a few flakes struck off an edge (as in the assemblage of Naiyena Engol 10, late Oldowan, ca. 1.8 Ma, Roche in Roche et al. 2005) and the tens of blocks from which up to 70 flakes have been extracted following a similar organized core reduction pattern (Lokalalei 2C, early Oldowan, 2.34 Ma; Delagnes and Roche 2005), the range of technical possibilities is wide which, as can be seen, does not correspond to the concept of evolution through complexification! In addition to the knapped material, the "pounded pieces" also seem to play a discriminating role which is only beginning to be evaluated. In West Turkana, for example, these pounded pieces are more numerous in the early Oldowan sites than in later ones – when they are not practically absent – whereas they are present in most of the Oldowan sites of Olduvai. However it may be, in the context of our field of research – in which time, distance and, like it or not, the hominins' taxonomy, the technical competence and needs of the groups, and idiosyncrasy play a considerable role – the observation of such differences is not surprising. That is why we have opted for the paradigm of variability and/or diversity in the Oldowan technical productions of West Turkana, this does not accord with the concept of stasis as it was first set out from the material of Gona (2.6/2.5 Ma) and in reference to that of Olduvai (Semaw 2000; Semaw et al. 1997), i.e. from 2.6 Ma to the emergence of the Acheulean, the Oldowan forms a unique, simple and homogenous ensemble. Now that it has been clarified that "...the word 'stasis' doesn't imply a lack of diversity but rather a state of equilibrium" (Stout et al. 2010:489), it would appear this disaccord ought to fade away. Pushing back the Plio-Pleistocene limit to 2.6 Ma (Leigh Mascarelli 2009) has deprived paleoanthropologists and prehistorians of a dividing

line that had considerable advantages, and the term "pre-Oldowan" (used for the period before 2 Ma) has doubtless quite rightly had its day. Nonetheless, we are not dealing here so much with a single uniform entity, but with Oldowans – and not just "early" and "late" – each with its own characteristics.

To Conclude

At the end of his 1984 article, Glynn wrote: "Reflexion shows that the cumulative differences in knowledge are subtle rather than sweeping. It is the state of inquiry more then the state of knowledge that has advanced." I should tend to think that as of now the cumulative differences in knowledge are still subtle, and I am not certain the state of inquiry has progressed that much. What new lines of research would he not have explored or envisaged himself?

On the level of state of knowledge, from the mid 1970s – even if it was not immediately accepted and/or recognized – it was already known (Corvinus 1976; Corvinus and Roche 1976; Roche and Tiercelin 1977, 1980) that the first evidence for very early knapping activity was located in Ethiopia at "Hadar" (now Gona) between 2.6 and 2.5 Ma, which was later on confirmed (Harris 1983; Semaw et al. 1997; and see Semaw et al. 2009 and Stout et al. 2010). The Omo sites had also been brought to light and constituted a first "Pliocene" landmark at 2.3 Ma on the Ethiopian part of the Lake Turkana basin (Chavaillon 1976; Merrick and Merrick 1976). Next was the important group of East Africa Pleistocene Oldowan sites – first and foremost Olduvai Bed 1 (Tanzania), fully published in 1971 (Leakey 1966), the KBS sites at Koobi Fora (Kenya; Isaac 1976; Isaac and Isaac 1997), and Chesowanja at Baringo (Kenya; Gowlett et al. 1981) for the earliest (1.9/1.6 Ma); and such as Melka Kunture in Ethiopia (Chavaillon et al. 1979), the Karari Escarpment sites at Koobi Fora (Harris 1976) and Peninj in Tanzania (Isaac 1965, 1967) for the later sites (from 1.6 Ma). Currently, and in spite of intensive research - especially at Gona - no discovery has been made that allows an earlier age than 2.6/2.5 Ma to be ascribed with certainty to a proven technical activity; highlighting two surface bone fragments, lying on sediments dated at 3.39 Ma and showing cutmarks (McPherron et al. 2010) is a too tenuous clue, out of context and with no certainty these marks were inflicted by an intentionally produced cutting tool. Lastly, very little is still known about the period 2.6/2.5 Ma - the excavated surfaces and unearthed material being so small in quantity (Semaw et al. 2009; Stout et al. 2010). On the other hand, the period between 2.4 and 2.3 Ma has gained a few sites to its credit - at Hadar in Ethiopia (Kimbel et al. 1996) and in West Turkana in Kenya (see references in this article) - but is still poorly represented. Two sites - Fejej in Ethiopia (Asfaw et al. 1991; de Lumley and Beyene 2004) and Kanjera near Lake Victoria in Kenya (Plummer et al. 1999) have filled in an empty period around 2 Ma. Lastly, the late Oldowan now has the benefit of more plentiful documentation, and not just in eastern Africa.

As for site formation, the models have multiplied (see for instance Plummer 2004 for a comprehensive review): from the central place foraging proposed and defended by Glynn himself (Isaac 1976, 1978, 1983), passing through the carnivore kill site and routed foraging model (Binford 1984) the favored place hypothesis (Schick 1987), the resource defense model (Rose and Marshall 1996), the dual unit foraging model (Oliver 1994) etc. - of which I am not sure their contributions are either new, or very convincing, any more than are their methodological aspects. I would say rather we should have learned from the last two decades of research that a single explanation model cannot account

for the complexity of evolutionary process. The diet of early man, on the other hand, has certainly favored new reflections, with lively discussions about carnivory and the hominins/carnivores relationship, as well as numerous models, sometimes very sophisticated (Aiello and Wheeler 1995). At least it can be said the study of hominin technical behaviour has slightly progressed and - as has been seen - provoked some controversy. But it definitely seems more that the debates appear to have shifted, or certain enquiries been taken further, rather than that really new lines of research have emerged - especially when the comparison is made with the decade Glynn synthesised in his article, and which was so fertile, largely owing to the impulse he gave to it.

Acknowledgements
No need to say that the work in West Turkana is the result of a team venture, that include, Mzalendo Kibunjia, Sonia Harmand, Pierre-Jean Texier and Sophie Clément for archaeology, Jean-Philip Brugal (paleontology), Sandrine Prat (paloanthropology), Gourguen Davtian (geomatic), Craig Feibel, Arnaud Lenoble, Chris Lepre, Rhonda Quin and Xavier Boes (geology); Mathieu Schuster and Jean-Jacques Tiercelin (geology) have also participated to the project. We are indebted to our Kenyan workers for their constant efforts and dedication to work, more particularly to Rafael M. Kioko, Christopher Kirua, John M. Kafuka, Bernard K. Mulwa; William Ekadil, Ekai Emekwi, Francis E. Emekwi, Alfred Kole, James Ekwiyeni, Sammy L. Lokodi, Ethekon L. Lokorodi, David Masika, Ali M. Mutisya, Frederic Mwanza, Akwam A. Nares, Longolei K. Ngolokerem, Leeyio P. Martin.

We thank the Office of the President and the Board of Governors of the National Museums of Kenya to allows to conduct archaeological research is West Turkana. The work of WTAP is funded by the French Foreign Affairs (Ministère des Affaires étrangères et européennes DGM/ATT/RECH Pôle Sciences Humaines et Sociales), and the CNRS (program ECLIPSE/INSU), and we thank Total Kenya for constant logistic support.

Endnotes

1   The elongated items, necessary for these measures, are unfortunately rare in our sites: by their nature for the lithic items and by absence or deterioration for the faunal remains.

2   "...artifacts came to be thought of not as a set of objects, but as drop-out of a dynamic system. This clearly implied that in order to understand them and to devise useful counts and measures one would have to understand in some details the processes by which they are produced, used, rejuvenated and finally discarded" (Isaac 1984).

References

Aiello, L. C., and P. Wheeler
1995 The expensive-tissue hypothesis. *Current Anthropology* 36(2):199-222.

Asfaw, B., Y. Beyene, S. Semaw, G. Suwa, T. D. White, and G. WoldeGabriel
1991 Fejej: A new paleoanthropological research area in Ethiopia. *Journal of Human Evolution* 21:137-143.

Binford, L. R.
1981 *Bones. Ancient Men and Modern Myths.* Academic Press, New York.

1984 *Faunal Remains from Klasies River Mouth.* Academic Press, Orlando, FL.

Braun, D., M. Rogers, J. Harris, and S. Walker
2008 Landscape-scale variation in hominin tool use: Evidence from the Developed Oldowan. *Journal of Human Evolution* 55(6):1053-1063.

Braun, D. R., T. Plummer, P. Ditchfield, J. V. Ferraro, D. Maina, L. C. Bishop, and R. Potts
2008 Oldowan behavior and raw material transport: Perspectives from the Kanjera Formation. *Journal of Archaeological Science* 35(8):2329-2345.

Brown, F. H,. and C. S. Feibel
1991 Stratigraphy, depositional environments and paleogeography of the Koobi Fora Formation. In *Koobi Fora Research Project, Volume 3, The fossil ungulates: Geology, fossil artiodactyls and paleoenvironments*, edited by J. M. Harris, pp. 1-30. Clarendon, Oxford.

Brown, F. H., and P. N. Gothogo
2002 Stratigraphic relation between Lokalalei 1A and Lokalalei 2C, Pliocene archaeological sites in West Turkana, Kenya. *Journal of Archaeological Science* 29:699-702.

Chavaillon, J.
1976 Evidence for the technical practices of early Pleistocene hominids, Shungura Formation, Lower Omo Valley, Ethiopia. In *Earliest Man and Environments in the Lake Rudolf Basin*, edited by Y. Coppens, F. C. Howell, G. L. Isaac, and R. E. Leakey, pp. 565-573. University of Chicago Press, Chicago.

Chavaillon, J., N. Chavaillon, F. Hours, and M. Piperno
1979 From the Oldowan to the Middle Stone Age at Melka Kunture (Ethiopia): Understanding cultural changes. *Quaternaria* 21:87-114.

Corvinus, G.
1976 Prehistoric exploration at Hadar, Ethiopia. *Nature* 261:571-572.

Corvinus, G., and H. Roche
1976 La préhistoire dans la région de Hadar (Bassin de l'Awash, Afar, Ethiopie): Premiers résultats. *l'Anthropologie* 80(2):315-324.

de la Torre, I.
2011 The early stone age lithic assemblages of Gadeb (T=Ethiopia) and the Developed Oldowan / early Acheulian in East Africa. *Journal of Human Evolution* 60:768-812.

de Lumley, H., and Y. Beyene (Editors)
2004 *Les sites préhistoriques de la region de Fejej, Sud-Omo, Ethiopie, dands leur contexte stratigraphique et paléontologique.* Association Pour la Diffusion de la Pensée Française, Paris, Paris.

Delagnes, A., and H. Roche
2005 Late Pliocene hominid knapping skills: The case of Lokalalei 2C, West Turkana, Kenya. *Journal of Human Evolution* 48:435-472.

Delagnes, A., J.-P. Brugal, S. Harmand, A. Lenoble, S. Prat, J.-J. Tiercelin, and H. Roche
2006 Interpreting pachyderm single carcass sites in the African Middle Pleistocene record: A multidisciplinary approach in the site of Nadung'a 4 (Kenya). *Journal of Anthropological Archaeology* 25:448-465.

Feibel, C., F. H. Brown, and I. McDougall
1989 Stratigraphic context of fossil hominids from the Omo group deposits: Northern Turkana Basin, Kenya and Ethiopia. *American Journal of Physical Anthropology* 78:595-622.

Feibel, C. S., J. M. Harris, and F. H. Brown
1991 Paleoenvironmental context for the Late Neogene of the Turkana Basin. In *Koobi Fora Research Project. Vol. 3: The Fossil Ungulates: Geology, Fossil Artiodactyls and Palaeoenvironments*, edited by J. M. Harris, pp. 321-370. Clarendon Press, Oxford UK.

Goldman-Neuman, T., and E. Hovers
2009 Methodological considerations in the study of Oldowan raw material selectivity: Insights from A.L. 894 (Hadar, Ethiopia). In *Interdisciplinary approaches to the Oldowan*, edited by E. Hovers and D. Braun, pp. 71-84. Springer, Berlin.

Gowlett, J. A. J., J. W. K. Harris, D. Walton, and B. A. Wood
1981 Early archaeological sites, hominid remains and traces of fire from Chesowanja, Kenya. *Nature* 294:125-129.

Harmand, S.
2005 *Matieres Premières Lithiques et Comportements Techno-économiques des Hominines Plio-pléistocène du Turkana Occidental, Kenya*. University Paris-X, Nanterre, France.

2009a Raw material and economic behaviours at Oldowan and Acheulean in the West Turkana region, Kenya. In *Lithic Materials and Paleolithic Societies*, edited by B. Adams and B. Blades, pp. 3-14. Blackwell Publishing, Oxford UK.

2009b Variability in raw material selectivity at the late Pliocene sites of Lokalalei, West Turkana, Kenya. In *Interdisciplinary Approaches to the Oldowan*, edited by E. Hovers and D. R. Braun, pp. 85-97. Springer, Berlin.

Harmand S., H. Roche, and C. Feibel
in prep Nasura 1, a new Late Pliocene Oldowan site in the Nachukui Formation, West Turkana, Kenya: Preliminary results.

Harris, J. M., F. H. Brown, and M. G. Leakey
1988 Stratigraphy and paleontology of pliocene and plestocene localities west of lake Turkana Kenya. *Contributions in Science*. Natural History Museum of Los Angeles County, Los Angeles CA.

Harris, J. W. K.
1983 Cultural beginnings: Plio-Pleistocene archaeological occurrences from the Afar, Ethiopia. *African Archaeological Review* 1:3-31.

Harris, J. W. K., and Isaac, G. Ll.
1976 The Karari industry: Early Pleistocene archaeological evidence from the terrain east of Lake Turkana, Kenya. *Nature* 262:102-107.

Isaac, B. (Editor)
1989 *The Archaeology of Human Origins. Papers by Glynn Isaac*. Cambridge University Press, Cambridge.

Isaac, G. L.
1965 The stratigraphy of the Peninj Beds and the provenance of the Natron Australopithecine mandible. *Quaternaria* 7:101-130.

1967 The stratigraphy of the Peninj Group: Early Middle Pleistocene formations west of Lake Natron, Tanzania. *Background to Evolution in Africa*. University of Chicago Press, Chicago.

1976 Plio-Pleistocene artifact assemblages from East Rudolf, Kenya. In *Earliest Man and Environments in the Lake Rudolf Basin: Stratigraphy, Paleoecology and Evolution*, edited by Y. Coppens, F. C. Howell, G. Ll. Isaac, and R. E. F. Leakey, pp. 552-564. Chicago University Press, Chicago.

1978 The food-sharing behavior of protohuman hominids. *Scientific American* (238):311-325.

1981 Stone age visiting cards: Approaches to the study of early land-use patterns. In *Patterns in the Past*, edited by I. Hodder, G. L. Isaac and N. Hammond, pp. 37-103. Cambridge University Press, Cambridge.

1983 Bones in contention: competing explanations for the juxtaposition of Early Pleistocene artifacts and faunal remains. In *Animals and Archaeology. I. Hunters and their Prey*, edited by J. Clutton-Brock, C. Grigson, vol. 3-19. British Archaeological Reports International series 163, Oxford.

1984 The archaeology of human origins: studies of the lower Pleistocene in East Africa 1971-1981. In *Advances in World Archaeology*, Vol. 3, edited by F. Wendorf, pp. 1-87. Academic Press, New York.

Isaac, G. L., and A. B. Isaac (Editors)
1997 *Plio-Pleistocene Archaeology, Vol. 5*. Clarendon Press, Oxford.

Kibunjia, M.
1994 Pliocene archaeological occurrences in the Lake Turkana basin. *Journal of Human Evolution* 27:159-171.

1998 Archaeological investigations of Lokalalei 1(GaJh5): A Late Pliocene site west of lake Turkana, Kenya. Ph.D. Dissertation, Anthropology, The State University of New Jersey, Rutgers, USA.

Kibunjia, M., H. Roche, F. H. Brown, and R. E. Leakey
1992 Pliocene and Pleistocene archaeological sites west of Lake Turkana, Kenya. *Journal of Human Evolution* 23(5):431-438.

Kimbel, W. H., R. C. Walter, D. C. Johanson, K. E. Reed, J. L. Aronson, Z. Assefa, C. W. Marean, G. C. Eck, R. Bobe, E. Hovers, Y. Rak, C. Vondra, T. Yemane, D. York, Y. Chen, N. M. Evensen, and P. E. Smith
1996 Late Pliocene *Homo* and Oldowan Tools from the Hadar Formation (Kada Hadar Member), Ethiopia. *Journal of Human Evolution* 31:549-561.

Leakey, M. D.
1966 A review of the Oldowan culture from Olduvai Gorge, Tanzania. *Nature* 210:462-466.

1971 *Olduvai Gorge, Vol. 3: Excavations in Beds I and II, 1960-1963*. Cambridge University Press, Cambridge.

Leigh Mascarelli, A.
2009 Quaternary geologists win timescale vote. *Nature* 624.

Lenoble, A., and P. Bertran
2004 Fabric of Palaeolithic levels: Methods and implications for site formation processes. *Journal of Archaeological Science* 31:457-469.

McPherron, S. P., Z. Alemseged, C. W. Marean, J. G. Wynn, D. Reed, D. Geraads, R. Bobe, and H. A. Béarat
2010 Evidence for stone-tool-assisted consumption of animal tissues before 3.39 million years ago at Dikika, Ethiopia. *Nature* 466(7308):857-860.

Merrick, H. V., and J. P. S. Merrick
1976 Archaeological occurences of Earlier Pleistocene age from the Shungura Formation. In *Earliest man and environments in the Lake Rudolf Basin*, edited by Y. Coppens, F. C. Howell, G. L. Isaac and R. E. Leakey, pp. 574-584. University of Chicago Press, Chicago.

Oliver, J. S.
1994 Estimates of hominid and carnivore involvement in the FLK Zinjanthropus fossil assemblage: Some socioecological implications. *Journal of Human Evolution* 27:267-294.

Plummer, T.
2004 Flaked stones and old bones: biological and cultural evolution at the dawn of technology. *Yearbook of Physical Anthropology* 47(118-164).

Plummer, T., L. C. Bishop, P. Ditchfield, and J. Hicks
1999 Research on late Pliocene Oldowan sites at Kanjera South, Kenya. *Journal of Human Evolution* 36(2):151-170.

Prat, C., J. Brugal, H. Roche, and P. Texier
2003 Nouvelle découvertes de dents d'hominidés dans le membre Kaitio de la formation de Nachukui (1, 65-1, 9 Ma), Ouest du Lac Turkana (Kenya). *Comptes Rendus Palevol.* 2:685-693.

Prat S., S. Harmand, J-J. Tiercelin, K. Kimeu, A. Lenoble, J-P Brugal, P-J Texier, and H. Roche
in prep New hominid specimens from the Kaitio Member (1.65-1.9 Myr) in West Turkana (Kenya).

Prat S, J-P. Brugal, J-J. Tiercelin, J-A. Barrat, M. Bohn, A. Delagnes, S. Harmand, K. Kimeu, M. Kibunjia, P-J. Texier, et al.
2005 First occurence of early *Homo* in the Nachukui Formation (West Turkana, Kenya) at 2.3-2.4 Myr. *Journal of Human Evolution* 49(2):230-240.

Roche, H., et al.

2005 *Rapport collectif 2005 du West Turkana Archaeological Project, Mission Préhistorique au Kenya.* Ministère des Affaires Étrangères.

2008 *Rapport collectif 2008 du West Turkana Archaeological Project, Mission Préhistorique au Kenya.* Ministère des Affaires Étrangères.

2009 *Rapport collectif 2009 du West Turkana Archaeological Project, Mission Préhistorique au Kenya.* Ministère des Affaires Étrangères.

Roche, H., and J. J. Tiercelin

1977 Découverte d'une industrie lithique ancienne *in situ* dans la formation d'Hadar, Afar central, Ethiopie. *Comptes Rendus de l'Academie des Sciences,* Paris D 284:187-194.

1980 Industries lithiques de la formation Plio-Pléistocene d'Hadar: campagne 1976. *Proceedings of the VIIIth Panafrican Congress of Prehistory and Quaternary Studies* 194-198. Nairobi.

Roche, H., A. Delagnes, J.-P. Brugal, C. S. Feibel, M. Kibunjia, V. Mourre, and P.-J. Texier

1999 Early hominid stone tool production and knapping skill 2.34 Myr ago in West Turkana. *Nature* 399:57-60.

Roche, H., J.-P. Brugal, A. Delagnes, C. S. Feibel, S. Harmand, M. Kibunjia, S. Prat, and P.-J. Texier

2003 Les sites archéologiques plio-pléistocènes de la Formation de Nachukui (Ouest Turkana, Kenya): Bilan préliminaire 1997-2000. *Comptes Rendus Palevol.* 2:663-731.

Rose, L., and F. Marshall

1996 Meat eating, hominid sociality, and home bases revisited. *Current Anthropology* 37:307-338.

Schick, K. D.

1987 Modeling the formation of Early Stone Age artifact concentrations. *Journal of Human Evolution* 16(7/8):789-808.

Schoeninger, M. J.

2010 Toward a $\delta^{13}C$ isoscape for primates. In *Isoscapes: understanding movement, pattern and process on earth through isotope mapping,* edited by J. B. West, G. Bowen, J., T. E. Dawson and K. P. Tu, pp. 319-333. Springer, Dordrecht.

Semaw, S.

2000 The world's oldest stone artifacts from Gona, Ethiopia: their implications for understanding stone technology and patterns of human evolution between 2.6-1.5 Million years ago. *Journal of Archaeological Science* 27:1197-1214.

Semaw, S., P. Renne, J. W. K. Harris, C. S. Feibel, R. L. Bernor, N. Fesseha, and K. Mowbray
1997 2.5-million-year-old stone tools from Gona, Ethiopia. *Nature* 385:333-336.

Semaw, S., M. J. Rogers, and D. Stout
2009 Insights into Late Pliocene lithic assemblage variability: The East Gona and Ounda Gona South Oldowan Archaeology (2.6 million years ago), Afar Ethiopia. In *The Cutting Edge: New Approaches to the Archaeology of Human Origins*, edited by A. Schick and N. Toth, pp. 211-246. Stone Age Institute Publication Series. Stone Age Institute Press, Gosport, IN.

Stout, D., J. Quade, S. Semaw, M. Rogers, and N. Levin
2005 Raw material selectivity of the earliest stone toolmakers at Gona, Afar, Ethiopia. *Journal of Human Evolution* 48(4):365-380.

Stout, D., S. Semaw, M. J. Rogers, and D. Cauche
2010 Technological variation in the earliest Oldowan from Gona, Afar, Ethiopia. *Journal of Human Evolution* 58(6):474-491.

Tiercelin, J.-J., M. Schuster, H. Roche, J.-P. Brugal, P. Thuo, S. Prat, S. Harmand, G. Davtian, J.-A. Barrat, and M. Bohn
2010 New considerations on the stratigraphy and environmental context of the oldest (2.34 Ma) Lokalalei archaeological site complex of the Nachukui Formation, West Turkana, northern Kenya Rift. *Journal of African Earth Sciences* 58(2):157-184.

Whiten, A., K. Schick, and N. Toth
2009 The evolution and cultural transmission of percussive technology: Integrating evidence from palaeoanthropology and primatology. *Journal of Human Evolution* 57(4):420-435.

6

THE EMPIRE OF THE ACHEULEAN STRIKES BACK

John A. J. Gowlett

The Acheulean was the premier archaeological tradition of Lower Paleolithic research in the 1950s. In the 1960s Glynn Isaac made seminal contributions to its study at Olorgesailie in Kenya, but interest in the Acheulean was becoming eclipsed by the remarkable discoveries which made the Oldowan the centerpiece of human origins research for the next generation. Isaac nevertheless laid foundations for a modernization of study which presaged many later developments. In recent years, the Acheulean has re-emerged as a focus of interest, with studies increasing in number, scope and theoretical impact - its empire has struck back. Its million and a half years are again recognized to be central to our understanding of human evolution, especially in its middle and later stages. This paper examines Glynn Isaac's contributions, which laid foundations for modern Acheulean research in terms of landscapes, artifact studies and chronological perspective, continuing to look at them in the perspective of developments of the subsequent 25 years.

The Acheulean is a grand archaeological tradition sitting square in the middle of the early archaeological record. It lasts for more than a million years. It has the first tools with clear design form (Figure 6.1). Its sameness and its variety are both well known to us. It preserves sites in many contexts. Its archaeology has enjoyed a renaissance in recent years, but where has it come from and where is it headed?

In this paper I shall try to examine the roots and course of this renaissance, noting that our hard-earned knowledge is controversial both in frame and in detail. Many scholars have played a part in the development of Acheulean archaeology since World War II. In Africa Desmond Clark laid especially significant foundations, through the vast investigations at Kalambo Falls, and explorations on many other sites, and his own very practical insights into the nature of hunter

Figure 6.1 A large well-finished hand-axe from Kilombe demonstrates that fine finish and symmetry are sometimes found in bifaces a million years old. The general symmetry is offset by a slight asymmetry of the tip - with one convex and one straight edge - illustrating a 'symmetry breaking' which occurs in many bifaces. It may relate to preferential use of one side of the tool. Scale in cm.

gatherer life and ecology. Mary Leakey also achieved a huge amount at Olduvai. In Europe we should acknowledge Bordes, Roe and Tuffreau, who set out new typologies and excavation techniques. It was Glynn Isaac, however, who hauled the Acheulean around the corner into the New Archaeology, and beyond. He was the modernizer. It would be tempting to claim that Isaac's work on the Acheulean has formed the foundations of the many developments of recent years. For Africanists there would be much truth in this idea, but in Europe and Asia often that has been less the case - many have made pronouncements and advanced hypotheses that derive from more local schools of Archaeology. That is not surprising, and indeed Isaac's chief fieldwork contributions to the Acheulean came from just six years work on just one site, Olorgesailie (Isaac 1968, 1977). Many of the insights gained by both Clark and Isaac, however, came from their knowledge of the deep record in Africa, and of the Acheulean in its totality - environment, geology, fauna, and the whole repertoire of its artifacts. This immersion was coupled with a robust knowledge of statistics and their straightforward application. Others have approached the Acheulean on a far narrower front, often concentrating just on its bifaces - perhaps surprisingly that has been more rather than less a tendency of recent years.

The different currents of research show that archaeology does not progress monolithically, but in a multi-stranded way. The Acheulean is unusual in having a sort of double or even treble function. It is local and regional archaeology; it is a tool for exploring human evolution; and it represents a stage in popular view, with the hand-axe as its iconic symbol. For some students and many of their (non-Paleolithic) teachers, the Acheulean is a few mantras, largely uncritically repeated and accepted.

The 1950s in Africa

The New Archaeology characterizes the century before 1960 as "the long sleep" (Renfrew 1982:6). That is unkind to African prehistorians in particular, as they confronted the post-war period with a new energy and organization. They made many advances in the period after WWII, centered on the meetings allowed by the Pan-African Conferences. It was during this period in the 1950s that Desmond Clark formulated his main ideas, set out in two general books (Clark 1959, 1970), but tested and proofed in his earlier excavations, and published in detail (Clark 1969, 2001).

These were focused on Kalambo Falls in southern Africa, where the actual finds led the way in forcing reinterpretations. Clark's voluminous writings had a huge effect - hand-axes were not just typological items - they came from "living surfaces" and served as part of a "toolkit" performing tasks within ecological constraints. Clark also aimed to date this past, but this was rather more difficult. Radiocarbon was the main tool, and it did not reach back far enough. It was left to Glynn Isaac to stretch the Acheulean back in time, conceptually as well as chronologically. He argued clearly for the 1.4 million years time depth, first established at Peninj (Brock and Isaac 1974; Isaac 1967a, 1969; Isaac and Curtis 1974).

The Isaac Contributions

When Glynn Isaac was working at Olorgesailie in the 1960s he was in a time of crucial changes. The Leakeys had just made their shattering discoveries at Olduvai, finding hominids and artifacts in close association (Leakey 1959, 1961). Potassium-Argon dating was reshaping the timescale; and the ideas of the New Archaeology were just beginning to appear (Evernden and Curtis 1965; Binford and Binford 1966; Isaac 1969). It is interesting to note that this was a very short phase or "window" - essentially, as the new ideas came in, the Acheulean itself was being eclipsed, and the Oldowan and early hominids were coming to dominate the scene of enquiry. Isaac himself moved back to the Oldowan and it became the best known theme of his research - note that Blumenschine's (1991) thorough and perceptive review of Isaac's ideas mentions the Acheulean only once or twice. For roughly the last quarter of the 20th century the Acheulean was headed for a tertiary role, in the shadows compared with both human beginnings and later human revolutions. Nevertheless, Isaac did make major contributions in this phase, and the current reawakening of ideas makes them of particular interest.

Two things were clear from the start in the 1960s - Olorgesailie was already well known as a large hand-axe site; and Merrick Posnansky had already described there a contemporary non-hand-axe site ascribed to the "Hope Fountain" (Posnansky 1959). This was a flake industry, and the implication of the name was that it belonged to a different "culture", much as Francois Bordes saw contemporary variants of the Mousterian, and as the Acheulean and "Clactonian" hung on in western Europe (Bordes 1972; White 2000). A further recent factor was the typological work of Maxine Kleindienst (1961, 1962), which had set the major site complexes of eastern Africa into a common framework, just as Bordes was doing for the European Lower/Middle Paleolithic (and before Mary Leakey published her own schemes in the 1960s; Leakey 1967). Isaac built from these rich but puzzling foundations.

What did Isaac achieve for the Acheulean? First, he promoted a landscape approach that tallied with the New Archaeology, but that was realized in practice because of the variety of occurrences demonstrably present on the ground. In Europe the Mousterian debate was taking shape (Binford and Binford 1966; Binford 1973; Bordes and Bordes 1971; Bordes 1972; Mellars 1969, 1996). The question there was, did variation stem from the co-presence of Mousterian tribes, as Bordes believed, or were explanations more functional as asserted by the Binfords, or even chronological as advocated by Paul Mellars? At Olorgesailie there were the large hand-axe accumulations such as DE89, but also localities dominated by scrapers, or by flakes or heavy-duty material, and there were also hints of an elephant butchery site. The last was not clear cut, but elsewhere as at Isimila there were indications of Acheulean hippo butchery, a pattern of involvement with large animal carcasses later confirmed on other sites (including Olorgesailie itself; Potts et al. 1999).

Then there were the taphonomic experiments - the sort that you could only do if you were living on a site as a warden, and had the time to garner materials as well as ideas. The problem of explaining accumulations was already in Isaac's mind (cf. Isaac 1967). One wonders why? - but he had to field the questions of tourists day after day, and above all they wish to know how so many hand-axes have ended up in one place. The link that Isaac construed between hand-axe concentrations and sandy channels was echoed later by observations at East Turkana. It took him to the work of Leopold, who had studied the dynamics of artifact movements in channels ("kinematic waves"), and it also led him to carry out the actualistic taphonomic experiments, which make Isaac one of the pioneers of this discipline alongside Bob Brain (e.g., Brain 1981; Leopold et al. 1966).

Isaac was equally interested in the bones - how did accumulations survive? His experience was that bones tended to disperse, raided by numerous carnivores. In his experiments, bones had to be put under metal meshes to slow down the movements (Isaac 1967b). As noted before (Gowlett 1989) there was a methodological irony here; Lewis Binford, equally interested in bones and their taphonomy, was interested in the way accumulations formed, and how through "middle range" theory you would tell whether the main agents were carnivores or hominids (e.g., Binford 1981). Isaac was at that time more interested in the dispersal, which also would clearly affect interpretations. This offset, the facing in different directions, seemed to play a significant part in shaping the Binford-Isaac debates.

Also here we find the derogatory label of "speculation" (Binford 1981). Binford believed in formalized hypothesis testing through rigorous application of a specific methodology (even if that sometimes turned out to be misapplied). Isaac was more concerned with making general models about the past, although he never forgot about the testing. This freer mode of thinking was part of the Cambridge way: for David Clarke "imagination" was one of the most important requirements in an archaeologist. The antithesis of approach is still with us - those who are very strictly tied to a sense of scientific methodology often feel that the "thinkers" are out of bounds.

Artifact variation remains at the heart of the Olorgesailie research, although Isaac like many others had mixed feelings. The Acheulean could mop up an enormous amount of detailed research, and give very little back in return (but "the puzzling variety... may be more than a nuisance, it may be an important source of information"; Isaac 1977:206). He was interested in issues of cultural stability and diffusion that have a modern ring (as worked on again by Steele 1994, or Lycett 2010). Isaac's Models paper of 1972 set out ideas of networks that are still relevant, as these have become a focal point of social brain studies. The point of interest now is how far

local "cultural speciation" takes place - and then how far this plays a role in social competition models as seen by Alexander (1989). We see the similar questions arising from recent studies of chimpanzee culture (Whiten et al. 1999, 2005).

Isaac appreciated - perhaps from Charles McBurney, one of his teachers at Cambridge - the difference between modern "classes" and ancient "types" (Gowlett 1999). The duty of the archaeologist is to seek out the latter - as conceived by their ancient makers. We could not simply pronounce them into existence by making arbitrary divisions of our material. Confusion on this point still pervades literature and opinions. Like McBurney, Isaac believed that types had to be demonstrated. If they might be part of a continuum, you had to try and test whether there are separate modes that confirm the separate existence of distinct groups (such as "hand-axes" and "cleavers" - in the particular case Isaac wondered whether they could be linked through a graded series of "chisel bit" bifaces; Isaac 1977:120, 123).

So Isaac eliminated all the subtypes of bifaces that permeate Kleindienst, Bordes (and to a lesser extent Roe's "pointed" and "ovate" traditions; Roe 1968). But eliminating subtypes, he increased measurements - these were designed to investigate flake scars, degree of finish, and symmetry with a higher level of objectivity (e.g., 1977:233). The emphasis on measurement for evaluating the validity of typology was something new and important. Isaac showed us the "false concreteness" of many arbitrary divisions which archaeologists make, but unfortunately these lessons have not always been learnt. He also came close to working out allometry (see below), making a graph of hand-axe breadth and length, partly from figures given by Gilead (1970). These show that as bifaces (in general) became smaller through the Pleistocene, so they also became relatively broader (Isaac 1977:139).

Strikingly, Isaac found time to tackle each major category of artifacts - and all this within a Ph.D. Beyond the landscapes and the bifaces he wrote thesis (and monograph) chapters on the heavy-duty artifacts; the flake scrapers; and the cores and flakes. For each he devised new sets of measurements, some of which were taken up and developed by others (for example Toth (1985) in the case of flakes - Toth's scheme has been adopted by many others).

Isaac was one of the first to look at the Acheulean variation with new analytical techniques - the Principal Components analysis (PCA) with Corruccini was a pioneering example of application of the technique, and it gave a graphic presentation in two dimensions of evidence reduced from as many as 12 variables (Isaac 1977:215). That was indeed ground breaking, especially as in the 21st century many accounts involve only the Acheulean bifaces. Although he pioneered new techniques in studying the bifaces, and they were clearly a great part of the Olorgesailie enterprise, Isaac did not in publication show a great interest in their place in cognition or origins of creativity. His approach is more matter of fact, and his interest was primarily in transmission of craft traditions (influenced by Binford and Deetz 1967, and Owen 1963 amongst others).

Where did Isaac's ideas come from? Much came from his own cast of mind, and ability to make new links. We can gain some clues from his bibliographies, aided by the time gap between his thesis (1968) and the final publication. The Olorgesailie monograph of 1977 has about 280 references - more of an achievement then than now, when electronic sources are readily available. They show a strong foundation of 1950s fieldwork; a select number of older references in each discipline; and an engagement with the new archaeology - but not quite what one might

expect. Binford figures prominently, and so do the hunters and gatherers (Isaac was at the Man the hunter symposium). S. R. Mitchell and especially R. A. Gould represent the ethnography. So also feature his Cambridge contemporaries such as Derek Roe and Wilfred Shawcross, but not his great friend David Clarke, or the other American new archaeologists. This suggests that the New Archaeology was less of a cohesive movement or club than it might look with hindsight; also that Isaac was highly focused in dipping into other disciplines in a select way gaining a maximal return for his effort. He added around 50 references in reworking the dissertation to produce the monograph - these give more insights, but many of them are updates of his own prolific writing or recent fieldwork. All together, the reading underlying Isaac's Acheulean was practical as much as theoretical, and often drew strongly on select sources.

The Acheulean Now - 25 Years More
Although Glynn Isaac largely finished with the Acheulean with his publication of 1977, he remained well up with its issues (and of course intended to return to its research at Peninj). But the truth is that he did die in 1985, and that he would be very disappointed with us if nothing had been achieved in the next following 25 years. He might well say, "What have you folks been up to in all this time?" Anyone entrusted by Isaac to run an excavation for a day or two will know that the scrutiny on return was good-humored and supportive but not uncritical.

So what have we done? We can approach the Acheulean under headings - its "big geometry" of space and time; its use of landscapes; the bifaces; and the rest - the other artifacts and their contexts. They run together increasingly as different studies interlock (cf. the "multifactorial" perspective of Machin 2009).

The "Big Geometry"

This first aspect highlights how far we have become a far more extensive and well integrated research community across continental boundaries. In Europe there are striking contributions from Spain, Britain, France, and Italy; and recently these have driven Acheulean dates back towards the one million year mark (see papers in Gamble and Porr 2005; Goren-Inbar and Sharon 2006a; Soressi and Dibble 2003, and below). The Middle East has been another strong focus (Bar-Yosef 2006; Goren-Inbar and Sharon 2006b), and the picture for the rest of Asia has (finally) been shaping up (Pappu and Akhilesh 2006; Paddaya et al. 2006, and references below). The hand-axes are universally acknowledged as the definitive marker for the Acheulean, although there is not much more than intuitive agreement as to "what" they map out and represent. For most scholars they seem to represent at least loosely a single long-lasting cultural tradition, and to be sufficient to bind it together into a whole (Lycett and Gowlett 2008), despite the puzzles which they still present (Bar-Yosef 2006; Wynn 1995) and the rare importance of material that goes beyond stone (Thieme 2005). Alternatively, there could be a set of ideas which are lost and reinvented many times over. But if we find something like hand-axes in the Andes 3,000 years ago, or in Australia 500 years ago, they would not be regarded as Acheulean. Our appreciation of the Acheulean thus shows some element of continuity. This domain is now broadly accepted to begin 1.7 million years ago in Africa, with key evidence coming from Kokiselei at West Turkana (Roche et al. 2003; Lepre et al. 2011), and other East African sites. As Semaw et al. (2009) point out, the chronology of the beginnings is still seen differently by different authors. Elsewhere in Africa dates are not much later, with the hand-axe idea seeming

to appear within the Developed Oldowan, and to be widespread by 1.4 million years. Peninj, Konso-Gardula, Chesowanja and Wonderwerk Cave show slightly different aspects of this phenomenon. They suggest that the early Acheulean may be truly no more than the Developed Oldowan plus a useful new tool - which gradually becomes applied in more and more contexts, but never in all of them.

It is widely accepted that the Acheulean spreads out into Europe, the Middle East and most of Asia at somewhat later dates, chiefly on the grounds that the ex-Africa dates we have are slightly younger, rather than from any theoretical argument. The dates have been getting older, though without very solid claims of exceeding one million years. Ubeidiya remains in the range 1.0-1.4 Ma (Bar-Yosef and Goren-Inbar 1993; Isampur possibly >1.0 (Paddaya et al. 2006); in Spain dates suddenly now approach one million (Scott and Gibert 2009); Happisburgh takes this timescale into northern Europe (Parfitt et al. 2010). The evidence from Spain suggests both that Acheulean bifaces appeared there far earlier than generally expected, and also that they may have assumed distinctive local forms fairly early on - a possibility perhaps supported by bifaces from Arago and more early dates from Italy (Barsky and de Lumley 2010; Lefèvre et al. 2010).

It is not yet conclusively shown that there was a million year old Acheulean extending across to the Far East, but Isampur and Bose begin to stake out a claim for it (Paddaya et al. 2006; Petraglia et al. 2005; Hou et al. 2000). The details are controversial - and the status of the Chinese and Korean finds particularly so (Norton et al. 2006; Petraglia and Shipton 2008; Lycett and Gowlett 2008; Lycett and Bae 2010). In these domains, beyond the traditional Movius line, there are bifaces, and picks, but in debate the question has been, are they "real" Acheulean? The question is difficult to answer because there is no real formal Acheulean definition, beyond the presence of bifacial artifacts with a long axis and a certain degree of bilateral symmetry (itself debated; McNabb et al. 2004; Underhill 2007). Some of the Chinese specimens, at least, fall within the range of the Acheulean as expressed further west (Petraglia and Shipton 2008). At the same time such a huge site as Zhoukoudian plainly does not have bifaces, and they remain very rare across China as a whole.

Here it is worth returning to some of Isaac's points. After all, he stressed assemblage variability - hand-axes need not be a major component of assemblages, as the work of Potts and colleagues has confirmed in the area of Olorgesailie (Potts et al. 1999). Several other major complexes show the extent of "internal" variability within a site complex - Ubeidiya, Notarchirico and the Somme, to name but three (Bar-Yosef and Goren-Inbar 1993; Piperno et al. 1998; Piperno and Tagliacozzo 2001; Tuffreau et al. 1997). In some whole regions there are scarcely any bifaces - not just in the east but in the west, as in the Britanny Colombanian (Monnier 1996). The presence and number or size and shape of the bifaces is not fixed by an archaeological diktat, but is to be mapped out carefully through local and comparative studies. Did the Acheulean makers then get to all areas? The argument is as pertinent in parts of Europe as in Asia, and it is compounded by microlithic puzzles (Burdukiewicz 2009). In eastern Europe, Israel and parts of Africa, precociously early small tool industries have appeared that must be factored into our explanations (Glaesslein 2009; Zaidner et al. 2003). In one of the many Acheulean paradoxes hand-axes themselves are seen to be both common and rare.

Dating of the Acheulean longue durée has

steadily improved. A preoccupation for Isaac, and for Mary Leakey, was whether the Acheulean changed during its central million years. Isaac (1977) espoused the concept of very gradual change surrounded by considerable "noise" (he attributed this partly to a random walk or stochastic variation in style especially in bifaces). Mary Leakey thought that there was no change through the Olduvai sequence. Gradually it is becoming clear that Isaac's explanation fits well - early industries have a certain hallmark, and so do late ones. In between, from about 1.2 million to about 400,000 years ago in Africa, there seems to be very little difference between industries, beyond somewhat random factors of raw materials and perhaps local style. After that date there is a consensus that new features appear, such as new modes of blank striking which seem sometimes to incorporate specific Levallois ideas, also found in flake production. The Victoria West of southern Africa is one of the more puzzling cases, because it may have deeper chronological roots (Chazan 2009; Lycett 2009; Porat et al. 2010), but in one way or other a change in pattern is evident also at Kapthurin, Qesem, on the Somme and elsewhere by around 300,000 years ago (Barkai et al. 2003, 2005; Johnson and McBrearty 2010; McBrearty 1999; Sharon and Beaumont 2006; Tuffreau et al. 1997; White and Ashton 2003).

Only after 200,000 it seems do African bifaces start to reduce in size - as seen along the Nile, in the Western Desert (e.g., Schildt and Wendorf 1977). In Europe and Asia patterns may be slightly different from zone to zone, giving support to Wynn and Tierson's (1990) findings of regional variation, and the "phylogenetic" studies recently made by Lycett (2010).

The end of the Acheulean raises its own set of questions. Authors provide many dates, from 300,000 to less than 100,000 years ago. When does the Acheulean "really" finish? It seems to trouble us that this is purely a matter of definition. If, however, we tie our definitions to a single artifact class - as we do - then the result is bound to be a particular kind of problem. Bifaces seem to reflect primarily a culturally-maintained set of functional solutions to everyday tasks which recur, and as such they are not well suited to our needs of exclusive classification, which would require some obvious complete artifact regime change between the Lower and Middle Paleolithic.

Landscape

Through the last generation, Acheulean landscapes have begun to emerge in a series of scales, from the grandest on continental scale, down to the local - paralleling perhaps Clive Gamble's ideas of regional and local hominid networks (Gamble 1996). At the largest there is the issue, we have seen, of "Acheulean or not?" On the smaller scale researchers have doggedly exploited variability in pursuit of understanding the mapping of artifacts onto landscapes, an extracting more general lessons about behavior.

Olorgesailie demonstrates one of the great Acheulean landscapes. Potts and colleagues have built much beyond the original foundations. Their work stresses the need for delimiting precise stratigraphic intervals, and using excavation rather than just surface scatters to get an accurate idea of the contents of a landscape (Potts et al. 1999). They show clearly how bifaces are just one aspect of the record, rare indeed in most of the Olorgesailie basin. The unusual concentrations of bifaces "appear to be correlated spatially with two factors: the proximity of lava outcrops and, to a lesser extent, the presence of channels". Test excavations and detailed surface survey through the basin indicate that the Site Museum area is highly unusual in having large aggregates of bifaces

(Potts et al. 1999:781). Similar messages come from the archaeology of the Somme, where Tuffreau and colleagues have carried out equally detailed work. Again, the percentages of bifaces vary markedly; they are concentrated in particular areas; and there is a similar sense of repeatedness and localization in the activities. To these landscapes we can add more - in Asia the patterns of bifaces in the Hunsgi-Baichbal Basin (Isampur) and elsewhere; in Arabia too artifact distributions extend along channels or are found by rocky outcrops (Chauhan 2009; Noll and Petraglia 2003; Petraglia et al. 2005; Petraglia 2003). There are different lessons but common themes that interrelate to a surprising degree.

Apart from large biface concentrations, there is another pole of variation in the usually sparse scatters of artifacts associated with individual large animal carcasses. Often small tools, these also include limited numbers of bifaces, sometimes specifically associated with butchery. Such evidence occurs as far apart as Olorgesailie I5, Aridos and Boxgrove (Potts et al. 1999; Villa 1990, 1996; Pope and Roberts 2005).

The Bifaces

These patterns in the production and discard of bifaces allow great opportunities for intercomparisons - that have been realized so far to a limited extent (notwithstanding Chauhan 2010; Crompton and Gowlett 1993; Gamble and Marshall 2001; Lycett and Gowlett 2008; Noll and Petraglia 2003; Wynn and Tierson 1990). Acheulean archaeology has far deeper historical roots than recent discoveries of chimpanzee cultures. Perhaps for that very reason we have been slower in making the big intercomparisons, since our methodologies themselves have diverged culturally to a far greater extent, making it difficult to achieve a single cohesive approach. A huge amount has been achieved in biface studies - but it probably remains fair to say that archaeology has more to do to match up form and function. A very obvious approach, finally becoming practically realizable is the 3D digitization of artifacts. It will not achieve full value without a new generation of better analyses. Shirley Crompton gave a pointer in her early 3D analyses of MSA points (Crompton 1997), which effectively harness techniques of solid modeling.

At the moment the commonality of traditional measurements often does not go far beyond Length, Breadth and Thickness, but these can be useful. We are fortunate to have the bifaces preserved, with their three-dimensional solid geometry, when so much else does not survive. Strangely, this hard reality seems not quite good enough, and some analytical techniques work from elaborate transformations (Costa 2010; Crompton and Gowlett 1993; Iovita 2010; Saragusti et al. 1998; Nowell et al. 2003). These, including a new morphometrics, can be justified in the sense that the Acheulean is so intractable that we need to take statistical hammers to crack nuts. But part of our luck is that the bifaces are tangible real world objects made for every day purposes, and our techniques must always return to this point.

We can look briefly at some of the main ideas. Symmetry is a well known feature of Acheulean bifaces that was of incidental interest to Isaac, who nevertheless developed measurements that could express symmetry/asymmetry. The ideas have been tackled by numerous authors since Toth's (1990) paper. McNabb and colleagues have recently developed an "eyeball" technique, and Hardaker his "flip test" (McNabb et al. 2004; Hardaker and Dunn 2005). The fascination with symmetry has to be backed up with theory - the symmetry is there, but why should it be important? Arguments have been based on sexual selection, aesthetics (evolution

of the visual cortex), function, and neutral drift (Hodgson 2009; Kohn and Mithen 1999; Lycett 2008, 2009; Machin et al. 2007; Machin 2009; McNabb et al. 2004). Kohn and Mithen argued for sexual selection, provoking interest, but also some later sceptical responses (e.g. Nowell and Chang 2009; Machin 2008). It seems fair to say that the arguments have not really got to grips with "what" in the biface gives the sexual spice. For example large bifaces are not always the most symmetrical - which would count for more, the size or the symmetry? What would be the significance of "small and beautiful"? Wynn (2002) has noted the importance of symmetry-breaking, which may represent a compromise between the simplest design solution and functional needs for a slight skew (Figure 6.1). A more fundamental problem, also prompting us to draw on theoretical reserves, is that any oval cobble selected to make a hand-axe will already possess a great deal of symmetry; our schemes would ideally distinguish between "pre-symmetry" and deliberately imposed symmetry.

Symmetry is just one aspect of the rules seemingly embodied in Acheulean bifaces. Pope et al. (2006) express a now dominant view that there is a deliberateness of imposition of shape in the Acheulean beyond the mere intuition of modern human observers. Is there a mental template? Numbers of writers have mused on this point. We owe it to Deetz (1967) in the first place. Chase (2008) sees it as close to a tautology - an inevitable consequence of the idea of an artifact. The problem with bifaces is their very variation - surely, the makers were not intending specifically to make all the forms which they eventually produced? Surely that implies the lack of a fixed template? It seems more likely that Acheulean toolmakers operated through instruction sets which were adjustable according to immediate practical needs. The rule systems can be isolated more objectively through studying large assemblages, a development of an idea used by Isaac in comparing Olorgesailie assemblages. How many features were important to the makers? In bifaces some measured variables have very high correlations, others low ones. It would seem that the makers were more focused on some features than others. In this they may have been managing cognitive load, with a system of priorities.

Allometry – size-related change – has proved another somewhat unexpected major area of development. Allometry studies show progress with two courses of development. McPherron in the tradition of Frison believes that retrimming accounts for much variability in Acheulean bifaces (McPherron 2000, 2006). In this view the retrimming will lead to a size-shape shift, as trimming focuses on the tip rather than the butt of the specimen. Crompton and Gowlett set out another approach which sees the allometry as largely functional (Crompton and Gowlett 1993; Gowlett et al. 2001). Their approach implies that the size range we see is intentional (i.e. each size of artifact made is appropriate for some task). Some support for this view is given by the discovery that modern artifact classes which have a broad size range show a similar allometry – screw-drivers offering a good example (Gowlett 2009). But the two approaches are far from incompatible. Logically, as has been suggested by Dibble (1988) for the French Mousterian, retrimming will take place more when raw material is scarce or difficult to replace. Most of the big African Acheulean sites demonstrate a superabundance of material, but in European contexts flint was often scarce and precious, and

retrimming is quite evident, as at Boxgrove (Pope et al. 2006; Pope and Roberts 2005).

The Rest

It would be good to find that research on all the other artifact categories deserves equal mention. Isaac knew the importance of other artifacts, and so do the other fieldworkers who find them. In fact far less has been done for most of them. Canadian archaeologists have achieved most for the heavy-duty artifacts (Kleindienst 1962; Willoughby 1987; Cormack 1997). Sharon (2009) has directed a new attention towards the Acheulean giant cores. The best this paper can do is lay down a marker. At least we can pose a question – why for a million years did the tool makers invest their formal rules so much in the hand-axes and so little in the cores, flakes etc.? Were they challenged cognitively over managing so many rule systems in their heads side by side? That is possible, but chimpanzees manage artifact sets. The answer might lie in wood, where a world of forms may have vanished. The changes from 400,000 are widely thought to have something to do with the development of hafting (Barham 2010). This in turn may have been linked with new cognitive and social capabilities.

Conclusions

Acheulean studies have not just gone forward a great deal in the last 25 years – with rise of a new research community the empire of the Acheulean has struck back. Its 1.5 million years are again central to our understanding of thinking about human evolution, as important as any earlier and later revolutions. Its studies have also, naturally, become more diffuse as they have multiplied. They are, however, recognizably still in the same basic picture. There is much more detail in the record, and in the approaches. Together these make the Acheulean more like other archaeologies. Glynn

Isaac undoubtedly contributed enormously to the foundations of modern study. His Acheulean research, we can now see, was actually the dry run for his remarkable contributions to the study of the earliest technology and earliest human adaptation. If we see the Acheulean in some sense as just the Oldowan plus hand-axes, then Isaac's journey easily makes sense, and the freshness and originality of his early contributions are the more apparent. The "plus" of the hand-axes is an enormous benefit to archaeologists, when they become involved with detailed cultural variation, cognitive evolution and the networks of sociality. Great strides have been made and a lot more is to come in these areas.

Acknowledgements

I owe a debt of gratitude to Glynn Isaac, who made possible my first experiences in Africa, responding to a kind intervention of Pat Carter. I salute colleagues from those days including Jack Harris and Rob Foley, Clive Gamble and Marcel Otte. I thank Clive Gamble for helping with a reference. My recent research has been supported by the British Academy, especially in the Lucy to Language project. I am grateful to Jeanne Sept and David Pilbeam for their work in commemorating Glynn's achievements.

References

Alexander, R. D.

1989 Evolution of the human psyche. In *The Human Revolution: Behavioural and Biological Perspectives on the Origins of Modern Humans*, edited by P. Mellars and C. Stringer, pp. 455-513. Edinburgh University Press, Edinburgh.

Barham, L.

2010 A technological fix for 'Dunbar's Dilemma'? In *Social Brain and Distributed Mind*, edited by R. Dunbar, C. Gamble, and J. Gowlett, pp. 367-389. British Academy, London.

Barkai, R., A. Gopher, and R. Shimelmit

2005 Middle Pleistocene blade production in the Levant: An Amudian assemblage from Qesem Cave, Israel. *Eurasian Prehistory* 3:39-74.

Barkai, R., A. Gopher, S. E. Lauritzen, and A. Frumkin

2003 Uranium series dates from Qesem Cave, Israel, and the end of the Lower Paleolithic. *Nature* 423:977-979.

Barsky, D., and H. de Lumley

2010 Early European Mode 2 and the stone industry from the Caune de l'Arago's archeostratigraphic levels "P". *Quaternary International* 223-224:71-86.

Bar-Yosef, O.

2006 The known and the unknown about the Acheulian. In *Axe Age: Acheulian Toolmaking from Quarry to Discard*, edited by N. Goren-Inbar and G. Sharon, pp. 479-494. Equinox, London.

Bar-Yosef, O., and N. Goren-Inbar

1993 The lithic assemblages of 'Ubeidiya: A Lower Palaeolithic site in the Jordan Valley. *Qedem* 34.

Binford, L. R.

1973 Interassemblage variability: The Mousterian and the 'functional' argument. In *The Explanation of Culture Change: Models in Prehistory*, edited by C. Renfrew, pp. 227-254. Duckworth, London.

1981 *Bones: Ancient Men and Modern Myths*. Academic Press, New York.

Binford, L. R., and S. R. Binford

1966 A preliminary analysis of functional variability in the Mousterian of Levallois facies. *American Anthropologist* 68:236-295.

Blumenschine, R. J.

1991 Breakfast at Olorgesailie: The natural history approach to Early Stone Age Archaeology. *Journal of Human Evolution* 21:307-327.

Bordes, F. H.

1972 *A Tale of Two Caves*. Harper and Row, New York.

Bordes, F. H., and D. Sonneville-Bordes

1971 What do Mousterian types represent? The significance of variability in Palaeolithic assemblages. *World Archaeology* 2:61-73.

Brain, C. K.

1981 *The Hunters or the Hunted? An Introduction to African Cave Taphonomy*. University of Chicago Press, Chicago.

Brock, A., and G. Ll. Isaac

1974 Palaeomagnetic stratigraphy and chronology of hominid-bearing sediments east of Lake Rudolf, Kenya. *Nature* 247:344-348.

Burdukiewicz, J. M.

2009 Lower Palaeolithic transitions in the northern latitudes of Eurasia. In *Sourcebook of Paleolithic Transitions*, edited by M. Camps and P. Chauhan, pp. 195-209. Springer, New York.

Chase, P. G.

2008 Form, function and mental templates in Paleolithic lithic analysis. Paper presented at the symposium *From the Pecos to the Paleolithic: Papers in Honor of Arthur J. Jelinek*, Society for American Anthropology Meetings, Vancouver, BC, 28 March 2008.

Chauhan, P. R.

2009 The South Asian Paleolithic record and its potential for transitions studies. In *Sourcebook of Paleolithic Transitions*, edited by M. Camps and P. Chauhan, pp. 121-139. Springer, New York.

2010 Metrical variability between South Asian handaxe assemblages: Preliminary observations. In *New Perspectives on Old Stones: Analytical Approaches to Paleolithic Technologies*, edited by S. J. Lycett and P. R. Chauhan, pp. 119-166. Springer, New York.

Chazan, M.

2009 Assessing the Lower to Middle Paleolithic transition. In *Sourcebook of Paleolithic Transitions*, edited by M. Camps and P. Chauhan, pp. 237-243. Springer, New York.

Clark, J. D.

1959 *The Prehistory of Southern Africa.* Harmondsworth, Penguin.

1969 *Kalambo Falls Prehistoric Site Vol. I. The Geology, Palaeontology and Detailed Stratigraphy of the Excavations.* Cambridge, Cambridge University Press.

1970 *The Prehistory of Africa.* London, Thames and Hudson.

Clark, J. D. (Editor)

2001 *Kalambo Falls, Vol. 3.* Cambridge University Press, Cambridge.

Cormack, J. L.

1997 Are the Sangoan and the Acheulean 'Industrial Complexes' Distinct? In *Archaeological Sciences 1995: Proceedings of a Conference on the Application of Scientific Techniques to the Study of Archaeology*, edited by A. Sinclair, E. Slater, and J. Gowlett, pp. 395-399. Oxbow Monograph 645, Oxford.

Costa, A. G.

2010 A geometric morphometric assessment of plan shape in bone and stone Acheulean bifaces from the Middle Pleistocene site of Castel di Guido, Latium, Italy. In *New perspectives on Old Stones: Analytical Approaches to Paleolithic Technologies*, edited by S. J. Lycett and P. R. Chauhan, pp. 23-41. Springer, New York.

Crompton, S. Y.

1997 Technology and morphology: Does one follow the other? In *Archaeological Sciences 1995: Proceedings of a Conference on the Application of Scientific Techniques to the Study of Archaeology*, edited by A. Sinclair, E. Slater, and J. Gowlett, pp. 413-419. Oxbow Monograph 645, Oxford.

Crompton, R. H., and J. A. J. Gowlett

1993 Allometry and multidimensional form in Acheulean bifaces from Kilombe, Kenya. *Journal of Human Evolution* 25:175-199.

Deetz, J.

1967 *Invitation to Archaeology*. Natural History Press, Garden City, New York.

Dibble, H. L.

1988 Typological aspects of reduction and intensity of utilization of lithic resources in the French Mousterian. In *The Upper Pleistocene Prehistory of Western Eurasia*, edited by H. L. Dibble and A. Montet-White, pp. 181-197. University of Philadelphia Press, Philadelphia.

Evernden, J. F., and G. H. Curtis

1965 Potassium-argon dating of late Cenozoic rocks in East Africa and Italy. *Current Anthropology* 6:343-385.

Gamble, C.

1996 Making tracks: Hominid networks and the evolution of the social landscape. In *The Archaeology of Human Ancestry*, edited by J. Steele and S. Shennan, pp. 253-277. Routledge, London.

Gamble, C. S., and G. D. Marshall

2001 The shape of handaxes, the structure of the Acheulian World. In *A Very Remote Period Indeed: Papers on the Palaeolithic Presented to Derek Roe*, edited by S. Milliken and J. Cook, pp. 19-27. Oxbow, Oxford.

Gamble, C. S., and M. Porr (Editors)

2005 *The Hominid Individual in Context: Archaeological Investigations of Lower and Middle Palaeolithic Landscapes, Locales and Artefacts*. Routledge, London.

Gilead, D.

1970 Handaxe industries in Israel and the Near East. *World Archaeology* 2,1:1-11.

Glaesslein, I.

2009 Patterns of choice and context in pre-neanderthal Europe. (Proc. UISPP Lisbon) *L'Anthropologie* 113:198-210.

Goren-Inbar, N., and G. Sharon

2006a *Axe Age: Acheulian Toolmaking from Quarry to Discard*. Equinox, London.

2006b Invisible handaxes and visible Acheulian biface technology at Gesher Benot Ya'aqov, Israel. In *Axe Age: Acheulian Toolmaking from Quarry to Discard*, edited by N. Goren-Inbar and G. Sharon, pp. 111-154. Equinox, London.

Gowlett, J. A. J.

1989 Introduction. *The Archaeology of Human Origins: Papers by Glynn Isaac*, edited by Barbara Isaac, pp. 1-10. Cambridge University Press, Cambridge.

1999 The work and influence of Charles McBurney. In *The Great Archaeologists*, edited by T. Murray, pp. 713-726. ABC-Clio, Oxford. Vol. II, Encyclopaedia of Archaeology.

2009 Artefacts of apes, humans and others: Towards comparative assessment. *Journal of Human Evolution* 57:401-410.

Gowlett, J. A. J., R. H. Crompton, and Li Yu
2001 Allometric comparisons between Acheulean and Sangoan large cutting tools at Kalambo Falls. In *Kalambo Falls Prehistoric Site, Vol 3: The Earlier Cultures: Middle and Earlier Stone Age*, edited by J. D. Clark, pp 612-619. Cambridge University Press, Cambridge.

Hardaker, T., and S. Dunn
2005 The Flip Test - A new statistical measure for quantifying symmetry in stone tools. *Antiquity Gallery, Antiquity* 306 (online).

Hodgson, D.
2009 Evolution of the visual cortex and the emergence of symmetry in the Acheulean techno-complex. *C.R. Palevol* 8:93-97.

Hou, Y., R. Potts, B. Yuan, Z. Guo, A. Deino, W. Wei, J. Clark, G. Xie, and W. Huang
2000 Mid-Pleistocene Acheulean-like stone technology f the Bose Basin, South China. *Science* 287:1622-1626.

Isaac, G. Ll.
1967a The stratigraphy of the Peninj Group - Early Middle Pleistocene formations west of Lake Natron, Tanzania. In *Background to Evolution in Africa*, edited by W. W. Bishop, and J. D. Clark, pp. 229-258. University of Chicago Press, Chicago.

1967b Towards the interpretation of occupation debris: Some experiments and observations. *Kroeber Anthropological Society Papers* 37:31-57; reprinted in *The Archaeology of Human Origins: Papers by Glynn Isaac*, edited by B. Isaac, pp. 191-205. Cambridge University Press, Cambridge.

1968 The Acheulian site complex at Olorgesailie: A contribution to the interpretation of Middle Pleistocene culture in East Africa. Unpublished Ph.D. Dissertation, University of Cambridge.

1969 Studies of early culture in East Africa. *World Archaeology* 1:1-28.

1972 Early phases of human behaviour: Models in Lower Palaeolithic archaeology. In *Models in Archaeology*, edited by D. L. Clarke, pp. 167-199. Methuen, London,

1977 *Olorgesailie: Archaeological Studies of a Middle Pleistocene Lake Basin*. University of Chicago Press, Chicago.

Isaac, G. Ll., and G. H. Curtis.
1974 Age of Early Acheulian industries from the Peninj Group, Tanzania. *Nature* 249:624-627.

Iovita, R.
2010 Comparing stone tool resharpening trajectories with the aid of elliptical Fourier analysis. In *New Perspectives on Old Stones: Analytical Approaches to Paleolithic Technologies*, edited by S. J. Lycett and P. R. Chauhan, pp. 235-253. Springer, New York.

Johnson, S. R., and S. McBrearty
2010 500,000 year-old blades from the Kapthurin Formation, Kenya. *Journal of Human Evolution* 58:193-200.

Kleindienst, M. R.
1961 Variability within the late Acheulean assemblage in eastern Africa. *South African Archaeological Bulletin* 16,62:35-52.

1962 Components of the East African Acheulian assemblage: An analytic approach. In *Actes du IVe Congres panafricain de Préhistoire et de l'étude du Quaternaire*, edited by C. Mortelmans and J. Nenquin, pp. 81-105. Tervuren, Belgium.

Kohn, M., and S. Mithen
1999 Handaxes: Products of sexual selection? *Antiquity* 73:518-26.

Leakey, L. S. B.
1959 A new fossil skull from Olduvai. *Nature* 184:491-493.

1961 New finds at Olduvai Gorge. *Nature* 189:649-650.

Leakey, M. D.
1967 Preliminary survey of the cultural material from Beds I and II, Olduvai Gorge, Tanzania. In *Background to Evolution in Africa*, edited by W. W. Bishop and J. D. Clark, pp. 417-446. University of Chicago Press, Chicago.

1975 Cultural patterns in the Olduvai sequence. In *After the Australopithecines*, edited by K. W. Butzer and G. Ll. Isaac, pp. 477-494. Mouton, The Hague.

Lefèvre, D., J.-P. Raynal, G. Vernet, G. Kieffer, and M. Piperno
2010 Tephro-stratigraphy and the age of the ancient Southern Italian Acheulean settlements: The sites of Loreto and Notarchirico (Venosa, Basilicata, Italy). *Quaternary International* 223-224:360-368.

Leopold, L. B., W. W. Emett, and R. M. Myrick
1966 Channel and hillslope processes in a semiarid area, New Mexico. *Geological Survey Professional Paper* 352-G, 193-253. Washington D.C.

Lepre, C. J., H. Roche, D. V. Kent, S. Harnand, R. L. Quinn, J.-P. Brugal, P.-J. Texier, and C. S. Feibel
2011 An earlier origin for the Acheulian. *Nature* 477:82-85.

Lycett, S. J.
2008 Acheulean variation and selection: does handaxe symmetry fit neutral expectations? *Journal of Archaeological Science* 35:2640-2648.

2009 Are Victoria West cores 'proto-Levallois'? A phylogenetic assessment. *Journal of Human Evolution* 56:175-191.

2010 Cultural transmission, genetic models and Palaeolithic variability: Integrative analytical approaches. In *New Perspectives on Old Stones: Analytical Approaches to Paleolithic Technologies*, edited by S. J. Lycett and P. R. Chauhan, pp. 207-234. Springer, New York.

Lycett, S. J., and C. J. Bae
2010 The Movius Line controversy: The state of the debate. *World Archaeology* 42:521-544.

Lycett, S. J., and J. A. J. Gowlett
2008 On questions surrounding the Acheulean 'tradition'. *World Archaeology* 40:295-315.

McBrearty, S.
1999 The Archaeology of the Kapthurin formation. In *Late Cenozoic Environments and Hominid Evolution: A Tribute to Bill Bishop*, edited by P. Andrews, and P. Banham, pp. 143-156. Geological Society, London,

McNabb, J., F. Binyon, and L. Hazelwood
2004 The Large Cutting Tools from the South African Acheulean and the question of social traditions. *Current Anthropology* 45:653-677.

McPherron, S. P.
2000 Handaxes as a measure of the mental capabilities of early hominids. *Journal of Archaeological Science* 27:655-663.

2006 What typology can tell us about Acheulian handaxe production. In *Axe Age: Acheulian Tool-making from Quarry to Discard*, edited by N. Goren-Inbar and G. Sharon, pp. 267-285. Equinox, London.

Machin, A.
2008 Why handaxes just aren't that sexy: A response to Kohn and Mithen (1999). *Antiquity* 82:761-769.

2009 The role of the individual agent in Acheulean biface variability: A multi-factorial model. *Journal of Social Archaeology* 9(1):35-58.

Machin, A. J., R. T. Hosfield, and S. J. Mithen
2007 Why are some handaxes symmetrical? Testing the influence of handaxe morphology on effectiveness. *Journal of Archaeological Science* 34:883-93.

Mellars, P. A.
1969 The chronology of Mousterian industries in the Perigord region. *Proceedings of the Prehistoric Society* 38:134-171.

1996 *The Neanderthal Legacy. An Archaeological Perspective from Western Europe.* Princeton University Press, Princeton.

Monnier, J. L.
1996 Acheuléen et industries archaïques dans le Nord-Ouest de la France. In *L'Acheuléen dans l'Ouest de l'Europe*, edited by A. Tuffreau, pp. 145-153. Publications du CERP, Université des Sciences et Technologies de Lille, 4.

Noll, M., and M. P. Petraglia
2003 Acheulean bifaces and early human behavioural patterns in East Africa and South India. In *Multiple Approaches to the Study of Bifacial Technologies*, edited by M. Soressi, and H. Dibble, pp. 31-53. Pennsylvania Museum of Archaeology and Anthropology, Philadelphia.

Norton, C. J., K. Bae, J. W. K. Harris, and H. Lee
2006 Middle Pleistocene handaxes from the Korean Peninsula. *Journal of Human Evolution* 51:527-536.

Nowell, A., and M. Chang
2009 The case against sexual selection as an explanation of handaxe morphology. *PaleoAnthropology* 2009:77-88.

Nowell, A., K. Park, D. Metaxas, and J. Park
2003 Deformation Modeling: A methodology for the analysis of handaxe morphology and variability. In *Multiple approaches to the study of bifacial technologies*, edited by Marie Soressi and Harold L. Dibble, pp. 193-208. University of Pennsylvania Museum of Archaeology and Anthropology, Philadelphia.

Owen, R. C.
1963 The patrilocal band: A linguistically and culturally hybrid social unit. *American Anthropologist* 67:675-690.

Paddaya, K., R. Jhaldiwal, and M. D. Petraglia
2006 The Acheulian quarry at Isampur, Lower Deccan, India. In *AxeAge: Acheulian Toolmaking from Quarry to Discard*, edited by N. Goren-Inbar and G. Sharon, pp. 45-73. Equinox, London.

Pappu, S., and K. Akhilesh
2006 Preliminary observations on the Acheulian assemblages from Attirampakkam, Tamil Nadu. In *Axe Age: Acheulian Toolmaking from Quarry to Discard*, edited by N. Goren-Inbar and G. Sharon, pp. 155-180. Equinox, London.

Parfitt, S. A., N. M. Ashton, S. G. Lewis, R. L. Abel, G. R. Coope, M. H. Field, R. Gale, P. G. Hoare, N. R. Larkin, M.D. Lewis, et al.
2010 Early Pleistocene human occupation at the edge of the boreal zone in northwest Europe. *Nature* 466:229-233.

Petraglia, M. D.
2003 The Lower Palaeolithic of the Arabian Peninsula: occupations, adaptations and dispersals. *Journal of World Prehistory* 17:141-179.

Petraglia, M. D., and C. Shipton
2008 Large cutting tool variation west and east of the Movius Line. *Journal of Human Evolution* 55:962-966.

Petraglia, M. D., C. Shipton, and K. Paddaya
2005 Life and mind in the Acheulean: a case study from India. In *The Hominid Individual in Context*, edited by C. Gamble and M. Porr, pp. 197-219. Routledge, London.

Piperno, M., and A. Tagliacozzo
2001 The Elephant Butchery Area at the Middle Pleistocene site of Notarchirico (Venosa, Basilicata, Italy). In *The World of Elephants, 1st International Congress, Rome 2001*, edited by G. Cavarretta, P. Gioia, M. Muss, and M. R. Palombo, pp. 230-236. Consiglio Nazionale delle Ricerche, Rome.

Piperno, M., D. Lefèvre, J.-P. Raynal, and A. Tagliacozzo
1998 Notarchirico, an early Middle Pleistocene site in the Venosa Basin. *Anthropologie* (Brno) 36:85-90.

Pope, M., and M. Roberts
2005 Observations on the relationship between Palaeolithic individuals and artifact scatters in the Middle Pleistocene site of Boxgrove, U.K. In *The Hominid Individual in Context*, edited by C. Gamble and M. Porr, pp. 81-97. Routledge, London.

Pope, M., K. Russel, and K. Watson
2006 Biface form and structured behaviour in the Acheulean. *Lithics* 27:44-57.

Porat, N., M. Chazan, R. Grün, M. Aubert, V. Eisenmann, and L. K. Horwitz
2010 New radiometric ages for the Fauresmith industry from Kathu Pan, southern Africa: Implications for the Earlier to Middle Stone Age transition. *Journal of Archaeological Science* 37:269-283.

Posnansky, M.
1959 A Hope Fountain site at Olorgesailie, Kenya Colony. *South African Archaeological Bulletin* 14:83-89.

Potts, R., A. K. Behrensmeyer, and P. Ditchfield
1999 Paleolandscape variation and Early Pleistocene hominid activities: Members 1 and 7, Olorgesailie Formation, Kenya. *Journal of Human Evolution* 37:747-788.

Renfrew, C.
1982 Explanation revisited. In *Theory and Explanation in Archaeology: The Southampton Conference*, edited by C. Renfrew, M. J. Rowlands, and B. A. Segraves, pp. 5-23. Academic Press, New York.

Roche, H., J.-P. Brugal, A. Delagnes, C. Feibel, S. Harmond, M. Kibunjia, S. Prat, and P.-J. Texier
2003 Les sites archéologiques plio-pléistocènes de la formation de Nachukui, Ouest-Turkana, Kenya: Bilan synthétique 1997-2001. *C.R. Palevol* 2:663-673.

Roe, D. A.
1968 British Lower and Middle Palaeolithic handaxe groups. *Proceedings of the Prehistoric Society* 34:1-82.

Saragusti, I., I. Sharon, O. Katzenelson, and D. Avnir
1998 Quantitative analysis of the symmetry of artefacts. Lower Paleolithic handaxes. *Journal of Archaeological Science* 25:817-825.

Schild, R. and F. Wendorf
1977 *The Prehistory of Dakhla Oasis and Adjacent Desert*. Wroclaw, Polska Akademia Nauk Instytut Historii Kultury Materialnej.

Scott, G. R., and L. Gibert
2009 The oldest hand-axes in Europe. *Nature* 461:82-85.

Semaw, S., M. Rogers, and D. Stout
2009 The Oldowan-Acheulian transition: Is there a 'Developed Oldowan' artifact tradition? In *Sourcebook of Paleolithic Transitions*, edited by M. Camps and P. Chauhan, pp. 173-193. Springer, New York.

Sharon, G.
2009 Acheulian giant core technology: A worldwide perspective. *Current Anthropology* 50:335-367.

Sharon, G., and P. Beaumont
2006 Victoria West: A highly standardized prepared core technology. In *Axe Age: Acheulian toolmaking from quarry to discard*, edited by N. Goren-Inbar and G. Sharon, pp. 181-199. Equinox, London.

Soressi, M., and H. Dibble (Editors)
2003 *Multiple Approaches to the Study of Bifacial Technologies*. Pennsylvania Museum of Archaeology and Anthropology, Philadelphia.

Steele, J.
1994 Communication networks and dispersal patterns in human evolution: A simple simulation model. *World Archaeology* 26:126-143.

Thieme, H.
2005 The Lower Palaeolithic art of hunting: The case of Schöningen 13 II-4, Lower Saxony, Germany. In *The Hominid Individual in Context*, edited by C. Gamble and M. Porr, pp. 115-132. Routledge, London.

Toth, N.
1985 The Oldowan reassessed: A close look at early stone artefacts. *Journal of Archaeological Science* 12:101-121.

1990 The prehistoric roots of a human concept of symmetry. *Symmetry: Culture and Science* 1, 3:257-281.

Tuffreau, A., A. Lamotte, and J.-L. Marcy
1997 Land-use and site function in Acheulean complexes of the Somme Valley. *World Archaeology* 29:225-241.

Underhill, D.
2007 Subjectivity inherent in by-eye symmetry judgements and the large cutting tools of the Cave of Hearths, Limpopo Province, South Africa. *Papers of the Institute of Archaeology* 18:103-113.

Villa, P.
1990 Torralba and Aridos: Elephant exploitation in Middle Pleistocene Spain. *Journal of Human Evolution* 19:299 309.

1996 Torralba et Aridos: Hhasse, charognage et depécage d'éléphants au Pleistocène moyen en Espagne. In *L'Acheuléen dans l'Ouest de l'Europe*, edited by A. Tuffreau, pp. 61-71. Publications du CERP, Université des Sciences et Technologies de Lille, Lille.

White, M. J.
2000 The Clactonian Question: On the Interpretation of Core-and-Flake Assemblages in the British Lower Paleolithic. *Journal of World Prehistory* 14:1-63.

White, M. J., and N. M. Ashton
2003 Lower Paleolithic Core Technology and the Origins of the Levallois Method in North-Western Europe. *Current Anthropology* 44:598-609.

Whiten A., J. Goodall, W. C. McGrew, T. Nishida, V. Reynolds, Y. Sugiyama, C. E. G. Tutin, R. W. Wrangham, and C. Boesch
1999 Cultures in chimpanzees. *Nature* 399:682-685.

Whiten, A., V. Horner, and F. B. M. de Waal

2005 Conformity to cultural norms of tool use in chimpanzees. *Nature* 437:737-740.

Willoughby, P. R.

1987 *Spheroids and Battered Stones in the African Early and Middle Stone Age.* Cambridge Monographs in African Archaeology 17.

Wynn, T.

1995 Handaxe enigmas. *World Archaeology* 27:10-24.

2002 Archaeology and cognitive evolution. *Behavioral and Brain Sciences* 25:389-438.

Wynn, T., and F. Tierson

1990 Regional comparison of the shapes of later Acheulean handaxes. *American Anthropologist* 92:73-84.

Zaidner, Y., A. Ronen, and J.-M. Burdukiewicz

2003 L'industrie microlithique de Paléolithique inférieur de Bizat Ruhama, Israël. *L'Anthropologie* 107:203-222.

7

THE DIET OF EARLY HUMANS: A SUMMARY OF THE CRITICAL ARGUMENTS 40 YEARS AFTER ISAAC'S INITIAL INSIGHTS

Manuel Domínguez-Rodrigo

Introduction

In 1971, Glynn Isaac published a paper "The diet of early man: Aspects of archaeological evidence from Lower and Middle Pleistocene sites in Africa" in which he critically discussed the evidence available for reconstructing human subsistence. He described archaeologists' efforts to make accurate interpretations in this regard as "...a little like navigating in the vicinity of an iceberg: more than four-fifths of what is of interest is not visible" (Isaac 1971:280). He noted a "potentially depressing review of factors which may distort the archaeological record of subsistence," while stressing at the same time that archaeological refuse constituted the "firmest document" available for addressing this question. Isaac cautiously interpreted some Early Pleistocene sites as the result of hunting or scavenging, where meat-eating was the main target in carcass exploitation, and for the Middle Pleistocene record he cited several sites as "spectacular evidence of success in hunting" (including in this observation some evidence we would now attribute to the early Pleistocene). According to him, human carnivory was at the heart of the behavior responsible for the emergence of the archaeological record, which Isaac (1978, 1981) interpreted within the food-sharing model, later re-defined as "central-place foraging" (Isaac 1983). Isaac's cautiousness regarding the strategies of carcass acquisition by hominins was an intellectually fresh approach in contrast with the unsupported inference that most sites represented episodes of human hunting.

The scavenging hypotheses proposed during the 1980s and 1990s, framed as replacing the old hunting paradigm, lacked the same type of scientific support: for a long time, they remained either untested or unsatisfactorily tested, lacking proper empirical and analogical support. The prevalent processual approach, which structured analogical reasoning for the past three decades, provided evidence that some African Early Pleistocene sites did not resemble modern foragers' home bases but rather carnivores' faunal assemblages (e.g., Binford's 1981 middle-range theory applied to this archaeological record). Using such a framework, a functional link between stone tools and bones was assumed at most sites, and a passive scavenging hypothesis was produced that inferred a carnivore-dominated process of assemblage formation: hominins were obtaining marginal carcass resources at carnivores' kills (Binford 1981).

Such a passive scavenging model needed a clearer articulation and more appropriate analogs to be considered scientifically rigorous. That is why Blumenschine (1986) designed an ecological study of scavenging opportunities in a modern savanna (Serengeti, Tanzania). Using modern Serengeti habitats as proxies for past habitats and communities, his work suggested that certain scavenging opportunities may have existed in the past: passive scavenging was feasible in riparian closed-vegetation habitats during the terminal part of the dry season (Blumenschine 1986). Furthermore, this actualistic approach also showed that passive scavenging generally did not enable access to meat, but rather that long bone marrow and head contents would have been the main targets of passive scavenging hominins (Blumenschine 1986). One question remained: whether this analogical scenario could be satisfactorily applied to the past, in the sense of being empirically testable. Tappen (2001) showed it could not, since she discovered a greater variability in the scavenging conditions of modern savannas. Unless a clear link could be established between the type of savanna and its scavenging opportunities and the discovery of similar paleocological conditions in specific sites in the early Pleistocene savannas in Africa, researchers could not take any modern scenario as a good proxy for prehistoric savannas and confidently apply it to understand the trophic dynamics and the probabilities of scavenging opportunities that those paleolandscapes could have afforded. From an epistemological point of view, these analogical frameworks were not scientific (*sensu* scientific realism; Bunge 1998), since they could not confidently establish an empirical link between present and past environments and their trophic dynamics.

In spite of this important shortcoming, even if such an epistemological bridge could be built, researchers should understand that scavenging opportunities do not automatically mean scavenging hominins. Scavenging as a strategy used by prehistoric hominins can be scientifically defended only if it created physical traces which potentially could be unambiguously interpreted. These traces should be found on the bones of the carcasses that hominins exploited.

Recently, a thorough review has been made of the history of the hunting and scavenging debate (Domínguez-Rodrigo 2002; Domínguez-Rodrigo and Pickering 2003; Domínguez-Rodrigo et al. 2007). In the initial stage of this debate, the use of skeletal part profiles produced equifinality scenarios: the long bone-dominated early Pleistocene sites could have been the result of hominin selection of the highest-yielding carcass portions (Bunn and Kroll 1986, 1988) or the marginal exploitation of the marrow-bearing bones abandoned at felid kills (Blumenschine 1986, 1987). Meat was important for the former model, and hardly-so for the latter.

A second phase opened in the 1990s, which switched focus from skeletal elements to bone surface modifications. Here, two approaches emerged to interpreting the order of access by hominins to carcasses: one focused on carnivores' tooth marks (Blumenschine 1995) and the other on hominin-imparted cut marks (whose study had been pioneered by Bunn and Kroll [1986, 1988]). The clash between these approaches (defending radically opposing interpretations) meant that by the 2000s some researchers became skeptical of both approaches, particularly the latter cut-mark-based approach, arguing that experimental samples were small (Blumenschine and Pobiner 2006; Blumenschine et al. 2007) or statistically undifferentiated (Lupo and O'Connell 2002), that cut marks were highly variable and, thus, not informative of butchery behaviors (Lyman 2006), and, that the ethological/experimental analogs used to understand the resources available for scavenging had failed to reproduced the array of resources that felid kills could yield (Pobiner 2007, 2008).

On the threshold of a new decade, almost 40 years after Isaac (1971) saw hunting and scavenging as complementary rather than opposing strategies of carcass exploitation (but with special emphasis on meat consumption), researchers dealing with the earliest archaeological record still seem confused and skeptical regarding our ability to disentangle the empirical threads of both hypotheses. Here I will show that ambiguity (defined as alternative hypotheses unable to provide a single best interpretation) has no place in scientific approaches (Bunge 1998) and that, at this point, there are only two possible outcomes of this debate, based on the assertion that only what is scientifically supported is valid: one of the competing hypotheses shows reduced heuristics (and, therefore, it is probably wrong) and archaeologists have the tools to falsify it; or, alternatively, we still lack enough evidence to assert either hypothesis and the hunting-scavenging debate (and thus the importance of meat in early human evolution) cannot be dealt with within the realm of science.

What does taphonomy say about early site formation?

My taphonomic review of most published Oldowan sites from 2.6 Ma to 1 Ma - focused on sites from Omo, Gona, Gadeb, Fejej and Hadar (Ethiopia), Olduvai and Peninj (Tanzania), Koobi Fora (Kenya) and Swartkrans Member 3 (South Africa) - showed that bone assemblages in most "Type C" sites (Isaac's original label for accumulations of stone tools and bones from several animals) were either naturally formed involving processes that did not imply hominid participation, or had such poor preservation that one could not make any functional link between hominid behavior and bone deposition at the site (Domínguez-Rodrigo 2009). This leaves archaeologists with a scanty evidence with which to reconstruct early hominid behavior during the late Pliocene and early Pleistocene.

The "Olduvai effect" caused by the taphonomic revision of the Olduvai sites also proves that accidental association of stone tools and bones was not only possible but also fairly common in the Plio-Pleistocene (Domínguez-Rodrigo et al. 2007a). The paleosurfaces where stone tools were discarded were exposed for prolonged time periods during which successive unrelated depositional events by non-hominid biotic agents might have generated the spatial association that archaeologists unearthed. This was previously stressed by Binford (1981). This is also especially true in a large portion of the Olduvai archaeological record from Bed I and Bed II, since materials appear vertically dispersed spanning large depositional periods. Therefore, no functional assumptions can be made based solely on spatial associations of bones and stone tools.

By using a multivariate approach for the analysis of these sites, including different variables of taphonomic value to discriminate human and carnivore behaviors in the accumulation and modification of bone assemblages, I think it becomes clear that an important proportion of the Early Pleistocene sites at Olduvai Gorge were formed mainly by carnivores, namely felids (Figure 7.1; Domínguez-Rodrigo et al. 2007).

This is of great relevance because it shows that if hominins were scavengers, they must have been actively involved in processing felid-abandoned carcasses in those spots, as suggested by Blumenschine (1986). However, taphonomic studies show that hominins systematically disregarded complete marrow-bearing bones at those sites (a substantial portion of them appear complete and percussion marks are virtually absent) and only hyenas were interested in exploiting them intermittently (Domínguez-Rodrigo et al. 2007a; Egeland and Domínguez-Rodrigo 2008). Furthermore, in those cases where some carcass exploitation by hominins is evident, meat and not marrow was the target of their exploitation,

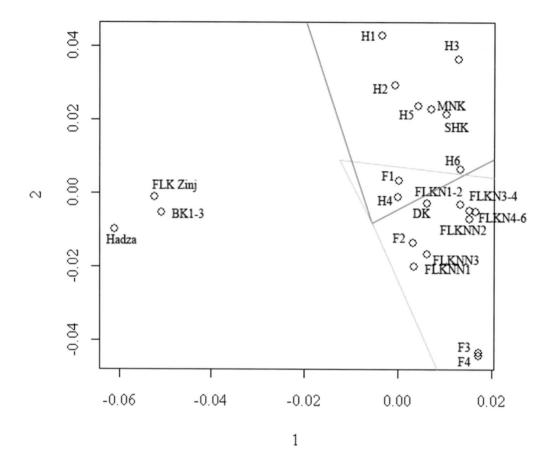

Figure 7.1 Multiple correspondence analysis performed in R, where a two-dimension solution of seven factors (taphonomic variables: Long bone profile, axial profile, tooth mark frequencies on long bone midshafts, percentages of complete long bones, frequency of digested bone, frequency of cut marks and percussion marks on long bones) compares spotted hyena bone assemblages (H1, Amboseli; H2, Maasai Mara den large animals; H3, Maasai Mara den small animals; H4, Syokimau den small animals; H5, Syokimau den large animals; H6, KND2 small animals) and felid-created assemblages (F1, Portsmut leopard lair; H2, Hakos river leopard lair; H3, WU/BA001; H4, lion-consumed carcasses from Maasai Mara and Tarangire, unpublished). Data were derived from personal study of the collections and Egeland et al. (2008), Prendergast and Domínguez-Rodrigo (2008), Brain (1981), de Ruiter and Berger (2005), Lupo and O'Connell (2002), Bunn (1993), Domínguez-Rodrigo et al. (2007a, b). Notice how most Olduvai sites cluster in between hyena and felid assemblages, with the exception of FLK Zinj and BK, which cluster close to an average of Hadza assemblages.

suggesting that those carcasses were independently obtained and consumed by hominins, resulting in independent-agent palimpsests (Domínguez-Rodrigo et al. 2007a, 2010).

A very small number of early Pleistocene sites have been taphonomically documented to be mostly or fully anthropogenic: FLK Zinj and BK (Olduvai Gorge), ST4 (Peninj), and maybe FwJj14 (Koobi Fora) (Domínguez-Rodrigo et al. 2007a; Pobiner et al. 2008). The information contained in them is of critical importance in understanding the diet and behavior of early humans. Actualistic research conducted with the goal of reproducing some taphonomic variables observed in those sites (related to the order of access to carcasses) is thus essential for providing archaeologists with analogical frameworks to understand the taphonomic signatures discovered in fossil assemblages.

On the Importance of How Scavenging Scenarios are Experimentally Modeled

The adequacy of analogies depends on how the conceptual premises of experiments are designed. If there is a hierarchy of principles that can be applied to the components of actualistic research, it can be argued that the most important one is the adequate articulation of premises (see Wylie 1988) in the elaboration of referential frameworks. Researchers create these analogs primarily to understand behaviors represented in and responsible for the archaeological record. The significance of analogy as a non-objective entity was initially stressed by Richter (1928). It entails a series of assumptions, some of them selected by the researcher, in a dynamic dialectic between the ideas that researchers try to test and the way the testing is eventually carried out (Domínguez-Rodrigo 2008).

The application of this important conceptual structure to the study of scavenging modeling requires that the experimental contexts are carefully selected. Resource availability in carcasses eaten by felids has been documented to tightly depend on the degree of competition (see critical review in Domínguez-Rodrigo 2010). When food is abundant because there are no competitors, lions abandon more carcass resources than when carnivores are more abundant and overlap in the use of the space. When modern humans have an impact on trophic dynamics, this also has an important effect on how carnivores behave; for instance, when humans chase lions, these may consume carcasses hurriedly and abandon them unfinished. This is why experiments modeling felid behavior in early Pleistocene savannas must be carried out in ecosystems with minimal or no human impact on how carnivores interact with one another and with their game. This renders experiments created in artificial contexts (e.g., in captivity or in ecologically-altered private properties) of doubtful validity for modeling resource availability and carcass modification by carnivores in the wild.

Observations of resources available after felid consumption of carcasses in natural parks or reserves have yielded similar results regarding the degree of thoroughness with which carcasses are consumed (see Blumenschine [1986] for Serengeti and Domínguez-Rodrigo et al. [2007] for Maasai Mara, Tsavo-Galana-Kulalu). This has recently received further support from similar ongoing studies in Tarangire National Park (Tanzania; work in progress). The only outlier in this analogical approach comes from Pobiner (2007). The reasons for this departure from the patterns documented in national parks have been argued to be the intrinsic variability of lion behavior. This would have important consequences for the way tooth mark and cut mark patterns have been modeled. However, Domínguez-Rodrigo (2010) argued that the reasons for Pobiner's

(2007) purported widening of felid behavioral variability can be explained by three fundamental experimental variables that have not received proper attention: sample size, sample control and the trophic dynamics of the context where the study was carried out.

Pobiner's (2007) small sample of nine lion-fed carcasses in the wild was obtained at a private ranch. This contrasts with the much larger samples found in previous studies. For instance, Blumenschine (1986) reported data from more than 250 carcasses in the Serengeti ecosystem. Domínguez-Rodrigo's (1999, 2009) lion-fed carcass sample is 35 (16 of them butchered) from national parks or reserves (Maasai Mara, Tsavo-Galana-Kulalu, Kenya). Dominguez-Rodrigo's lion-fed carcass sample was documented with full control (the observation of lions consuming carcasses was carried out until lions abandoned their prey and bones were subsequently collected) and it can be guaranteed that no other carnivore processed the carcasses. Pobiner's (2007) lion-fed carcasses in the wild were inferred to have been exclusively eaten by lions but this was never fully documented, because carcasses were recovered the morning following their nocturnal undocumented consumption. Therefore, it is possible that other carnivores also intervened in carcass consumption. This sample lacks control and, thus, is questionable.

Most importantly, Pobiner's sample in the wild was obtained on a private property where hyenas have been killed and removed from the trophic dynamics and lions are also frequently killed. The lack of competition of lions and their fear of humans account for the pattern of more incomplete carcass consumption, as documented by Pobiner, compared to the patterns observed in Maasai Mara, Tsavo, Galana, Kulalu, and Serengeti. Given that Pobiner's pattern can be explained by contextual variables and that the resources available after felid consumption of carcasses are fairly similar when documented in national parks and reserves, what follows will be based on the studies carried out in anthropogenically undisturbed ecological contexts, which are the only ones that can be used as more reliable proxies for what carnivores did in early Pleistocene savannas.

What can cut marks tell us about early Pleistocene hominin carcass exploitation strategies?

Domínguez-Rodrigo et al. (1997a) argued that primary access to carcasses could be differentiated from secondary access by the following criteria: cut mark frequencies per element and bone section types, and frequency of cut-marked midshaft meat-bearing bones relative to all long bone cut-marked specimens. Criticisms of the validity of cut marks to infer butchery behaviors have been based on the use of mere frequencies of cut marks per specific long bone portions and element type (Lyman 2006) or of combining global frequencies and appendicular anatomical sections (e.g., upper, intermediate and lower limb bones) using non-parametric tests (Lupo and O'Connell 2002). I also criticized the use of non-parametric tests on a sample of experiments (Domínguez-Rodrigo 2003) on the grounds of incomparability of the samples derived by different researchers because frequencies were tallied differently and following diverse incomparable methods. Furthermore non-parametric tests are not as powerful as parametric approaches.

By using all the variables together with the experimental data on butchery, including the ethnographic data (Lupo and O'Connell 2002), one could overcome the commonly small sample sizes that archaeologists have used in their experiments and the shortcomings of non-parametric tests by applying robust statistics. This can be

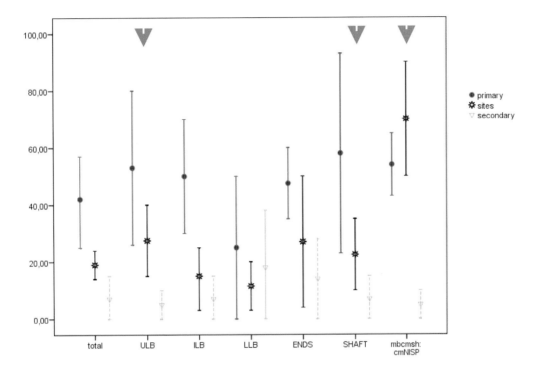

Figure 7.2 Analysis of 95% C.I. for the frequency of cut-marked long bone specimens from medium-sized carcasses from modern experiments replicating primary and secondary access to carcasses and early Pleistocene anthropogenic assemblages (sites). The confidence intervals were calculated using the t distribution, where t0·025 is the critical value of t with n-1 degrees of freedom. The analysis was derived using robust estimators by bootstrapping the sample 1000 times and then obtaining robust confidence intervals by applying a Monte Carlo technique both on quantiles and on the sampling error estimator. To calculate the two-tail sampling error in the bootstrapped distribution, a nested bootstrapping approach was performed. This consisted of generating 100 bootstrapped samples from the bootstrapped data derived from the original vector used to estimate the 0.025 and 0.975 quantiles, and then those nested samples were applied to estimate the sampling error. The total number of bootstrapped samples used to create a 95% confidence interval equals 10,000. The statistical analysis was performed in R by using the Robust library. Data for the analysis are from Domínguez-Rodrigo et al. (2002, 2007a), Domínguez-Rodrigo (1997a, 1997b), Lupo and O'Connell (2002), and Pobiner et al. (2008).

done by bootstrapping the samples and using robust estimators of mean and confidence intervals (Figure 7.2; Wilcox 2005). When doing this, the fossil sample shows some overlap with both primary and secondary access experimental scenarios (e.g., lower limb bones and ends), which can be explained by the fact that the fossil samples underwent taphonomic processes that did not affect the experimental samples. This can clearly be seen in the global percentage of cut marks on limb bones, where the fossil sample basically does not overlap with either experimental sample (Figure 7.2). Several factors may be responsible, among which bone surface preservation, the impact of dry fragmentation and different specimen size representation are probably the most important ones. Despite that, the fossil sample of sites considered taphonomically anthropogenic (FLK Zinj, BK1, BK2, BK3 and BK4 (Olduvai Gorge); ST4 (Peninj); FwJj14a,b and GaJi14 (Koobi Fora)) shows a significant departure from secondary access experimental scenarios when considering frequencies of cut-marked specimens from upper long bones and shafts. These become some of the most diagnostic taphonomic features for interpreting primary access to carcasses by hominins when considering both variables together as upper limb bone shafts (Domínguez-Rodrigo 1997a, 1997b; Domínguez-Rodrigo et al. 2007; Lupo and O'Connell 2002). Fossil shafts seem to be cut-marked at lower rates than in primary access experimental scenarios only because the BK samples used presented the raw cut mark estimates (11.9% for BK1, 13.2% for BK2, 10.3% for BK3 and 5.7% for BK4) without taking into account the intense dry breakage and the surface preservation. Once that is accounted for, the estimates are significantly higher (18.5% for BK1, 20.5% for BK2, 17.9% for BK3 and 11% for BK4). The fossil sample including these corrected data when bootstrapped produced a range between 15% and 29%, which is above the upper margin of the range of variation of secondary access experiments.

However, the intermediate ranges of cut-marked bone produced in the fossil sample, because of different properties of the fossil specimens when compared to the experimental ones, can be overcome if using variables that are less sensitive to different specimen size representation (fragmentation) and bone surface preservation. A good example is the ratio of cut-marked meat-bearing mid-shaft specimens when compared to the total sample of long bone

cut-marked specimens. Taphonomic processes are supposed to affect the cut-marked sample equally, irrespective of element type and portion type. This is why when considering the identified cut-marked portion of the sample, the distribution of cut-marked specimens can be useful in discriminating primary from secondary access. In Figure 7.2, it can be seen that the proportion of cut marks on meaty bones best discriminates between primary and secondary access.

As further support for inferring primary access to carcasses by hominins during the FLK Zinj (1.8 Ma) and BK (1.2 Ma) times, the study of the exact location of cut marks on each bone portion shows an important fraction of cut-marked specimens occurred on locations where no scraps survive after carnivores' initial carcass defleshing. For a more detailed description see Domínguez-Rodrigo et al. (2007) and Domínguez-Rodrigo et al. (2009).

What can tooth marks tell us about early Pleistocene hominin carcass exploitation strategies?

Passive scavenging from carnivores, as currently conceived, should be modeled in the following terms: felids were the primary consumers of carcasses, exploiting most of their flesh; hominids followed them by focusing mostly on marrow extraction; and hyenas intervened lastly by deleting grease-bearing bones (Blumenschine 1986). From a conceptual point of view, all the carnivore-only experiments carried out by Blumenschine (1988) that inspired the carnivore-hominid-carnivore model (Selvaggio 1994; Blumenschine 1995) were carried out using hyenas (Blumenschine 1995; Capaldo 1995, 1997) or a mix involving felids, hyenids and canids and tallying the resulting tooth mark frequencies together when comparing them to FLK Zinj (Selvaggio 1994). Tooth marks on mid-shafts at FLK Zinj were estimated to be high and, hence, carnivores were assumed to have initially defleshed the carcasses exploited by hominins (Blumenschine 1995; Capaldo 1995; Selvaggio 1994).

Some authors argue that carnivore initial consumption of carcasses should yield high frequencies of tooth-marked long bone mid-shafts (Blumenschine 1995; Capaldo 1995). However, the few carnivore-first(only) experiments carried out with hyenas fail to reproduce the range of tooth marking produced by hyena consumption of bone remains, as can be documented in spotted hyena dens, which is usually substantially lower than is reported experimentally (Domínguez-Rodrigo and Pickering 2010). Shafts from hyena dens are tooth-marked on average about 50% less than is reported in carnivore-first(only) experiments. There are even spotted hyena den assemblages with as low as 29% tooth-marked specimens (in contrast with >70% reported in experiments; Kuhn et al. 2009). In addition, it should be stressed that the use of "carnivore" models that do not pay attention to the type of bone-modifying agent are inadequate, since tooth mark frequencies depend on the type of carnivore involved in carcass consumption (see discussion in Domínguez-Rodrigo and Pickering 2010). For instance, suids (omnivores) can produce similar frequencies of tooth-marked bone to those of hyenas both in primary (carnivore-first) or secondary (carnivore-last) experimental models (Domínguez-Solera and Domínguez-Rodrigo 2009). Furthermore, in contrast to the moderate to high range of tooth marks produced by hyenas, felids have been shown to produce very low frequencies of tooth-marked bone. Long bones from carcasses consumed by felids and subsequently broken either by them or by hammerstones produce average frequencies of

tooth-marked shafts under 12% (Domínguez-Rodrigo et al. 2007b).

Not surprisingly, the application of the carnivore-first(only) experimental samples to the fossil record produced contradictory results. A recent study of tooth mark frequencies at FLK Zinj shows that the percentage previously identified by Blumenschine and colleagues is extremely inflated and results from misidentifying biochemical marks, caused by fungi and bacteria in combination with root etching, as carnivore-imparted tooth marks (Domínguez-Rodrigo and Barba 2006, 2007). The actual frequency of tooth marks on mid-shafts from all carcasses at FLK Zinj is <20%, much lower than any carnivore-first experimental model produced by Blumenschine (1988) or Selvaggio (1994). According to Blumenschine (1988, 1995), this would be indicative of primary access to fleshed carcasses by hominins.

A lingering issue regarding order of access to carcasses by hominins and carnivores is the overlap of cut and tooth marks. Potts and Shipman (1981) documented some specimens bearing cut marks overlying previous tooth marks at the Olduvai sites. This overlap was used as an argument to support hominin scavenging from previously tooth-marked carcass remains, both in the early Pleistocene and, recently, also for the middle Pleistocene (Blasco and Rosell 2009). In the most recent taphonomic reassessment of all the Olduvai Bed I sites, we did not find any case of overlap of cut marks over tooth marks (Domínguez-Rodrigo et al. 2007a). We identified a few cases of trampling marks overlying tooth marks and biochemical marks, abrasion marks caused by roots on biochemical marks, and some tooth marks overlying cut marks. However, even if we assume that cut marks overlying tooth marks may exist in the fossil record (see some examples in Blasco and Rosell 2009), one common assumption is to attribute the tooth marks directly to carnivores. White and Toth (2007) have remarked that this assumption is unfounded. Hominins may have been as likely a tooth-marking agent as other carnivores. A large number of modern foragers usually combine tools and teeth to deflesh carcass remains. This has been documented from African (Landt 2007) to Amazonian foragers (Martínez 2009). Although chewing occurs mostly on small carcass remains (e.g., some pastoralists have produced extreme damage on some goat bones [in Fisher 1995]), considerable damage to bones from larger prey has also been documented among the Hadza, where long bone ends of medium-sized animals are chewed after marrow extraction (Oliver 1993). White and Toth (2007) show that Pleistocene hominins were much better equipped to chew on bone than modern humans. Chimpanzees have also been reported to chew on bones (Pickering and Wallis 1997). Marks produced by humans in different contexts mimic the damage pattern documented among small to medium-sized carnivores (Oliver 1993; Elkin and Mondini 2001; Landt 2007; Martínez 2009). Humans can tooth-mark 35% of all bone specimens in their assemblages (Martínez 2009).

Therefore, the null hypothesis is to assume that tooth marks on fossil assemblages can be the result of hominins as well as other carnivores. I collected a bone assemblage of goats chewed by Maasai and have a few examples of tooth marks overlaid by cut marks and vice versa (Figure 7.3). A cut mark overlying a tooth mark is no more proof of hominin scavenging than it is of hominins using their teeth and tools to deflesh (and even chew on the grease-bearing portions of) bones.

The Inextricable Evidence of Plant Consumption

Hominin diet has been inferred from its most visible refuse (bones), but ethological and ethnographic analogies clearly document that hominins probably acquired the greatest diversity of minerals, carbohydrates, fatty acids and vitamins from plant foods (some of which even contained high amounts of protein). Isaac (1971, 1984) emphasized that early hominin diet was probably dominated by plant foods. The lack of preservation of this important part of hominin diet necessarily biases our reconstruction of hominin subsistence towards the most visible part – animal foods. Sept (1984, 1994, 2001) studied plant food availability and created what is still the most complete analogical framework to understanding plant distribution and potential plant food types available to hominins. This good foundation requires going one step further and creating an interpretive link between what was potentially used by hominins and what they actually consumed.

Some advances have been achieved in this regard in the past few years. One of involves analyses of plant phytoliths in natural (Bamford et al. 2008; Barboni et al. 1999, 2010) and archaeological soils (Ashley et al. 2010b). Some phytoliths have been documented on stone tool active edges (Domínguez-Rodrigo et al. 2001). Another discovery suggesting plant consumption by hominins is the increasing body of evidence showing that at some archaeological localities, such as most Olduvai Bed I sites, battering activities linked to the exploitation of non-animal resources have been documented (Diez-Martin et al. 2010) and seem to predominate over the tool-cutting activities that would result from animal butchery (de la Torre and Mora 2005). It is surely a matter of little time until a clear link between these battering activities

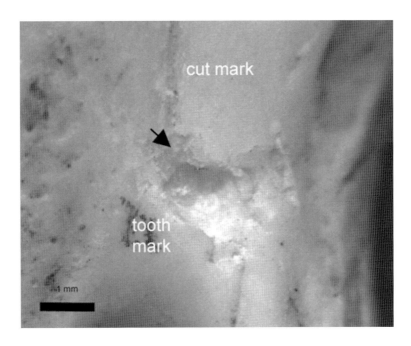

Figure 7.3 Cut mark overlapping a human tooth mark on an ethnoarchaeological specimen which was documented to have been eaten by a Maasai by using both a knife and his teeth. The widening of the cut mark as it enters the tooth pit and the striation inside the groove extending as far as the inside of the rim of the pit (arrow) can be observed.

and the exploitation of specific plant resources can be made.

What do we know of "early man's diet" that Isaac did not know?

In one of Isaac's last papers, he argued that the issue that hominids were eating substantial quantities of meat from large animals at some sites could be treated as settled (Isaac 1984). He argued that after the early years of the revisionist approach to the early archaeological record, the cumulative differences in knowledge were subtle rather than sweeping when compared with the previous decade. Isaac also stressed that it was the state of inquiry more than the state of knowledge that had advanced. Something along these lines could be said about the bulk of our knowledge of early hominin

subsistence behavior now if compared to 25 years ago. In the big picture, we can provide many more convincing arguments to support an interpretation of early Pleistocene hominin behavior based on the consumption of meat from relatively large animals. However, it is in the cumulative progression of small details that our knowledge now enables us to go a few steps ahead of Isaac's (1984) agnosticism over whether meat was obtained through hunting or scavenging. Taphonomists in the past three decades have been able to provide powerful analogical frameworks for differentiating various types of scavenging: passive versus confrontational, and human scavenging from one carnivore type versus another. It is through the application of these analogical frameworks to the archaeological record, and through a varied set of analytical techniques, that we now know that anthropogenic archaeofaunal assemblages are much less common than was believed during Isaac's time, that the few anthropogenic sites from the early Pleistocene preserve compelling evidence of primary access to small and medium-sized carcass resources, and that we are at the threshold of being able to differentiate whether this interpretation of primary access is the result of hunting or an alternative dynamic opportunistic strategy. I believe that Isaac would have been thrilled to see the outcome of a research program that he initiated through his students and collaborators. Given the evidence available at his time, his interpretations in this regard have been remarkably well reinforced by the progression of research over the years. He knew that much more could be achieved when he claimed that we could go further than we had yet done and that archaeologists would obviously have fun responding to the challenge (Isaac 1984).

Acknowledgements

I would like to express my appreciation to J. Sept and D. Pilbeam for their kind invitation to be part of this tribute to Glynn Isaac and for their comments to an earlier draft of this work. B. Isaac kindly shared some unpublished information from Isaac's work in Lake Natron, which helped me carry out my research at Peninj and reinforce his interpretation of the importance of meat in early Pleistocene hominin behavior.

References

Ashley, G., D. Barboni, M. Domínguez-Rodrigo, A. Mabulla, H. T. Bunn, F. Diez-Martin, and E. Baquedano
2010 A spring and wooded habitat at FLK Zinj and their relevance to origins of human behavior. *Quaternary Research* (in press).

Bamford, M. K., I. G. Stanistreet, H. Stollhofen, and R. M. Albert
2008 Late Pliocene grassland from Olduvai Gorge, Tanzania. *Palaeogeography Palaeoclimatology Palaeoecology* 257:280-293.

Barboni, D., R. Bonnefille, A. Alexandre, and J. D. Meunier
1999 Phytoliths as paleoenvironmental indicators, West Side Middle Awash Valley, Ethiopia. *Palaeogeography Palaeoclimatology Palaeoecology* 152:87-100

Barboni, D., G. Ashley, M. Domínguez-Rodrigo, A. Mabulla, H. T. Bunn, and E. Baquedano
2010 Phytolith-inferred paleo-vegetation at FLK North and surrounding localities during upper Bed I time, Olduvai Gorge, Tanzania. *Quaternary Research* (in press).

Binford, L.R.
1981 *Bones: Ancient Men and Modern Myths*. Academic Press, New York.

Blasco, R., and J. Rosell

2009 Who was the first? An experimental application of carnivore and hominin overlapping marks at the Pleistocene archaeological sites. *Comptes Rendus Palevol* 8:579-592.

Blumenschine, R. J.

1986 Early hominid scavenging opportunities: Implications of carcass availability in the Serengeti and Ngorongoro ecosystems. *British Archaeological Reports, International Series* 283.

1987 Characteristics of an early hominid scavenging niche. *Current Anthropology* 28:383-407.

1988 An experimental model of the timing of hominid and carnivore influence on archaeological bone assemblages. *Journal of Archaeological Science* 15:483-502.

1995 Percussion marks, tooth marks and the experimental determinations of the timing of hominid and carnivore access to long bones at FLK Zinjanthropus, Olduvai Gorge, Tanzania. *Journal of Human Evolution* 29:21-51.

Blumenschine, R. J., and B. Pobiner

2006 Zooarchaeology and the ecology of oldowan hominid carnivory. In *Evolution of the Human Diet: The Known, the Unknown and the Unknowable*, edited by P. Ungar, pp. 167-90. Oxford University Press, Oxford.

Blumenschine, R. J., C. R. Peters, S. D. Capaldo, P. Andrews, J. K. Njau, and B. L. Pobiner

2007 Vertebrate taphonomic perspectives on Oldowan hominin land use in the Plio-Pleistocene Olduvai Basin, Tanzania. In *Breathing Life into Fossils: Taphonomic Studies in Honor of C.K. (Bob) Brain*, edited by T. R. Pickering, K. Schick, and N. Toth, pp. 161-179. Stone Age Institute Press, Gosport, Indiana.

Brain, C. K.

1981 *The Hunters or the Hunted? An Introduction to African Cave Taphonomy*. Chicago, University of Chicago Press.

Bunge, M.

1998 *Philosophy of Science*. Transaction Publishers, Londres.

Bunn, H. T.

1993 Bone assemblages at base camps: A further consideration of carcass transport and bone destruction by the Hadza. In *From Bones to Behavior. Ethnoarchaeological and Experimental Contributions to the Interpretations of Faunal Remains, Center for Archaeological Investigations*, edited by J. Hudson, pp. 156-168. Southern Illinois University.

Bunn, H. T., and E. M. Kroll

1986 Systematic butchery by Plio-Pleistocene hominids at Olduvai Gorge, Tanzania. *Current Anthropology* 27(5):431-442.

Bunn, H. T., and E. M. Kroll
1988 Fact and fiction about the FLK Zinjanthropus floor: Data, arguments, and interpretations. (reply to L. R. Binford). *Current Anthropology* 29(1):135-149.

Capaldo, S. D.
1995 Inferring hominid and carnivore behavior from dual-patterned archaeological assemblages. Ph.D. Thesis. Rutgers University, New Brunswick.

1997 Experimental determinations of carcass processing by Plio-Pleistocene hominids and carnivores at FLK 22 (Zinjanthropus), Olduvai Gorge, Tanzania. *Journal of Human Evolution* 33:555-597.

de Ruiter, D., and L. Berger
2005 Leopards as taphonomic agents in dolomitic caves: Implications for bone accumulations in the hominid-bearing deposits of South Africa. *Journal of Archaeological Science* 27:665-684.

Diez-Martin, F., P. Sanchez, M. Domínguez-Rodrigo, A. Mabulla, H. T. Bunn, R. Barba, and E. Baquedano
2010 A place to crush: New insights into hominin lithic activities at FLK North Bed I (Olduvai Gorge, Tanzania). *Quaternary Research*.

Domínguez-Rodrigo, M.
1997a Meat-eating by early hominids at the FLK 22 Zinjanthropus site, Olduvai Gorge, Tanzania: An experimental approach using cut mark data. *Journal of Human Evolution* 33:669-690.

1997b A reassessment of the study of cut mark patterns to infer hominid manipulation of fleshed carcasses at the FLK Zinj 22 site, Olduvai Gorge, Tanzania. *Trabajos de Prehistoria* 54:29-42.

1999 Flesh availability and bone modification in carcasses consumed by lions. *Palaeogeography, Palaeoclimatology and Palaeoecology* 149:373-388.

2002 Hunting and scavenging by early humans: The state of the debate. *Journal of World Prehistory* 16:1-54.

2003 Bone surface modifications, power scavenging and the "display" model at early archaeological sites: A critical review. *Journal of Human Evolution* 45:411-416.

2008 Conceptual premises in experimental design and their bearing on the use of analogy: An example from experiments on cut marks. *World Archaeology* 40:67-82.

2009 Are all Oldowan sites palimpsests? If so, what can they tell us about hominid carnivory? In *Interdisciplinary Approaches to Understanding the Oldowan*, edited by E. Hovers, and D. Braun. Springer, New York.

2010 Distinguishing between apples and oranges again: A response to Pobiner (2008) on the importance of differentiating between epistemically trivial and non-trivial analogies. *Journal of Taphonomy*.

Domínguez-Rodrigo, M., and R. Barba

2006 New estimates of tooth marks and percussion marks from FLK Zinj, Olduvai Gorge (Tanzania): The carnivore-hominid-carnivore hypothesis falsified. *Journal of Human Evolution* 50:170-194.

2007 Five more arguments to invalidate the passive scavenging version of the carnivore-hominid-carnivore model: A reply to Blumenschine et al. (2007a). *Journal of Human Evolution* 53:427-433.

Domínguez-Rodrigo, M., and T. R. Pickering

2003 Early hominids, hunting and scavenging: A summary of the discussion. *Evolutionary Anthropology* 12:275-282.

2010 A multivariate approach for discriminating bone accumulations created by spotted hyenas and leopards: Harnessing actualistic data from East and Southern Africa. *Journal of Taphonomy* 8.

Domínguez-Rodrigo, M., C. P. Egeland, and R. Barba

2007a *Deconstructing Olduvai*. New York, Springer.

Domínguez-Rodrigo, M., C. P. Egeland, and T. R. Pickering

2007b Models of passive scavenging by early hominids: Problems arising from equifinality in carnivore tooth mark frequencies and the extended concept of archaeological palimpsests. In *African Taphonomy: A Tribute to the Career of C. K. "Bob" Brain*, edited by T. Pickering, K. Schick, and N. Toth, pp. 255-268. CRAFT Press (Indiana University), Bloomington.

Domínguez-Rodrigo, M., J. Serrallonga, J. Juan-Treserras, L. Alcalá, and L. Luque

2001 Woodworking activities by early humans: A plant residue analysis on Acheulian stone tools from Peninj, Tanzania. *Journal of Human Evolution* 39:421-436.

Domínguez-Rodrigo, M., A. Mabulla, H. T. Bunn, R. Barba, F. Diez-Martin, C. P. Egeland, E. Espilez, A. Egeland, J. Yravedra, and P. Sanchez

2009. Unraveling hominin behavior at another anthropogenic site from Olduvai Gorge (Tanzania): New archaeological and taphonomic research at BK, Upper Bed II. *Journal of Human Evolution* 57:260-283.

Domínguez-Rodrigo, M., H. T. Bunn, A. Mabulla, F. Diez-Martin, D. Barboni, S. Domínguez-Solera, R. Barba, P. Sánchez, G. Ashley, and F. Baquedano

2010 The FLK North (Olduvai Gorge, Tanzania) Bed I sequence revisited: Disentangling hominin and carnivore activities at a spring. *Quaternary Research*.

Domínguez-Solera, S., and M. Domínguez-Rodrigo

2009 A taphonomic study of bone modification and tooth-mark patterns on long limb bone portions by suids. *International Journal of Osteoarchaeology* 19:345-363.

Egeland, C.P., and M. Domínguez-Rodrigo

2008 Taphonomic perspectives on hominid site use and foraging strategies during Bed II times at Olduvai Gorge, Tanzania. *Journal of Human Evolution* 55:1031-1052.

Egeland, A., C. P. Egeland, and H. T. Bunn

2008 Taphonomic analysis of a modern spotted hyena den (*Crocuta crocuta*) from Nairobi, Kenya. *Journal of Taphonomy* 6:275-300.

Elkin, D., and M. Mondini

2001 Human and small carnivore gnawing damage on bones: An exploratory study and its archaeological implications. In *Ethnoarchaeology of Andean South America. Contributions to Archaeological Method and Theory*, edited by L. A. Kuznar, pp. 255-256. International Monographs in Prehistory, Ethnoarchaeological Series 4.

Fisher, J. W.

1995 Bone surface modification in zooarchaeology. *Journal of Archaeological Method and Theory* 1:7-65.

Isaac, G. L.

1971 The diet of early man: Aspects of archaeological evidence from Lower and Middle Pleistocene sites in Africa. *World Archaeology* 2:278-299.

1978 The food-sharing behavior of protohuman hominids. *Scientific American* 238:90-108.

1983 Bones in contention: Competing explanations for the juxtaposition of early Pleistocene artifacts and faunal remains. In *Animals and Archaeology: Hunters and their Prey*, edited by J. Clutton-Brock, and C. Grigson, pp. 3-19. British Archaeological Reports, International Series 163.

1984 The archaeology of human origins: Studies of the Lower Pleistocene in East Africa 1971-1981. In *Advances in Old World Archaeology*, edited by F. Wendorf, and A. Close, pp. 1-87. Academic Press, New York.

Kuhn, B. F., L. R. Berger, and J. D. Skinner

2009 Variation in tooth mark frequencies on long bones from the assemblages of all three-extant bone collecting hyaenids. *Journal of Archaeological Science* 36:297-307.

Landt, M. J.

2007 Tooth marks and human consumption: Ethnoarchaeological mastication research among foragers in the Central African Republic. *Journal of Archaeological Science* 34:1629-1640.

Lyman, R.

2006 Analyzing cut marks: Lessons from artiodactyl remains in the northwestern United States. *Journal of Archaeological Science* 32:1722-1732.

Lupo, K. D., and J. F. O'Connell

2002 Cut and tooth mark distributions on large animal bones: ethnoarchaeological data from the Hadza and their implications for current ideas about early human carnivory. *Journal of Archaeological Science* 29:85-109.

Martínez, G.

2009 Human chewing bone surface modification and processing of small and medium prey amongst the Nukak (foragers of the Colombian Amazon). *Journal of Taphonomy* 7:1-20.

Oliver, J. S.

1993 Carcass processing by the Hadza: Bone breakage from butchery to consumption. In *From Bones to Behavior. Ethnoarchaeological and Experimental Contributions to the Interpretations of Faunal Remains, Center for Archaeological Investigations*, edited by J. Hudson, pp. 200-227. Southern Illinois University.

Pickering, T. R., and J. Wallis,

1997 Bone modifications resulting from captive chimpanzee mastication: Implications for the interpretation of Pliocene archaeological faunas. *Journal of Archaeological Science* 24:1115-1127.

Pobiner, B.

2007 Hominin-carnivore interactions: Evidence from modern carnivore bone modification and early Pleistocene archaeofaunas (Koobi Fora, Kenya; Olduvai Gorge, Tanzania). Ph.D. Dissertation, Department of Anthropology, Rutgers University, New Brunswick.

2008 Apples and oranges again: Comment on 'Conceptual premises in experimental design and their bearing on the use of analogy: An example from experiments on cut marks'. *World Archaeology* 40:466-479.

Pobiner, B. L., M. Rogers, C. Monahan, and J. W. K. Harris

2008 New evidence for hominin carcass processing strategies at 1.5 Ma, Koobi Fora, Kenya. *Journal of Human Evolution* 55:103-130.

Potts, R., and P. Shipman

1981 Cutmarks made by stone tools from Olduvai Gorge, Tanzania. *Nature* 291:577-580.

Prendergast, M., and M. Domínguez-Rodrigo

2008 Taphonomic analyses of a hyena den and a natural-death assemblage near Lake Eyasi (Tanzania). *Journal of Taphonomy* 6:301-336.

Richter, R.

1928 Aktuopaläontologie und Paläobiologie; eine Abgrenzung. *Senkenbergiana* 19.

Selvaggio, M. M.

1994 Identifying the timing and sequence of hominid and carnivore involvement with Plio-Pleistocene bone assemblages from carnivore tooth marks and stone-tool butchery marks on bone surfaces, Ph.D. Dissertation. Rutgers University, New Brunswick.

Sept, J. M.

1984 Plants and early hominids in East Africa: A study of vegetation in situations comparable to early archaeological site locations. Ph.D. Dissertation, Department of Anthropology, University of Berkeley, California.

1994 Beyond bones: Archaeological sites, early hominid subsistence and the costs and benefits of exploiting wild plant foods in east African riverine landscapes. *Journal of Human Evolution* 27:25-45.

2001 Modeling the edible landscape. In *Meat-eating and Human Evolution*, edited by G. B. Stanford, and H. T. Bunn, pp. 73-98. Oxford University Press, Oxford.

Tappen, M.

2001 Deconstructing the Serengeti. In *Meat-Eating and Human Evolution*, edited by G. B. Stanford, and H. T. Bunn, pp. 13-32. Oxford University Press, Oxford.

Torre de la, I., Mora, R.,

 2005 *Technological strategies in the Lower Pleistocene at Olduvai Beds I and II.* University of Liege Press, Eraul 112, Brussels.

White, T., and N. Toth

2007. Carnivora and carnivory: Assessing hominid toothmarks in zooarchaeology. In *Breathing Life into Fossils: Taphonomic Studies in Honor of C.K. (Bob) Brain*, edited by T. R. Pickering, K. Schick, and N. Toth, pp. 281-296. Stone Age Institute Press, Gosport, Indiana.

Wilcox, R. R.

2005 *Introduction to Robust Estimation and Hypothesis Testing.* London, Elsevier Academic Press.

Wylie, A.

1988 "Simple" analogy and the role of relevance assumptions: Implications of archaeological practice. *International Studies in the Philosophy of Science* 2:134-150.

8

GLYNN ISAAC:
MEMORIES AND TRIBUTE

John D. Speth

Introduction

Glynn Isaac was one of the greats in archaeology. The profession certainly recognizes him as such, and this volume is further testimony to his lasting contribution to studies of human beginnings. But he is also one of the greats in my own life, and what he gave me early on in my fledgling career will last my lifetime. I beg the readers' indulgence if I begin with a bit of personal reminiscence, to show just how wonderful and important his gift to me was when I was just starting out in the profession and then, in the second half of the paper, I will turn to a more substantive contribution, but one which still follows a path that Glynn helped me get started on so many years ago. First, the personal side.

The Personal Side - The 1977
SAAs in New Orleans

Fresh out of graduate school and in my first academic job at Hunter College in New York City, the ink still not completely dry on my Ph.D. diploma, I wrote a paper on the seasonality of the occupations in Olduvai Gorge and published it in *Science* (Speth and Davis 1976). It was one of those papers that should never have been published anywhere, let alone in *Science*, but I didn't know it at the time. What was wrong with it? The conclusions of the paper were based heavily on patterning in the fauna from a number of the early archaeological localities in the Gorge, including FLK-Zinj and several others, but I knew absolutely nothing about fauna and,

worse yet, I had never heard the word "taphonomy". My immersion in fauna didn't begin for another year, in 1977, when I began digging a bison kill in New Mexico and had no choice but to deal with bones (Speth 1983). Prior to that I was a hard-core "lithics" person, specializing in the Middle Paleolithic of the Near East. After a failed attempt to get a dissertation out of a porcupine-infested Middle Paleolithic cave in Iran (a site I affectionately dubbed my "Pleistocene milkshake") I finally finished my degree with a hasty, library-based look at fracture mechanics in order to understand how and why flint broke the way it did (Speth 1972). So my Olduvai paper was a detour - a major one at the time - into a topic that I was, to put it politely, "ill-equipped" to deal with. In ordinary parlance, I didn't know a bone from the proverbial hole in the ground and I hadn't yet heard about taphonomy, a brand new and very hot topic in the 1970s for those who were working with the animal bones from early hominin sites (e.g., Brain 1969, 1975; Behrensmeyer 1975).

But I had heard about Glynn Isaac; I avidly read his papers as they began to appear in the late 1960s and early 1970s and regularly talked about his ideas in my classes at Hunter College (e.g., Isaac 1969, 1971, 1972a, 1972b, 1972c; Isaac et al. 1971). In 1977, shortly after the *Science* paper appeared, I ventured off to the 42nd Annual Meeting of the Society for American Archaeology in New Orleans (I hope my memory of these details serves me correctly here). To my utter surprise and amazement, I was approached by Glynn Isaac, who very politely asked if we could meet at 5:00 p.m. at the hotel registration desk and go somewhere to chat about my Olduvai paper. (Those who have been to an SAA meeting know all too well what a madhouse that area of the hotel is after the sessions let out.) But, sure enough, Glynn appeared right on schedule and off we headed to one of the bars in the hotel, Glynn leading the way and me, tagging along behind, hands sweating and knees shaking. As we entered the bar, there was Lew Binford holding forth with a bunch of other well-known figures in the profession. Seeing Glynn, Binford and the others invited him to join them at the table, which I think many another archaeologist would have done without a second thought, a few inviting the neophyte to come along, many others simply leaving the uncomfortable nobody stranded high and dry. But Glynn did neither; instead, he excused himself and said that he had promised to talk to me, so we headed off to a table at the back. Needless-to-say, as a total nobody I was absolutely floored by this unexpected and extremely kind gesture. So we sat down and began talking.

What came next was even more amazing, and has profoundly influenced me ever since. Glynn could have creamed my *Science* paper, and me in the process; and, in what had seemed like an interminable "upstream swim" through the mob scene from the hotel registration desk to the bar, I desperately tried to steel myself for what I was sure was coming. But he didn't. Instead, Glynn proceeded to compliment me on the paper, telling me how I had raised some very interesting issues, and agreeing, with complete evident sincerity, that seasonality was something important that we as archaeologists needed to know more about. Needless-to-say, I began to relax and my sagging self-confidence got a much

needed boost. Only after my ego was fully bolstered, did he then, ever so gently, slip in the "...but there are some issues that you should still consider...". That's when I first heard that there was something called taphonomy. Yes, criticism came, but not only was it gentle, it was positive and always constructive. At that point I didn't mind being criticized, because he had left my ego and self-confidence intact.

But the meeting in the bar wasn't over yet. What came next was truly extraordinary; Glynn invited me to come to Koobi Fora, as he put it, "to see for myself". And I did. In 1979, I joined his project as that season's intellectual "gadfly" (my expression, not Glynn's). He ran the field work, which that year was mostly at FxJj50 on the Karari Escarpment (Bunn et al. 1980), as a bona fide graduate course (he was still on the faculty at Berkeley at the time) and part of that course was a weekly seminar, held under an acacia tree in front of our tents. He wanted his students to be exposed, not just to his way of thinking, but to the way others, from other intellectual traditions, might approach the same issues. So each year he brought a different person in from the outside to participate in the seminar – 1979 was my year. And it was a truly fabulous experience in every imaginable way. And, yes, by the end of that three-month field season I had a pretty good idea what taphonomy was all about.

Needless-to-say, I have never forgotten that evening in the bar in New Orleans, and the life-transforming experiences that grew out of that one little event. Of course, the opportunity to go to Kenya and participate in the Koobi Fora project profoundly influenced me professionally. But to me the most important thing I came away with from that wonderful evening in New Orleans was a pedagogical model that, over the intervening years, I hope I have been able to at least approximate in my own interactions with both students and professionals alike. When critiquing someone's work, whether a student paper, a thesis, a grant proposal, or a professional publication, always begin with the positive in what people have done. It's always there, somewhere, though sometimes it may take a little digging to find it. Then move on to criticisms, if any are warranted, but do it gently, and always in as positive and constructive a light as possible.

The Personal Side - Things Nutritional

Life works in strange ways. Between the appearance of that ill-conceived publication in *Science* and my wonderful three-month stint in Koobi Fora, I became a North Americanist. Not by choice. The City University of New York - in fact the entire city of New York - went belly-up in the mid-1970s. Our teaching and research environment at Hunter College, to put it mildly, "deteriorated" and I began seriously looking for another job. Michigan, my graduate *alma mater*, came through with a job offer I couldn't refuse - but it came with one major, and very daunting, string attached - I had to become a North American archaeologist. I could work anywhere on the continent and in any time period. The only problem was that I had never had a graduate-level course on North American prehistory and knew next to nothing about it. But I gratefully accepted, gave up my newly-minted tenure at Hunter College, and went to work to establish an active program of field research and publication in North America.

Since I thrive best in hot climates, the southwest was the obvious choice for me as a field area. And I had done my undergraduate studies at the University of New Mexico, so the southwest felt like home. Moreover, the closest thing to the Paleolithic in North America was Paleoindian. So in the summer of 1976, the same summer in fact that the *Science* paper on

Olduvai appeared, I found myself out in New Mexico looking for the "perfect" Paleoindian site to dig. Not an easy task; one doesn't generally waltz into an area, totally unknown and green, and have someone just hand you a Blackwater Draw or a Lindenmeier. Nonetheless, with a lot of help from a number of very generous people, I finally did find a site, a bison kill near Roswell, New Mexico - the Garnsey Site - and spent the next two summers (1977 and 1978) digging there (Speth 1983). But my little black cloud was obviously still with me, the one I hold responsible for the "Pleistocene milkshake" in Iran and the bankruptcy of the City University of New York, because my "Paleoindian" site, despite the fact that parts of it were buried up to four meters below the surface, turned out to date to the mid-15th century AD!

Nevertheless, faced with a bison kill, Paleoindian or not, I had no choice but to begin learning about bones in a big hurry. I read everything I could get my hands on that dealt with bison, faunal analysis more generally, and taphonomy. Then in 1978, Binford's *Nunamiut Ethnoarchaeology* appeared, but I didn't have time to read it in Ann Arbor, so I schlepped it with me to Koobi Fora and it became my "leisure"-time reading on days off. I would lie in the sand in the middle of a broad, dry river bed at Kampi Ya Simba, our base on the Karari Escarpment, cooking myself in the sun, now and then watching a dik-dik whose territory included my tent, and began working my way through Nunamiut. As I read about caribou and utility indices, I began to get ideas how I might analyze the bison bones from Garnsey. I was also struck by how similar Garnsey was to FxJj50, so I began to wonder if one might be able to look at early hominin fauna from the utility perspective as well. Both sites were in semi-arid environments with very seasonal rainfall; both sites were situated next to an ephemeral stream channel; both sites involved butchered carcasses of large bovids; and the stone tools at both were equally "uninspiring". Glynn was intrigued, and as always encouraging, and we had many wonderful discussions along those lines, though he frequently "brought me back to earth" by reminding me that FxJj50 was 1.5 million years earlier than Garnsey, on a different continent, and made by a pre-modern hominin. Details.... During that field season we talked about many other things as well – about hunting and scavenging and how one might tell them apart, about taphonomy, modern hunter-gatherers like the San (Bushmen) and Hadza, and many other related, and unrelated, topics. By the time I returned to Ann Arbor, my head was swimming with new ideas and things I wanted to try.

In that formative year, 1979, my interests began to coalesce around "things nutritional". Glynn, and my experiences at Koobi Fora, were major catalysts in my move from lithics to bones, and then to more general nutritional issues. The Garnsey bison bones were stored in an abandoned wind tunnel on the Michigan campus (archaeologists end up in all sorts of strange places on college campuses). So, soon after my return from Kenya, I squirreled myself away in the tunnel, with no phone or windows to the outside world, and began cranking through the material. To be absolutely honest, when I launched into the analysis I had no idea where it was all going, but it didn't take long for some striking anomalies in the data to begin appearing. For example, the bison that were killed at Garnsey were mostly adult males, and the kills took place in the spring, in striking contrast to the widely known Northern Plains pattern documented by George Frison and his students and colleagues in which females were most often the prime targets and most events occurred in the fall or winter (Frison 1978).

The anomalies went further. The body-part representations for males and females were very different. Males had obviously been the principal targets for the Garnsey hunters, but they had killed some females as well. However, among the bones that the hunters discarded at the kill, female parts were often way overrepresented. Moreover, bones of females, as well as those of juveniles, weren't processed as thoroughly as those of adult males. Little by little it became apparent that there had been something the Indians had considered less desirable about the females and juveniles than the males. No doubt, it was partly attributable to the obvious differences in size. But something else was also involved in their decision-making. For quite some time I couldn't nail down what it was, but the key finally came when I began reading the travel accounts of 19th-century explorers like Lewis and Clark (Coues 1893) and Randolph Marcy (1863). Over and over again these accounts talked about running low on food in the winter or spring, sending out hunting parties, finding and successfully killing numerous elk or bison, but still returning empty-handed because the animals were "too lean for use". To these explorers, lack of fat seemed to be the critical variable, for they repeatedly talked about it. Not long afterwards, I stumbled upon Vihljalmur Stefansson's (1944) marvelous, *Arctic Manual*, which clearly spelled out the deleterious effects of a diet based heavily on lean meat, a condition common in the arctic that Stefansson referred to as "rabbit starvation".

Was the patterning I was beginning to tease out of the Garnsey bones a reflection of the same concern about fat that was so evident in the explorer accounts? In order to better understand what had structured the activities that took place at Garnsey, I began to peruse the wildlife literature, looking particularly at the many studies that dealt with the amount and distribution of fat in the carcasses of male and female ungulates that died or were shot at different times of year. The match between what was described in the wildlife studies and what I was seeing at Garnsey was striking.

So the next step was to find out why fat was so important. I had always assumed that if a person ate more meat than he or she needed to fulfill one's protein requirements, the rest would simply show up as unwanted "love-handles". It turned out that the real culprit was too much protein. Intake of protein above a certain threshold (typically about 250-300 g) exceeds the capacity of enzymes in the liver to deaminize the protein and excrete the resulting nitrogen. Continued intakes that exceed this threshold may lead to serious health issues and ultimately death. The explorers had been well aware of the problem, and it seems that the Garnsey hunters were aware of it too. These ideas finally made their way to print in 1983, first in an article co-authored with Kate Spielmann (Speth and Spielmann 1983), and then in a book on the Garnsey kill site (Speth 1983).

Glynn entered the picture again that same year with an invitation to participate in a Gordon Research Conference on "Diet and Human Evolution" to be held in Ventura, California early in 1984 (Anonymous 1983). Co-organized by Michael DeNiro, Alan Walker, and Glynn, this conference, which explicitly eschewed publication of the proceedings in order to encourage participants to "go out on a limb" with new and untested ideas, provided an ideal context for me to begin to explore implications of the "excess protein" arguments in other contexts, such as the increasingly popular view at the time of early hominins as dedicated passive scavengers (Binford 1981).

And that interest, once again catalyzed by Glynn at the Gordon Conference, continues to

this day. But it has also changed, shifting away from its original narrow focus on the nutritional problems that can arise from heavy reliance on lean meat, regardless of prey size, to a look at the broader array of factors - social and political, not just nutritional - that motivate both modern and ancient foragers to focus their efforts on hunting big game (Speth 2010; Speth et al. 2010). And this brings us to the perfect point to segue into the second half of the paper - why in fact do our classic African hunters and gatherers - the Bushmen and Hadza - devote so much of their time and effort pursuing big game when most of their prey, throughout the year, are among the leanest animals on the planet (Speth 2010)?

A New Look at an Old Topic - Big Game Hunting by the Bushmen and Hadza

In the nutritional literature the safe upper limit to daily per capita protein intake, where it is discussed at all, is typically expressed as a percentage of total calories - usually in the range of about 25% to 35% (e.g., Bilsborough and Mann 2006; Cordain et al. 2000). This leads to the impression that so long as adequate supplies of fat or carbohydrates are at hand, one can increase one's protein intake over a wide range of values without any negative health consequences. This would mean that excess protein intakes would only pose a serious problem during those seasons of the year when plant foods are in short supply - i.e., during the late winter/early spring in more northerly latitudes, or during the late dry season/early rainy season in more southerly latitudes, such as in much of sub-Saharan Africa. From this perspective hunting big game in Africa would seem to make eminent economic and nutritional sense over a substantial part of the year. Moreover, the meat from big game animals is obviously a very rich source of "high-quality" protein - in other words, not only is it abundant, it also has the ideal mix of amino acids in just the right proportions. Thus, to many paleoanthropologists and archaeologists a lot of human evolution can be seen as the story of how our ancestors became increasingly proficient in taking high-quality protein (meat) in larger and larger packages - in other words, big game hunting was at the very heart of what humans were all about.

Needless-to-say, this is an old idea, one that goes back to Charles Darwin's (1871) classic "feedback model". Nearly a century later it was still alive and well, forming the core of the "Man the Hunter" symposium, with Washburn and Lancaster (1968) its leading spokespersons. Nothing could epitomize the big game hunting perspective better than the famous giraffe hunt in John Marshall's classic movie "The Hunters", a film that was shown to countless introductory anthropology classes throughout the US and Canada during the 1970s and 1980s (and, no, the giraffe was not shot with a rifle at the end, but, yes, the movie was in fact a commingling of several different hunting episodes involving several different giraffes; (see Henley 2003:47-48). The "Man the Hunter" perspective is incredibly resilient, and still dominates much of our thinking today, as seen, for example, in recent popular papers explaining why humans excel at endurance running (Bramble and Lieberman 2004; Lieberman et al. 2009). And of course the archaeological evidence seems to support it - I could see it with my own eyes at FxJj50 during that wonderful field season at Koobi Fora - and arguments from optimal foraging theory have provided an elegant and very compelling logic.

In reality, however, the upper limit to safe protein intake is a (more or less) fixed ceiling somewhere in the neighborhood of 250-300 g (see Speth 2010:76-79; the actual value varies with body weight and some degree of adaptation to sustained elevated protein intakes).

Once a person exceeds the threshold the deleterious health consequences begin to set in, regardless of how much additional fat or carbohydrate the person might consume. Thus, without long-term storage many of the large-bodied animals that are hunted by the Bushmen and Hadza (e.g., giraffe) yield far more protein than the hunter or his family can safely use. One can't help but wonder, then, why they devote so much time and effort hunting the big ones?

While I was in Koobi Fora, another doubt about the nature and role of big game hunting began to gestate in my mind, although it took many years for it to materialize into something palpable and coherent. It first made its presence known during one of our many seminar discussions under the acacia tree. Glynn and I frequently talked about how one should, or shouldn't, use modern hunter-gatherer data in our attempts to generate ideas about the distant past. We often came at these issues from rather different perspectives. Glynn, following David Clarke's (1968) lead, felt that the most productive avenue of inquiry would be to approach early hominins by means of models and theory developed largely or entirely within the archaeological realm, whereas I felt that we had no choice but to work from the living - especially from ethnohistory and ethnography. These divergent philosophies made the exchanges all the more interesting and stimulating for me, because in order to hold my own I had to think through a lot of things that up to that point had been implicit or very poorly formulated in my mind.

The niggling doubt that began slowly working its way to the surface concerned the nature of the Bushman bow. If hunting, and especially big game hunting, really was so vital to the success and well-being of the foraging adaptation, why did anthropology's quintessential hunters use a bow that seemed to me to be little more than a toy, together with a poison that was very slow-acting? The combination almost guaranteed that after a successful hit, which apparently didn't happen all that often, the hunters would have to spend hours searching for their slowly dying, but often quite mobile and uncooperative quarry (witness the giraffe(s) in John Marshall's film). Only recently, however, did I finally decide to look into the matter by going to a local sporting goods franchise and asking a salesperson what sort of draw weight would be suitable for a child like my son, who was only about 11 at the time. The response was "not more than about 20 lbs". That's the typical draw weight of a Bushman bow! Why would such intrepid hunters, who had come to epitomize the hunting way of life, use a bow whose draw weight was recommended for children? Far more powerful bows were widespread in southern Africa, including recurved and backed forms that were genuine shock weapons (e.g., Noli 1992). The Bushmen must certainly have been aware of their existence. We anthropologists have spent so much time in awe of the Bushmen's uncanny ability to track animals that we lost sight of a more fundamental question - why did they use a weapon system that was so ineffective in killing or at least disabling their prey?

As these various threads began to gel and come together - e.g., the existence of a limit to the amount of protein that one can safely consume in a day; the fact that African big game animals produce periodic massive inputs of protein that exceed what a hunter and his family can possibly use; the Bushmen's reliance on an ineffectual bow and a very slow-acting poison; and the lengthy and frequently unsuccessful pursuits that often followed a successful hit - I finally decided to take a renewed look at the Bushman and Hadza literature, focusing specifically on how big game hunting actually fit into to their

overall subsistence system (Speth 2010). Although I began this foray with little hope that I would arrive at anything new or particularly helpful, the outcome turned out to be quite a surprise, and certainly very different from what I would have expected when Glynn and I first began talking about these issues at Koobi Fora in 1979. What follows is a brief summary of the results of this inquiry, beginning with the Bushmen, and then turning to Hadza.

Bushman hunters have very low success rates, particularly for large game (Hitchcock et al. 1996:175). For example, Richard Lee documented the hunting activities of Ju/'hoansi (!Kung San) over 28 days in July-August 1964. During that period, seven men put in a total of 78 person-days of hunting, successfully killing animals on only 23% of those days (Lee 1979:267; Hitchcock et al. 1996:182). A few years later, in 1968, John Yellen observed the Ju/'hoansi for a period of 80 days. During this period, men made no attempt to hunt on 14 days and failed to procure anything on an additional 25 days, indicating that on nearly 50% of the days the hunters made no successful kills. Moreover, most of what they caught were small animals, especially porcupines and springhare, as well as a number of birds (Hitchcock et al. 1996:175). If one considers only the ungulates, their success rate was much lower.

Not only do the San frequently fail in their attempts to kill big game, but, as already alluded to, the way they go about it is truly perplexing. When I was a graduate student in the late 1960s, publication of the *Man the Hunter* symposium volume was like a shot of adrenalin for those of us who were interested in hunters and gatherers (Lee and DeVore 1968). Almost overnight the Ju/'hoansi San (in those days referred to as the !Kung Bushmen) became the gold standard by which we viewed and interpreted the hunter-gatherer past.

Almost everything came to be seen "through Bushman eyes". Thus, the way the Ju/'hoansi hunted was the way all good hunters, past or present, must have hunted - with uncanny stealth (Stander et al. 1996) and amazing skill as trackers (Liebenberg 1990).

Yet, at more or less the same time, I was influenced by a fellow graduate student, George Frison, an expert on hunting, whose insights stemmed from vast archaeological expertise combined with years of experience as a hunter in his own right. According to Frison (1978:366), big game hunters whose livelihood depended on the outcome of the hunt would leave as little to chance as possible: "there was careful consideration...as to where animals were killed; nothing was killed where the effort of recovery exceeded the value of the meat...". A good hunter would also choose a weapon that was appropriate for the behavior and size of the prey, and direct a shot at the animal that would either kill it on the spot or at least immobilize it (Frison 1998:14579).

Contrast Frison's observations with Bushman hunting. The San use tiny bows, not shock weapons (Bartram 1997:325; Silberbauer 1981:206; Thomas 2006:128), and arrows smeared with a deadly but very slow-acting poison (Lee 1979:219; Thomas 2006:126), a combination that almost guarantees the hunters will have to track their prey (Hitchcock and Bleed 1997:354). How do Bushman bows stack up against what modern archers would consider appropriate for a hunting weapon? Sparano (2000:692) in *The Complete Outdoors Encyclopedia* recommends a draw weight of 9 kg (20 lb), or less, for children between the ages of six and 12, echoing what I had been told by a local sporting goods dealer, and at least 23 kg (50 lb) for deer hunting. In other words, the draw weight of Bushman bows falls within the range of weights that modern bowyers recommend for children. This is hardly a weapon designed to deliver a lethal or immobilizing shot to a large animal. As a result, Bushman hunters have to invest inordinate amounts of time tracking wounded animals across the landscape, animals that they frequently can't find or lose to other predators (e.g., Hitchcock and Bleed 1997:354; Hitchcock et al. 1996:185).

In other words, in stark contrast to Frison's perspective of what an economically-motivated hunter must do in order to make ends meet, the San leave a great deal to chance, fail frequently, and invest a huge amount of time and effort doing so. This is hardly a strategy designed to maximize returns of calories, protein, or fat, nor is it an effective way to minimize time, or opportunity costs, unless the payoff lies in some domain other than food.

The Bushman strategy becomes even more perplexing if one considers the number of years (actually, decades) that it takes a hunter to master the skills needed to successfully locate, pursue, and kill large game. According to Walker et al. (2002:639), hunters don't attain their peak level of performance until they are well into their thirties or even later, long after they pass their physical prime. If, as Kristen Hawkes (2000:65) so aptly put it, "in the long run, big game hunting is inferior to available alternative strategies for provisioning families," one has to wonder why hunters invest so much of their life, starting already as children, honing these particular skills?

Curiously, Hadza hunters in Tanzania, though using a much more powerful bow than the one typically used by the Bushmen, don't fare much better in their hunting endeavors. Even though Hadza bows have draw weights of 45 kg (100 lb) or more (Bartram 1997; Woodburn 1970), and arrow points smeared with poison, "individual hunters...fail to kill (or scavenge) large game on 97% of all hunting days" (Hawkes et al. 1997:573).

Even more eye-opening is Kristen Hawkes's (2000:64–65) comparison of Hadza hunting success with the return rates that hunters might expect if they instead devoted their efforts to other subsistence pursuits. Hadza men, on average, devote more than four hours per day to hunting, and yet take home only about 0.12 kg/hr of meat. Hawkes concludes that, in terms of caloric returns, adult Hadza men would enjoy higher return rates by gathering.

The tremendous day-to-day variance in hunting success, in which an "average hunter can expect a full month of failures for every day he scores," would be devastating as a family provisioning strategy, especially for children (Hawkes 2000:65). O'Connell et al. (2002:836) take this line of reasoning a step further, concluding that prestige, rather than nutrition, underlies the Hadza's focus on big game hunting. Thus, just in terms of success rates, big game hunting by the Hadza, as was the case for the Bushmen, seems like a very inefficient and unreliable way of putting food on the table.

Given the rather dismal and unreliable returns of Bushman and Hadza big game hunting, let us take a look at some of the alternative foods that are available to these hunter-gatherers and the times of year when these alternatives are available. In the interest of space, I have omitted many of the citations documenting the nutritional details of these alternative foods; the complete list of citations may be found in Speth (2010). Let's begin with the San.

Through Richard Lee's (1968, 1973:320, 1979) seminal work among the Bushmen in the 1960s, we know that the Ju/'hoansi relied heavily on mongongo nuts. In the area where Lee did his field work, groves of mongongo trees were extremely productive in most years, and

according to his input-output studies provided, on average, about 40% or more of the Ju/'hoansi's daily energy intake. The actual percentage varied seasonally from a low of about 10% in the late summer rainy season to as high as 90% in the fall and early winter (dry season) months when the fruits ripened and dropped to the ground. Mongongo nuts contain very high levels of fat and are also very rich in protein. Lee (1979:270) observed that Ju/'hoansi during the harvest season ate about 300 nuts per person per day, which according to Duke (2000:258) would contain "the caloric equivalent of 1,134 g of cooked rice and the protein equivalent of 396.9 g lean beef".

Particularly intriguing about mongongos in the context of the present discussion is the fact that their peak abundance occurs in the months of April, May, and June, precisely the months when the Ju/'hoansi also bring the most meat into camp (Hitchcock et al. 1996:201). Tsin beans (marama), another very important source of protein and fat for many San groups, ripen at this same time of year. In other words, Ju/'hoansi hunting activities peak at more or less the same time that the return rates from harvesting and processing mongongo nuts and tsin beans also peak (~1,300 kcal/hour for the mongongo according to Sih and Milton 1985:399). If there were any time of the year when the Ju/'hoansi would not need to hunt large game for either fat or protein, this would be it.

The same pattern also holds for the G/wi in the Central Kalahari and the Tyua in northern Botswana. While the G/wi and Tyua don't have mongongo nuts, they do have other protein- and fat-rich plant foods, such as tsin beans, marula nuts, and baobab fruits, which also reach their maximum availability during the rainy season, at more or less the same time that their hunting returns peak (Speth 2010:95–107; Wiessner 1981).

The Tyua have another important resource as well, in this case an insect, the mopane worm, that becomes available at this same time of year, and often in prodigious quantities. The mopane worm, the caterpillar or larval stage of the Emperor moth, has two "outbreaks" each year, the principal one occurring in the early months of the rainy season between November or December and January, and a second more minor one between March or April and May (Speth 2010:95-107). Mopane worms are rich in both protein and fat, providing about 450 kcal/100 g.

Thus, it would appear that Bushman groups throughout the Kalahari have access to a number of nutrient-dense and often quite abundant plant, animal, and insect resources that become available at more or less the same time of year that they undertake much of their big game hunting. This intriguing temporal convergence raises the possibility that the Bushmen hunt these animals for reasons other than fat or protein. Perhaps, instead, it is precisely because of the reliability and high fat and protein content of mongongos, baobabs, tsin beans, marula nuts, mopane worms, and others that Bushman hunters are able to afford the "luxury" of engaging in such a time-consuming, failure-prone, and costly activity. In other words, an explanation for their hunting behavior may well lie beyond the strictly nutritional realm.

Now let us shift our focus to the Hadza. For these East African foragers baobab fruits and seeds assume much the same role that mongongo nuts do for the Ju/'hoansi. Baobab seeds contain almost 30% fat by dry weight, and a similar or even higher concentration of protein (up to 36%). In fact, baobab seeds contain substantially more protein than agricultural plants like sorghum (11.4%), millet (11.9%), and manioc (0.9%). According to Murray et al. (2001:9),

baobab seeds in fact yield about the same amount of energy per 100 g dry weight as honey.

Baobab fruits ripen during the late dry season and/or early wet season, but remain edible for several months after they form (Marlowe 2006:363). According to Murray et al. (2001:12), Hadza "women consistently returned with dozens of baobab fruits or with significant quantities of seeds over the majority of months of the year either through direct fruit collection or through collection of seeds in baboon dung piles".

How does Hadza hunting covary with baobab availability? The Hadza engage in two principal types of hunting - intercept hunting at night from blinds during the late dry season when animals are concentrated close to major waterholes, and daytime encounter hunting which occurs throughout the year whenever hunters are out of camp. Curiously, despite their much more powerful bows, the Hadza, like the San, rely on slow-acting poison and frequently have to track their prey for several hours or longer after they have been wounded (O'Connell et al. 1992; Hawkes et al. 1991, 2001). In the studies conducted by Kristen Hawkes and colleagues in the mid- to late 1980s, the Hadza made 52 kills just in the last three months of the dry season - August, September, and October - compared to a total of only 19 kills during the remaining seasons (O'Connell et al. 1992:320-321). And the late dry season is precisely when the baobabs come into fruit. It is therefore tempting to conclude that Hadza big game hunting, like Bushman big game hunting, was only possible because it was underwritten by the availability of other productive, dependable, and cost-effective food sources. With baobabs, mongongos, tsin beans, marula nuts, and mopane worms as staples, the Hadza and Bushmen certainly don't need to hunt big game for either protein or fat. And in the Hadza case the late dry season also happens to be the worst possible time of year to pursue African ungulates for fat (Speth 2010).

Conclusions - Why Hunt Big Game?

Since its inception, paleoanthropology has been closely wedded to the idea that big game hunting by hunter-gatherers like the Bushmen and Hadza, and by our hominin ancestors in Africa, first and foremost, served as a means for acquiring energy and vital nutrients, especially protein. From such a perspective, it is not surprising that a significant part of the human story has been seen as a record of our ever-increasing prowess and sophistication at taking large prey. The assumption that big game hunting was primarily motivated by food needs has rarely been questioned. After all, few things in human evolution have seemed so intuitively obvious - meat is a nutrient-rich food, full of protein with the ideal array of essential amino acids, all in the right proportions; and big animals provide meat in large convenient packages, making them, hands down, the preferred target for foragers who have the organizational and technical means to kill them. Add in a "dash" of prestige for good measure, and you have the classic model of "Man the Hunter". But what seems so obvious and compelling at face value turns out to be less so when looked at more closely. The present endeavor has been an attempt to offer an alternative to the traditional view. I suggest instead that, for the Bushmen and Hadza, and for our forebears of the African Pleistocene, big game hunting was first and foremost about social, reproductive, and political goals, prestige among them, and that the nutritional component was actually the "dash" that got added in.

If Glynn were with us today, he might well disagree with my conclusions; but I am sure, whether he agreed or not, he would relish discussing them. And what would ensue would be a lively exchange, probably lasting long into the night, in which he would scrutinize every piece of the argument from one end to the other. And I would have my job cut out for me, trying to keep up with his insightful probing and questioning. In the end, however, when we both felt it was time to call it quits, I would retire knowing that I had just been treated to one of the most stimulating and rewarding evenings imaginable. I would have learned a lot in the process, and, who knows, I might even find myself setting off in a wholly new direction yet again. Glynn is sorely missed, both professionally and personally....

Acknowledgments

I am delighted to be able to participate in this Memorial to Glynn. I particularly appreciate the chance to say "thank you" to someone who played such an important role in my life. I spent three glorious months in Kenya in 1979, thanks to an invitation from Glynn, and those months on the Karari Escarpment had a tremendous impact on virtually everything I've done since. It's also wonderful as part of this same celebration to be able to say a special "thank you" to Barbara Isaac, who played such a huge role in making that field season a truly wonderful experience. There were lots of memorable moments at Kampi Ya Simba, far too many to mention here, but I often still think about that season and it brings an instant smile - even though Barbara frequently managed to best me in the evening when we gambled our daily ration of rum (two "tots" or capfuls) to see if we could get enough in our glass to actually taste it. I also had the pleasure of getting to know a number of Glynn's graduate students, many of whom have gone on to make their own valuable contributions to the field. I remember the seminars, and the many trips together on days off, with Henry Bunn, Zefe Kaufulu, Ellen Kroll, Fiona Marshall, Francis

Musonda, Kathy Schick, Jeanne Sept, Nick Toth, and Anne Vincent (I hope I haven't forgotten anyone). Those were great times. Finally, I am extremely grateful to Jeanne Sept and David Pilbeam for making this celebration of Glynn's life and many contributions a reality.

References

Anonymous
1983 Announcement. *Human Ecology* 11(3):363.

Bartram, L. E.
1997 A comparison of Kua (Botswana) and Hadza (Tanzania) bow and arrow hunting. In *Projectile Technology*, edited by H. Knecht, pp. 321-343. Plenum, New York.

Behrensmeyer, A. K.
1975 *The Taphonomy and Paleoecology of Plio-Pleistocene Vertebrate Assemblages East of Lake Rudolf, Kenya*. Volume 146(10):473-578. Harvard University, Museum of Comparative Zoology, Cambridge.

Bilsborough, S., and N. Mann
2006 A review of issues of dietary protein intake in humans. *International Journal of Sport Nutrition and Exercise Metabolism* 16(2):129-152.

Binford, L. R.
1978 *Nunamiut Ethnoarchaeology*. Academic Press, New York.

1981 *Bones: Ancient Men and Modern Myths*. Academic Press, New York.

Brain, C. K.
1969 The contribution of Namib Desert Hottentots to an understanding of Australopithecine bone accumulations. *Namib Desert Research Station Scientific Paper* 39:13-22.

1975 An introduction to the South African Australopithecine bone accumulations. In *Archaeozoological Studies*, edited by A. T. Clason, pp. 109-119. North-Holland Publishing Company, Amsterdam.

Bramble, D. M., and D. E. Lieberman
2004 Endurance running and the evolution of *Homo*. *Nature* 432(7015):345-352.

Bunn, H. T., J. W. K. Harris, G. L. Isaac, Z. Kaufulu, E. Kroll, K. Schick, N. Toth, and A. K. Behrensmeyer
1980 FxJj50: An Early Pleistocene site in northern Kenya. *World Archaeology* 12(2):109-136.

Clarke, D. L.

1968 *Analytical Archaeology*. Methuen, London.

Cordain, L., J. Brand Miller, S. B. Eaton, N. Mann, S. H. A. Holt, and J. D. Speth

2000 Plant-animal subsistence ratios and macronutrient energy estimations in worldwide hunter-gatherer diets. *American Journal of Clinical Nutrition* 71(3):682–692.

Coues, E. (Editor)

1893 *The History of the Lewis and Clark Expedition*, Vol. 1. Francis P. Harper, New York.

Darwin, C. R.

1871 *The Descent of Man, and Selection in Relation to Sex*. D. Appleton and Company, New York.

Duke, J. A.

2000 *Handbook of Nuts: Herbal Reference Library*. Herbal Reference Library Volume 4. CRC Press, Boca Raton.

Frison, G. C.

1978 *Prehistoric Hunters of the High Plains*. Academic Press, New York.

1998 Paleoindian large mammal hunters on the Plains of North America. *Proceedings of the National Academy of Sciences* 95(24):14576–14583.

Hawkes, K.

2000 Hunting and the evolution of egalitarian societies: Lessons from the Hadza. In *Hierarchies in Action: Cui Bono?*, edited by M. W. Diehl, pp. 59–83. Occasional Paper 27. Southern Illinois University, Center for Archaeological Investigations, Carbondale.

Hawkes, K., J. F. O'Connell, and N. G. Blurton Jones

1991 Hunting income patterns among the Hadza: Big game, common goods, foraging goals, and the evolution of the human diet. *Philosophical Transactions of the Royal Society (London), Series B. Biological Sciences* 334B(1270):243–251.

1997 Hadza women's time allocation, offspring provisioning, and the evolution of long postmenopausal life spans. *Current Anthropology* 38(4):551–577.

2001 Hadza meat sharing. *Evolution and Human Behavior* 22(2):113–142.

Henley, P.

2003 Film-making and ethnographic research. In *Image-based Research: A Sourcebook for Qualitative Researchers*, edited by J. Prosser, pp. 42–59. Routledge Falmer Press, London.

Hitchcock, R. K., and P. Bleed
1997 Each according to need and fashion: Spear and arrow use among San hunters of the Kalahari. In *Projectile Technology*, edited by H. Knecht, pp. 345-368. Plenum Press, New York.

Hitchcock, R. K., J. E. Yellen, D. J. Gelburd, A. J. Osborn, and A. L. Crowell
1996 Subsistence hunting and resource management among the Ju/'hoansi of northwestern Botswana. *African Study Monographs* 17(4):153-220.

Isaac, G. L.
1969 Studies of early culture in East Africa. *World Archaeology* 1(1):1-28.

1971 The diet of early man: Aspects of archaeological evidence from Lower and Middle Pleistocene sites in Africa. *World Archaeology* 2(3):278-298.

1972a Chronology and the tempo of cultural change during the Pleistocene. In *Calibration of Hominoid Evolution: Recent Advances in Isotopic and Other Dating Methods Applicable to the Origin of Man*, edited by W. W. Bishop and J. A. Miller, pp. 381-430. Scottish Academic Press, Edinburgh.

1972b Comparative studies of Pleistocene site locations in East Africa. In *Man, Settlement and Urbanism*, edited by P. J. Ucko, R. Tringham and G. W. Dimbleby, pp. 165-176. Duckworth, London.

1972c Early phases of human behaviour: Models in Lower Palaeolithic archaeology. In *Models in Archaeology*, edited by D. L. Clarke, pp. 167-199. Methuen, London.

Isaac, G. L., R. E. F. Leakey, and A. K. Behrensmeyer
1971 Archeological traces of early hominid activities, east of Lake Rudolf, Kenya. *Science* 173(4002):1129-1134.

Lee, R. B.
1968 What hunters do for a living, or, how to make out on scarce resources. In *Man the Hunter*, edited by R. B. Lee and I. DeVore, pp. 30-48. Aldine, Chicago.

1973 Mongongo: The ethnography of a major wild food resource. *Ecology of Food and Nutrition* 2(4):307-321.

1979 *The !Kung San: Men, Women, and Work in a Foraging Society*. Cambridge University Press, Cambridge.

Lee, R. B., and I. DeVore (Editors)
1968 *Man the Hunter*. Aldine, Chicago.

Liebenberg, L.
1990 *The Art of Tracking: The Origin of Science*. David Philip Publishers, Claremont, South Africa.

Lieberman, D. E., D. M. Bramble, D. A. Raichlen, and J. J. Shea
2009 Brains, brawn, and the evolution of human endurance running capabilities. In *The First Humans: Origin and Early Evolution of the Genus* Homo, edited by F. E. Grine, J. G. Fleagle and R. E. F. Leakey, pp. 77-92. Springer, Dordrecht.

Marcy, R. B.
1863 *The Prairie Traveler: A Handbook for Overland Expeditions.* Trubner, London.

Marlowe, F. W.
2006 Central place provisioning: The Hadza as an example. In *Feeding Ecology in Apes and Other Primates: Ecological, Physical and Behavioral Aspects*, edited by G. Hohmann, M. M. Robbins and C. Boesch, pp. 359-377. Cambridge University Press, Cambridge.

Murray, S. S., M. J. Schoeninger, H. T. Bunn, T. R. Pickering, and J. A. Marlett
2001 Nutritional composition of some wild plant foods and honey used by Hadza foragers of Tanzania. *Journal of Food Composition and Analysis* 14(1):3-13.

Noli, H. D.
1992 Archery in Southern Africa: The Evidence from the Past, Unpublished Ph.D. Dissertation, Department of Archaeology, University of Cape Town, Rondebosch, South Africa.

O'Connell, J. F., K. Hawkes, and N. G. Blurton Jones
1992 Patterns in the distribution, site structure and assemblage composition of Hadza kill-butchering sites. *Journal of Archaeological Science* 19(3):319-345.

O'Connell, J. F., K. Hawkes, K. D. Lupo, and N. G. Blurton Jones
2002 Male strategies and Plio-Pleistocene archaeology. *Journal of Human Evolution* 43(6):831-872.

Sih, A., and K. A. Milton
1985 Optimal diet theory: Should the !Kung eat mongongos. *American Anthropologist* 87(2):395-401.

Silberbauer, G. B.
1981 *Hunter and Habitat in the Central Kalahari Desert.* Cambridge University Press, Cambridge.

Sparano, V. T.
2000 *The Complete Outdoors Encyclopedia.* Macmillan, New York.

Speth, J. D.
1972 Mechanical basis of percussion flaking. *American Antiquity* 37(1):34-60.

1983 *Bison Kills and Bone Counts: Decision Making by Ancient Hunters.* University of Chicago Press, Chicago.

2010 *The Paleoanthropology and Archaeology of Big Game Hunting: Protein, Fat or Politics?* Springer, New York.

Speth, J. D., and D. D. Davis
1976 Seasonal variability in early hominid predation. *Science* 192(4238):441-445.

Speth, J. D., K. Newlander, A. A. White, A. K. Lemke, and L. E. Anderson
2010 Early Paleoindian big game hunting in North America: Provisioning or politics? *Quaternary International.*

Speth, J. D., and K. A. Spielmann
1983 Energy source, protein metabolism, and hunter-gatherer subsistence strategies. *Journal of Anthropological Archaeology* 2(1):1-31.

Stander, P. E., X. Ghau, D. Tsisaba, and X. Txoma
1996 A new method of darting: Stepping back in time. *African Journal of Ecology* 34(1):48-53.

Stefansson, V.
1944 *Arctic Manual.* Macmillan, New York.

Thomas, E. M.
2006 The lion/Bushman relationship in Nyae Nyae in the 1950s: A relationship crafted in the old way. In *The Politics of Egalitarianism: Theory and Practice,* edited by J. S. Solway, pp. 119-129. Berghahn Books, New York.

Walker, R. S., K. Hill, H. Kaplan, and G. McMillan
2002 Age-dependency in hunting ability among the Ache of eastern Paraguay. *Journal of Human Evolution* 42(6):639-657.

Washburn, S. L., and C. S. Lancaster
1968 The evolution of hunting. In *Man the Hunter,* edited by R. B. Lee and I. DeVore, pp. 293-303. Aldine, New York.

Wiessner, P.
1981 Measuring the impact of social ties on nutritional status among the !Kung San. *Social Science Information* 20(4-5):641-678. SAGE, London and Beverly Hills.

Woodburn, J.
1970 *Hunters and Gatherers: The Material Culture of the Nomadic Hadza.* British Museum, London.

9

HONEY AND FIRE
IN HUMAN EVOLUTION

Richard W. Wrangham

In his essay "Casting the net wide" Glynn Isaac (1978) drew attention to the importance in evolutionary studies of understanding the influences of all likely major food types. To achieve this, he wrote, "we need to make careful problem-oriented studies of environments that in specific ways are analogues of environments inhabited by evolving hominids. ... Such a broadening of scope will draw many different kinds of scientist into the argument about early man..." (Isaac 1978:249).

With Isaac's enjoinment in mind my aim here is to evaluate the role of honey in human evolution. Skinner (1991) previously reviewed the potential availability of honey for savanna hominids, but unfortunately the premise for his interest proved faulty. Skinner had suggested that possible signs of excessive dietary Vitamin A in *Homo erectus*, indicated by abnormal bone growth in KNM-ER 1808, might be due to eating excessive bee brood (as opposed to too much carnivore liver, which Walker et al. (1982) had proposed); but contrary to prior reports Skinner et al. (1995) found that bee brood contained very low levels of Vitamin A. Yaws is now an alternative explanation for the bone deposits in KNM-ER 1808 (Rothschild et al. 1995).

In the absence of paleo-pathological evidence honey has become archaeologically invisible: Saharan and South African rock paintings are the oldest direct indication of honey-eating in the human lineage (< 20,000 bp; Isack and Reyer 1989). Furthermore a review of paleo-diets concluded that honey "represented a relatively minor dietary component over the course of a year", representing around 3% of caloric intake (Cordain et al. 2005:341). However Cordain et al.'s review used data from only two forager groups, both living in continents without any naturally occurring honey bees (Ache (Paraguay) and Anbarra (Australia)). In contrast to Cordain et al.'s inference, honey can be a major component in the diets of African hunter-gatherers (below), for whom it vies with meat in its desirability and as an influence on nutrition, tool-using and social relationships. Honey is also a preferred food for chimpanzees, suggesting that it could have been important in pre-human diets. I therefore review current data from various "different kinds of scientists" including behavioral ecologists, ethnologists, and primatologists in an attempt to evaluate the possible significance of honey as a dietary item for Pliocene and Pleistocene hominids.

Evidence for a Long Evolutionary Relationship between Humans and Honey

Evidence that the foraging of human ancestors has included honey as a major component prior to 20,000 ya comes from the behavior of the greater honeyguide (*Indicator indicator*; Figure 9.1). This solitary-living bird habitually guides humans to honey. It is found throughout savannas of sub-Saharan Africa except for the northeastern and southwestern deserts, and is the only species in the family Indicatoridae known

Figure 9.1 Adult male greater honeyguide,
Indicator indicator. ©*Warwick Tarboton.*

to have the guiding habit, even though others also share its habits of feeding on beeswax (Short and Horne 2001). The greater honeyguide's relationship with humans has been characterized as "possibly the most advanced bird-mammal relationship in the world" (Dean et al. 1990). The behavior has been reviewed by Friedmann (1955) and described in most detail by Isack and Reyer (1989), who studied its interactions with Boran pastoralists of northern Kenya.

According to Friedmann (1955) and Isack and Reyer (1989), the greater honeyguide (henceforth "honeyguide") is attracted to many kinds of human activity, including walking, whistling, chopping wood and smoke, and nowadays motor vehicles and boats. A honeyguide that has recently visited an intact bee hive of *Apis*

mellifera appears to look for humans. On finding people it flies close, hops among nearby perches, fans its tail, arches and ruffles its wings, gives a characteristic two-note churring or chattering call and leads in the direction of the hive. The hive may be 1 km away or sometimes further. When a person comes within 5–15 m of the hive tree the bird waits in a neighboring tree, churring loudly until the person again nears the tree. People normally use axes to enlarge the hive entrance and smoke to subdue the bees. The bird waits patiently during the process, which may take more than an hour-and-a-half. After humans have left with their honeycomb the bird flies down to feed on larvae and wax in the fragments of comb that remain (often deliberately left for the bird by the honey-gatherers), thereby supplementing its regular diet of insects (Figure 9.2a–c). Like other *Indicator* species greater honeyguides have an unusual adaptation enabling them to digest wax, which is a preferred food on which at least one species (lesser honeyguide, *Indicator minor*) can survive for weeks at a time (Ambrose 1978; Diamond and Place 1988; Downs et al. 2002).

Honeyguides and humans each benefit from their relationship. Marlowe (2010) found that honeyguides led Hadza men only to *Apis* honey, never to the less valuable hives of stingless bees. Isaak and Reyer (1989) found that 96% of the hives to which honeyguides led Boran men would have been inaccessible to the birds until the men had opened them (cf. Short and Horne 2001, who estimate 50%). Furthermore the bird's chances of being stung were reduced by the smoke. Honeyguides have unusually thick skin and upraised, hard-edged slit nostrils, both possibly adapted as a defense against bee stings (Short and Horne 2001). Nevertheless greater honeyguides have been killed by being stung too often (Short and Horne 2001). With regard to the benefits for humans, Boran men in search of honey

Figure 9.2a–c. Sequence leading to an adult male greater honeyguide collecting honeycomb left on the ground by Hadza men following their extraction of honey from a wild Apis mellifera hive, east of Lake Eyasi, northern Tanzania. This bird had found four men in the process of examining an old hive (which proved to be unproductive). On arriving near the honeyhunters the honeyguide immediately started its display-call, and one of the men left to follow it. The honeyguide flew ca. 250 m away where it drew attention to a small tree containing an active hive. The Hadza follower rated the chances of finding a good supply of honey as being high, and duly returned to his companions to bring them for honey extraction. After calming the bees by inserting a smoking brand into the hive entrance, the Hadza men extracted at least eight combs of honey by hand, while the bird perched within 100 m, occasionally calling. After the honey-hunters had eaten (Figure 9.2a) they abandoned the site and left some discarded pieces of comb containing little or no honey (Figure 9.2b). Two filmers remained at ca. 25 m distance from the honey tree. During the next hour the bird made four visits to collect pieces of comb, initially after about 15 minutes (Figure 9.2c). He flew off each time carrying the comb up to 100 m away. Credits: Figure 9.2a–b: R. Wrangham; Figure 9.2c: Screen-capture from film by Bob Poole, courtesy of Benenson Productions Inc. Used with permission.

would give a loud specific whistle in order to attract a honeyguide. This led to birds arriving twice as often as without a whistle. When led by a honeyguide Boran honey-gatherers took an average of 3.2 hours to find a hive, compared to 8.9 hours on their own. Friedmann (1955) indicates shorter times.

Honeyguides and humans thus have a highly developed and mutually beneficial relationship for getting honey and hive products. Until recently this relationship was thought to result from humans taking advantage of guiding behavior that had co-evolved between the honeyguide and honey-badgers *Mellivora capensis*. However the notion that honeyguides and honey badgers have a co-evolved mutualistic relationship has been convincingly challenged.

The supposition of a close mutualism between honeyguides and honey badger began with Sparrman (1785-1786) and was demolished by Dean and Macdonald (1981) and Dean et al. (1990). Honey badgers are mostly nocturnal and terrestrial, although they can be active by day and sometimes climb trees. But Dean et al. (1990) note that no naturalist has ever claimed to see honeyguides leading honey badgers to honey, or even attempting to do so. The entire evidence for the claimed mutualism consists of a secondary citation from Sparrman (1785-1786) and various incomplete eyewitness reports cited by Friedmann (1955). Since 1955 no indications have emerged that honeyguides and honey badgers are routinely interested in each other's behavior, even though there has been sufficient study to reveal a previously unsuspected association between honey badgers and another bird (pale chanting goshawk, *Melierax canorus*; Dean et al. 1990). Informal playbacks of honeyguide calls produced no signs of interest in honey badgers. Short and Horne (2001) note that honey badgers have an excellent sense of smell and hearing,

apparently sufficient to enable them to find honey unaided. For instance honey badgers often raid apiaries at night. Smith and Horne (2001) suggest that possible reports of honeyguides leading honey badgers may have been confused with leading of the human observer. Unlike humans, honey badgers do not offer smoke to help quieten bees. In sum, even if honeyguides prove to occasionally guide honey badgers to honey the frequency is clearly too low to justify the concept of a co-evolved relationship.

Dean et al.'s (1990) elimination of honey badgers as a co-evolved mutualistic partner raises the question of why honeyguides are adapted to being guides. The habit certainly appears to be an evolved feature (rather than being socially learned), given that the honeyguide is a brood parasite (Spottiswoode and Colebrook-Robjent 2007). This means that the young has no opportunity to learn from either parent, so the adult's guiding behavior must rest on innate predispositions.

Accordingly the honeyguide's leading behavior must have evolved in partnership with at least one other species. There are three kinds of candidate. The partners could have been nonhuman honey-eaters other than honey badgers; extinct species of honey-eater; or hominids. Humans have been found to be the most severe contemporary predators of *Apis* honey in Africa compared to all other species (Caron 1978; Skinner 1991; Kajobe and Roubik 2006). Chimpanzees are a distant second but there are no records of honeyguides leading chimpanzees (or any other nonhuman) to honey. No extinct species other than hominids has been proposed or is likely to have been a major honey-eater. By contrast hominids are a plausible answer given the high rate of honey-collecting by humans and the likely preference for honey by ancestors of *Homo sapiens*. Hoesch (1937, cited by Dean et al. 1990) was the first to propose that hominids were responsible for the evolution of guiding behavior in honeyguides. Dean et al. (1990) concurred.

The Importance of Honey for Hunter-Gatherers

The implication of honeyguide behavior being adapted to hominid honey-gathering is clear. As Dean et al. argued (1990:100), the "selective advantages of the guiding behavior must have been considerable for this complex symbiosis to have evolved". Hominid ancestors must have extracted substantial amounts of honey on a regular basis. Data on hunter-gatherers support this idea by showing that honey has been important in recent African forager diets.

First, honey is a preferred and valuable food. African honey comes from three main sources, i.e., *Apis* honey bees, various stingless bees (Meliponini), and *Xylocopa* carpenter bees (Roubik 1989). Only *Apis* live as large colonies of relatively big bees, such that hives can contain several kilograms of honey; their honey is therefore the principal type exploited by humans. Stingless bees tend to produce amounts of honey per hive closer to a handful, while the honey (strictly 'bee-bread') of a *Xylocopa* bee would typically occupy less than 5 cm^3. Using questionnaires given to Hadza (Tanzania), Berbesque and Marlowe (2009) found that honey was the top-ranked food for both men and women, preferred even to meat, with *Apis* honey the most highly prized type. Honey was likewise reported to be the preferred food of Mbuti (Democratic Republic of Congo). After a man found a hive, whether close to the ground or as high as 30 m, he would invariably exploit it (Ichikawa 1981). The high preference makes sense. Wild honey is even more nutritious than marketed honey, since it tends to contain more larvae. Hadza honey bee honey had 403-439 cals/100 g dry weight, including 87.7-96.0% sugar (Murray et al. 2001).

It is also a good source of B-vitamins, vitamin C, and choline (Finke 2005). Honey is normally eaten rapidly but one of its advantages is that it can be stored for long periods (Skinner 1991).

Second, *Apis* bees are ubiquitous in African savanna environments and honey is often abundant. Kajobe and Roubik (2006) suggested a conservative mean density in tropical habitats of six *Apis* hives per km^2, but much higher levels have been recorded (e.g., 58 hives per km^2 in Zambia, and >12 in 100 yards in South Africa, Skinner 1991). Based on six hives per km^2 Skinner (1991) showed that hunter-gatherers would have 1,884 hives within a 10 km radius of camp, enough to warrant regular searching.

Third, as expected from the high preference for a food that is often abundant, tropical and subtropical hunter-gatherers worldwide regularly include substantial amounts of honey in their diets. All African hunter-gatherers appear to have eaten large amounts of honey, including Mbuti (Ichikawa 1981), Efe (Bailey 1991), Twa (Kajobe and Roubik 2006), Mbendjele Yaka (Lewis 2002), !Kung San (Lee 1979), Dorobo (Queeny 1952), and Hadza (Marlowe 2010). Over the year honey can be more important than meat. In the Ituri Forest (DR Congo) more calories were brought in by Efe men as honey than from any other food class, including meat or plants (Bailey 1991). On average they obtained 3.15 kg of honey per hive, for 1.7 hours of extracting (Bailey 1991). Hadza men brought back more calories of honey per hour foraging than any other food type. Though honey contributed only 8% by weight of food brought to camp over the year, compared to 28% for berries, 27% for meat, and 25% for tubers its contribution to the diet of men was probably much higher than 8%, since it was frequently eaten in large quantities out of camp (Figure 9.2a; Berbesque and Marlowe 2009; Marlowe 2010).

Apis honey production is normally seasonal, with the specific season depending on local variation in nectar production. In savannas of northern and western Tanzania people obtained *Apis* honey mostly in the wet seasons (six months per year; Berbesque and Marlowe 2009; Takeda 1976). Among the forest-living Efe and Mbuti the *Apis* honey seasons were mostly in dry periods from May to July (Ichikawa 1981; Bailey 1991). Ichikawa (1981) found that during the Mbuti honey season 80% of total energy intake (i.e., 1,900 calories per person per day) came from honey. At this time men largely abandoned hunting mammals. Ichikawa (1981) recorded 229 kg of honey consumed in one Mbuti camp (averaging 23 individuals) in 12 days, coming from 45 hives of *Apis* and four of stingless bees. The honey seasons were particularly comfortable for Mbuti. Mbuti men cut a maximum of four hives per day, rarely working more than six-to-seven hours to get honey. Honey season evenings were a time of relaxation when people sat around a big fire and sang honey songs.

Honey in Great Ape Diets

The fact that African hunter-gatherers regularly obtain more calories from honey than from meat draws attention to the importance of understanding how honey-eating evolved. Great apes offer an opportunity to speculate about honey in australopithecine diets.

The digestive systems of the great apes are rather uniform (Milton 1987), so data on the extant great apes can help with reconstructing australopithecine diets. All great apes show a strong appetite for honey. Bonobos (*Pan paniscus*) eat honey from at least three genera of Meliponine (stingless) bee in Wamba, DR Congo (Kano 1992), and from at least two genera of Meliponine bees in Lilungu, DR Congo (Bermejo et al. 1994). Chimpanzees, gorillas

(*Gorilla gorilla*), and orangutans (*Pongo pygmaeus*) all enjoy honey in captivity, so much so that researchers have found it to be a highly motivating food item in behavioral and cognitive experiments (Hirata and Celli 2003; Lonsdorf et al. 2009).

In the wild, however, there is considerable variation in the amount of honey-eating by great apes, despite the apparently ubiquitous presence of *Apis* honey bees in their habitats. Chimpanzees eat more *Apis* honey than other apes: they have eaten it in all populations where researchers have studied chimpanzees for more than a few months (at least 20 sites – McGrew et al. 2007; Watts 2008; Boesch et al. 2009; Gruber et al. 2009; Sanz and Morgan 2009). However, the total amounts eaten are always less than those of hunter-gatherers. Honey-eating appears to have been especially frequent among the savanna chimpanzees of Mt. Assirik, Senegal, where 23% of feces contained parts of *Apis* (McGrew 1983). Chimpanzees in the forests of Kahuzi-Biega (DR Congo) had beeswax in 3.1% of their feces, eating *Apis* honey almost entirely in the dry season (Yamagiwa and Basabose 2006). In Ruhija (Bwindi), a montane forest site in Uganda, 3% of feces or food remains included honey or bee products (Stanford 2006).

There is less consistency in honey-eating by other African apes. Honey has been recorded occasionally in the diet of one gorilla population (Kahuzi-Biega; Yamagiwa and Basabose 2006), but unlike sympatric chimpanzees, gorillas were not recorded eating *Apis* honey in Lope (Gabon) or Bwindi (Uganda; Tutin and Fernandez 1992; Stanford 2006), nor in the Virunga Mountains (Schaller 1963).

Finally despite bonobos eating stingless bee honey there is no published record of their eating *Apis* honey (McGrew et al. 2007). Observers confirm that bonobos have not been recorded eating *Apis* honey in Wamba (T. Furuichi pers. comm.), Lui Kotal (M. Surbeck, G. Hohmann and A. Fowler pers. comm.), or the short-term site of Kokolopori (A. Georgiev pers. comm.). This failure to exploit *Apis* honey is surprising since bonobos like Meliponine honey, and *Apis* bees are present throughout these forests. In Lui Kotal *Apis* bees have been recorded living at particularly high densities ("countless numbers daily"; McGrew et al. 2007).

No explanations have been proposed for variation in the intensity of *Apis* honey-eating among great apes. Hive density and honey quality could in theory vary, though it seems unlikely to differ systematically between chimpanzee and bonobo rainforest habitats given the proximity, similarity and latitudinal overlaps of the geographical ranges of these two species. Furthermore among hunter-gatherers high levels of honey-gathering occur in both savanna and forest habitats (Berbesque and Marlowe 2009; Ichikawa 1981). Other factors therefore seem likely to apply.

In theory motivation to eat honey may vary. However not only do captive apes work readily for honey and bonobos eat stingless bee honey, but variation in preference seems unlikely to be a major contributor since all great apes would benefit from high-energy food items, and the high sugar content (around 90%) means that honey is easily digested.

Second, great apes vary in how readily they climb. Gorillas' lower frequency of arboreality may help explain why they eat less *Apis* honey less than sympatric chimpanzees, given that most *Apis* hives are found in trees.

Third, chimpanzees tend to have a longer day-range than gorillas. They can therefore be expected to have a higher encounter rate with productive hives.

Fourth, chimpanzees might have a higher willingness to tolerate bee stings than other great apes. Certainly *Apis* bees are normally effective defenders. Within a few seconds of a chimpanzee attempting to obtain honey, scores or hundreds of bees attack the closest apes (and any humans who happen to be observing). Chimpanzees do not like to be stung, evidenced by the fact that they usually grab handfuls of honeycomb quickly before retreating rapidly to the ground and running to escape the bees. On rare occasions, however, bees do not defend their hives intensely. Under these unusual conditions chimpanzees can stay nearby and pull out honeycomb for periods as long as 20 minutes (Watts 2008; P. Bertolani pers. comm., Wrangham pers. obs.). Given that chimpanzees experience more aggression in social life (e.g., more frequent and severe beatings from each other) than bonobos, gorillas or orangutans do, they could in theory be more tolerant of the physical pain that normally accompanies extraction of *Apis* honey. Against this, Lui Kotal bonobos sometimes eat the brood of stinging wasps (M. Surbeck, G. Hohmann, and A. Fowler pers. comm.). If the relative disinterest of bonobos in extracting *Apis* honey is indeed associated with their greater aversion to pain, it may prove to be a paedomorphic trait (cf. Wobber et al. 2010).

Fifth, the cognitive abilities required to extract *Apis* honey may be more highly developed in chimpanzees than in other great apes. In favor of this idea chimpanzees frequently use tools to obtain honey, including specifically *Apis* honey in at least eleven wild populations (Sanz and Morgan 2009). Against it, however, chimpanzees in some populations obtain honey (of both *Apis* and stingless bees) entirely without tools (Budongo, Uganda - Gruber et al. 2009). Furthermore in populations where chimpanzees do use tools when extracting honey, they tend to use them more for Meliponine and *Xylocopa* than for *Apis* honey (e.g., Goualougo, Congo Republic - Sanz and Morgan 2009; Tai, Ivory Coast - Boesch et al. 2009). For example chimpanzees in Kanyawara (Kibale, Uganda) typically extract the 'bee bread' of *Xylocopa* bees using a stick, but rarely use tools when getting honey from *Apis* hives (pers. obs.). Finally the tools that chimpanzees use in the context of eating *Apis* honey are not generally designed to achieve access to the hole. Instead they serve to extract small amounts of honey (on dipping sticks, compared to pulling out chunks of honeycomb by hand) or to wave bees away (with leaves; Sanz and Morgan 2009). Neither of these activities produces much honey.

By contrast to *Apis* honey, the honey of stingless bees tends to induce extensive and complex tool use by chimpanzees including up to five different types of tool such as hammer, chisel, perforator, bodkin, brush-stick and dip-stick (Bai Hokou and Ndakan, Central African Republic - Fay and Carroll 1994; Lossi Forest, Congo Republic - Bermejo and Illera 1999; Ngotto et al. 2005; Dja, Cameroon - Deblauwe et al. 2006; Goualougo, Congo Republic - Sanz and Morgan 2009; Loango, Gabon - Boesch et al. 2009). Chimpanzees have been observed hammering with large pieces of wood to break into bee hives of stingless bees in five sites: Gombe, Tanzania (Goodall 1986), Bai Hokou, Central African Republic (Fay and Carroll 1994), Lossi, Congo Republic (Bermejo and Illera 1999), Goualougo (Sanz and Morgan 2009), and Loango, Gabon (Boesch et al. 2009).

Since *Apis* bees typically produce much larger quantities of honey than Meliponine stingless bees do (several kilograms compared to a handful), the lesser use of tools at *Apis* hives is striking. The obvious explanation for chimpanzees' low frequency of hammering at *Apis* hives is that they

are not willing to endure the painful stinging that would undoubtedly follow their hammering at the entry hole.

While these factors responsible for chimpanzees eating more honey than other apes are speculative, they suggest that although australopithecines could have exploited *Apis* honey at least as much as chimpanzees do, honey-eating was unlikely to have been substantially more frequent in australopithecines than in chimpanzees. Specifically, australopithecines probably climbed readily and had long day-ranges, giving them regular access to bee hives. Their relative brain size exceeded that of chimpanzees, suggesting a superior problem-solving ability, but among great apes tool use does not appear to be critical in facilitating access to *Apis* honey. The australopithecines' tolerance for physical pain is unknown, but given that the tools used by chimpanzees do not quieten stinging bees, it is hard to imagine that australopithecines would have stayed much longer than chimpanzees at a defended hive.

In short, the best guess about australopithecines is that like chimpanzees they lived in honey-rich environments, liked honey, and were able to obtain it regularly. Australopithecine behavioral adaptations to honey-collecting probably include tool-sets and multiple function tools (Boesch et al. 2009). However unless they used smoke, which is unlikely, australopithecines would not have been able to reduce the stinging rate of *Apis* bees. To judge from the high speed at which chimpanzees flee from aroused honey bees, the inability of australopithecines to quieten honey bees would have been the critical constraint on their collecting large amounts of honey. This conclusion is supported by the fact that no other large mammals are known to be successful in overcoming *Apis* defenses. For example among the eight species of bears, honey is a traditionally favorite food but only two are reported to include honey in their "main diet" and even for those species field studies show that honey is a minor item (Christiansen 2007). Both Bornean sun bears (*Helarctos malayanus*) and Peruvian spectacled bears (*Tremarctos ornatus*) supplement a predominantly insectivorous and frugivorous diet with occasional honey, mainly from stingless bees (Fredriksson et al. 2006). Sun bears, like chimpanzees, have been seen rapidly escaping from the defensive attack of *Apis* bees (G. Fredriksson pers. comm.). Only a few small mammals appear to be relatively impervious to honey bee stings (e.g., skunks (*Mephitis*) and opossums (*Didelphis*); Caron 1978).

About ten species of wax-eating honeyguides occur throughout Africa's tropical rainforests sympatric with chimpanzees, but none of them has developed a co-evolutionary relationship for obtaining honey (Friedmann 1955). This suggests that chimpanzees do not eat enough honey to allow the evolution of a symbiotic relationship with any species of honeyguide. Therefore if australopithecines obtained *Apis* honey at rates similar to those of chimpanzees, they too are unlikely to have developed a co-evolutionary relationship with honeyguides. This conclusion predicts that the guiding behavior of greater honeyguides began after the Pliocene.

Honey in the Ancestral *Homo* Diet

African honey-hunters concentrate on *Apis mellifera* honey, since this produces much more honey per hive than any other bees (e.g., Short and Horne 2001). *Apis mellifera* is the only African species of *Apis*. Its evolutionary history is uncertain. Mitochondrial DNA evidence once suggested that *Apis mellifera* arrived in northeastern Africa from the Middle East about one million years ago (Garnery et al. 1992), but genetic data

now suggest an African origin as early as 6–8 mya (Whitfield et al. 2006). Thus current evidence indicates that *Apis* honey has been available throughout the Pliocene.

Hunter-gatherers indicate the constraints that would have shaped Pleistocene gathering of *Apis* honey. Honey-hunters are always men, sometimes alone but often in a small group. By contrast, both men and women collect stingless bee honey among the Hadza (Marlowe 2010). Typically a single man climbs a honey-bearing tree carrying both an ax (to open the hole) and a smoldering log or leaves (to subdue the bees). In the Ituri Forest Bailey (1991) recorded a mean hive height of 19.1 m above ground, with a maximum of 51.8 m. Some hives were accessed only by using an elaborate construction of halters and bridges made from vines to traverse from one tree to another. Savanna trees are lower and easier, but can still be difficult (Figure 9.3).

After the honey-hunter extracts each piece of honeycomb from the hole he may keep it in a basket or drop it for companions to retrieve. By reducing bees' olfactory sensitivity, including to alarm pheromones (Visscher et al. 1995), smoke quietens bees enough that the honey-collector can empty the hive of its contents without receiving an insufferable number of stings. Even so, honey-collecting is a dangerous occupation. In the words of Coon (1972), "Men who collect honey from the tall trees or the faces of cliffs are faced with two principal problems, how to reach the hives and how to keep from being stung. ... It is the only way of collecting food that is at all comparable in danger to hunting elephants or whales". Without the control of fire it would seem impossible for a species vulnerable to honeybee stings to obtain large amounts of honey (Figure 9.4).

Homo erectus is the earliest hominid that plausibly controlled fire, possibly as early as 1.8–1.9 mya based on biological evidence (Wrangham

Figure 9.3 Using an axe to enlarge an arboreal entrance hole in a Zambian woodland. Honey-collecting is a dangerous occupation. ©Claire Spottiswoode.

scrapers. Presumably these tools would have been so inefficient that hole enlargement would have been slow and clumsy, and aggressive defense by bees therefore powerful. However Acheulian assemblages were also early. They were documented by Glynn Isaac at Peninj at 1.5-1.4 mya, including large cutting tools. Although the Peninj dates have been questioned, Acheulian tools are known in Gona (Ethiopia) and West Turkana (Kenya) prior to 1.6 mya (see de la Torre et al. 2008 for review). Heavy duty tools such as hand axes would clearly have facilitated hole enlargement. At Peninj they have been found with woody phytoliths embedded on their working surfaces, possibly from *Acacia* or *Salvadora* (Domínguez-Rodrigo et al. 2001). Containers could have included tortoise shells, bark fragments and large leaves, similar to but more elaborate than those occasionally used by chimpanzees (McGrew 1992). Equipped with ax, smoke and containers, *Homo erectus* is thus the oldest species whose tool-kit could in theory have allowed them to extract significantly more honey than chimpanzees obtain from *Apis* hives. They are therefore the earliest candidate for being a mutualistic partner with greater honeyguides.

Greater honeyguides were evidently already present when *Homo erectus* evolved, given that they include two highly divergent mitochondrial DNA lineages estimated to have split at least 2 to 3 million years ago (C. Spottiswoode, M. D. Sorenson and K. F. Stryjewski, pers. comm.). Accordingly *Homo* could in theory have started exploiting *Apis* honey with smoke, and developing a mutualistic relationship with greater honeyguides, shortly after the origin of *Homo erectus*.

Honey and Meat Compared

To judge from contemporary hunter-gatherers, once *Homo* successfully used smoke to quell *Apis*

and Carmody 2010). By 0.25-0.79 mya, archaeological evidence likewise indicates the control of fire (Wrangham 2009). Fire was not the only requisite for successful honey-collecting: axes were needed also, and containers (though not vital) would allow transporting. Initially (around 1.9-1.8 mya) *Homo erectus* could have chopped at a hive entrance with an Oldowan tool-kit of cores, large flakes, spheroids and

Figure 9.4 Collecting Apis mellifera *honey from a terrestrial hive, Zambia.*
Smoke is necessary to subdue the bees. ©Claire Spottiswoode.

aggression they could have eaten honey as their predominant source of calories for several weeks or months of the year. This raises the question of whether honey-eating would have had important implications for biological or social adaptations. Meat or other animal source products such as marrow are normally considered to be the most biologically and behaviorally influential of hunter-gatherers' preferred foods. However several authors have recently argued that honey can also have important effects (Boesch et al. 2009; Berbesque and Marlowe 2009). Since meat and other animal source foods (ASF) and honey are the only two strongly preferred classes of food that can be obtained by tropical savanna hunter-gatherers in large amounts, I briefly review their potential influences.

ASF and honey have much in common aside from being capable of seasonally dominating the diet. Compared to underground storage organs, fruits, leafy vegetables and even social insects ASF and honey are hard to find. Both incur relatively high risk not only in the sense of a forager possibly failing to obtain any, but also because foragers can be wounded in the attempt. Compensating for risk, the two food-classes are much appreciated partly because the items taste good, are nutrient-rich and can be found in large amounts. Both require considerable cognitive skills to obtain them, including using various kinds of tools.

Furthermore fitting the generalization for high-risk, high-gain classes of food, ASF and honey are obtained mainly by adult males (Bird 1999). Among the Hadza, men who are reputed to be skilled at hunting are reputed to be skilled honey-getters also (Marlowe 2010). Success in the search leads to men obtaining more than they can personally consume, which fosters a system of food distribution with social benefits for the producer and/or the opportunity for him to invest in his family (e.g., honey - Ichikawa 1981; Lewis 2002; meat - Wood 2006; Marlowe 2007). Cultures vary with respect to such aspects as how much of the food is eaten by the producer, who else benefits and who is responsible for distribution. ASFs and honey thus both tend to be influential foods that attract much social attention, can dominate the diet, present creative challenges for physical and social cognition, and are controlled (at least in their production) by adult males. Both may have therefore contributed to the social division of labor (Berbesque and Marlowe 2009).

There are important differences also. Unlike ASFs, honey is easily divisible without tools, is easily chewed, and requires no pounding or cooking before it is eaten. Honey cannot run away, which may help explain why Hadza men in the process of getting honey will switch to hunting if they see potential prey (Marlowe 2010). Berbesque and Marlowe (2009) found that Hadza men preferred meat to honey, whereas Hadza women preferred honey to meat, and suggested that the lower protein concentration of honey could therefore be important in food choices. The production of ASFs appears to be less variable over the year than honey, though more information is required. Whereas meat can be eaten by hominids who do not control fire, large amounts of honey probably cannot be obtained without the use of smoke. But once it is

obtained, it becomes an easy food for the weak-jawed young or the toothless old.

These differences imply parallel differences in the effects on hominid evolution. Mostly they suggest that honey has had less impact than meat, since honey demands a lower level of specialization to prepare and eat, and is unlikely to have been a major part of the diet until fire was controlled. But honey-eating has been studied relatively little, and Ichikawa's (1981) research shows that the social exchanges around honey can be complex and significant for camp life, including promoting labor differentiation among men (cf. Lewis 2002). Furthermore honey can be important even before it is collected. Among the !Kung a bee hive belongs to the man who finds it, a cultural fact that led to the capital punishment of a honey-thief (Lee 1979). The fact that ASFs are more archaeologically prominent whereas honey-eating is invisible has distorted the record of the relative importance of these two important food types. Further data on the distribution, abundance and productivity of *Apis* honey and its seasonal variation will contribute valuably to exploring the range of social implications from honey-gathering, and thus to completing Isaac's goal of a full accounting of the critical foods in human evolution.

Meanwhile we can reasonably propose that the annual cycle of Pleistocene African *Homo* included a regular few weeks or months, perhaps during the wet season, when adult males tended to abandon the hunt in order to get honey. This conclusion lies in contrast to the assumption that meat has always been more important than honey. We can now imagine that African *Homo* experienced honey seasons as periods of plenty during which individuals sometimes consumed such enormous amounts of glucose in short periods that there were possible implications for the evolution of the insulin system. Whereas a sustained period of high meat production has traditionally been viewed as an optimal time for human ancestors, the best time of all may have been the honey seasons.

Summary

1) The evidence of a co-evolved relationship between greater honeyguides and humans indicates that human ancestors heavily exploited *Apis mellifera* honey for long enough to affect natural selection on honeyguides;

2) contemporary African hunter-gatherers in both savanna and forest often have annual honey seasons lasting several weeks during which honey is the predominant preferred food;

3) chimpanzees everywhere eat honey of both stingless bees and *Apis*, but the amounts obtained and consumed are small compared to hunter-gatherers;

4) tool use by chimpanzees in eating honey is more developed for extracting small amounts of honey from the hives of stingless bees than for extracting large amount of honey from *Apis* hives. The danger of being stung by *Apis* bees is apparently an effective deterrent to prolonged extraction attempts;

5) current data suggest that although bonobos eat stingless bee honey, they make few or no attempts to obtain *Apis mellifera* honey;

6) australopithecines are unlikely to have eaten much more honey than chimpanzees because they did not have a successful way of defending themselves against *Apis* stings;

7) smoke is the only method used by primates to control aggressive defense by *Apis*. This means that *Homo erectus* is the earliest hominid that is likely to have obtained honey in large amounts; and,

8) after African *Homo* had become efficient at extracting *Apis* honey they would have experienced annual high-productivity seasons during which social relationships would have been importantly shaped by competition and cooperation in the honey quest. Physiology may have been affected also.

Acknowledgments

Glynn Isaac was a kind, inspiring and brilliant anthropologist. Thanks to David Pilbeam and Jeanne Sept for their invitation, to Rachel Carmody, Gabriella Fredricksson, Victoria Ling, Dave Morgan, Martin Muller, Crickette Sanz, Claire Spottiswoode and the editors for helpful comments, and to Claire Spottiswoode, Warwick Tarboton, Bob Poole and Benenson Productions Inc. for their photographs. For information about bonobo honey-eating I thank Alexander Georgiev, Andrew Fowler, Takeshi Furuichi, Gottfried Hohmann and Martin Surbeck. Bill Benenson and George Mavroudis kindly gave me the opportunity to witness the relationship between Hadza and greater honeyguides. I am grateful to Rob Foley and Marta Lahr for hosting me at the Leverhulme Center for Human Evolutionary Studies, University of Cambridge.

References

Ambrose, J. T.
1978 Birds. In *Honey Bee Pests, Predators and Diseases*, edited by R. A. Morse, pp. 215-226. Cornell University Press, Ithaca, NY.

Bailey, R. C.

1991 *The Behavioral Ecology of Efe Pygmy Men in the Ituri Forest, Zaïre.* University of Michigan Press, Ann Arbor, MI.

Berbesque, J. C., and F. W. Marlowe

2009 Sex differences in food preferences of Hadza hunter-gatherers. *Evolutionary Psychology* 7(4):601–616.

Bermejo, M., and G. Illera

1999 Tool-set for termite-fishing and honey extraction by wild chimpanzees in the Lossi Forest, Congo. *Primates* 40(4):619–627.

Bermejo, M., G. lllera, and J. Sabater Pi

1994 Animals and mushrooms consumed by bonobos (*Pan paniscus*): New records from Lilungu (Ikela), Zaire. *International Journal of Primatology* 15(6):879–898.

Bird, R.

1999 Cooperation and conflict: The behavioral ecology of the sexual division of labor. *Evolutionary Anthropology* 8:65–75.

Boesch, C., J. Head, and M. M. Robbins

2009 Complex tool sets for honey extraction among chimpanzees in Loango National Park, Gabon. *Journal of Human Evolution* 56:560–569.

Caron, D. M.

1978 Marsupials and mammals. In *Honey Bee Pests, Predators and Diseases*, edited by R. A. Morse, pp. 227–256. Cornell University Press, Ithaca, NY.

Christiansen, P.

2007 Evolutionary implications of bite mechanics and feeding ecology in bears. *Journal of Zoology* 272(4):423–443.

Coon, C. S.

1972 *The Hunting Peoples.* Jonathan Cape, London.

Cordain, L., S. B. Eaton, A. Sebastian, N. Mann, S. Lindeberg, B. A. Watkins, J. H. O'Keefe, and J. Brand-Miller

2005 Origins and evolution of the Western diet: Health implications for the 21st century. *American Journal Clinical Nutrition* 81:341–354.

Dean, W. R. J., and I. A. W. Macdonald

1981 A review of African birds feeding in association with mammals. *Ostrich* 52:135–155.

Dean, W. R. J., W. R. Siegfried, and I. A. W. Macdonald

1990 The fallacy, fact, and fate of guiding behavior in the greater honeyguide. *Conservation Biology* 4(1):99-101.

Deblauwe, I.

2006 New evidence of honey-stick use by chimpanzees in southeast Cameroon. *Pan African News* 13:2-4.

Diamond, J. W., and A. R. Place

1988 Wax digestion by black-throated honeyguides *Indicator indicator*. *Ibis* 130:557-560.

Domínguez-Rodrigo, M., J. Serrallonga, J. Juan-Tresserras, L. Alcalá, and L. Luque

2001 Woodworking activities by early humans: A plant residue analysis on Acheulian stone tools from Peninj (Tanzania). *Journal of Human Evolution* 40:289-299.

Downs, C. T., R. J. van Dyk, and P. Iji

2002 Wax digestion by the lesser honeyguide Indicator minor. *Comparative Biochemistry and Physiology Part A* 133:125-134.

Fay, J. M., and R. W. Carroll

1994 Chimpanzee tool use for honey and termite extraction in Central Africa. *American Journal of Primatology* 34(4):309-317.

Finke, M. D.

2005 Nutrient composition of bee brood and its potential as human food. *Ecology of Food and Nutrition* 44:257-270.

Fredriksson, G. M., S. A. Wich, and Trisno

2006 Frugivory in sun bears (*Helarctos malayanus*) is linked to El Nino-related fluctuations in fruiting phenology, East Kalimantan, Indonesia. *Biological Journal of the Linnaean Society* 89:489-508.

Friedmann, H.

1955 *The Honeyguides*. United States National Museum Bulletin 208. Smithsonian Institution, Washington, DC.

Goodall, J.

1986 *The Chimpanzees of Gombe: Patterns of Behavior*. Harvard University Press, Cambridge, MA.

Gruber, T., M. N. Muller, P. Strimling, R. W. Wrangham, and K. Zuberbühler

2009 Wild chimpanzees rely on cultural knowledge to solve an experimental honey acquisition task. *Current Biology* 19:1806-1810.

Hicks, T. C., R. S. Fouts, and D. H. Fouts
2005 Chimpanzee (*Pan troglodytes troglodytes*) tool use in the Ngotto Forest, Central African Republic. *American Journal of Primatology* 65(3):221-237.

Hirata, S., and M. L. Celli
2003 Role of mothers in the acquisition of tool-use behaviours by captive infant chimpanzees. *Animal Cognition* 6:235-244.

Hoesch, W.
1937 Ueber das "Honiganzeigen" von Indicator. *Journal fur Ornithologie* 85:201-205.

Ichikawa, M.
1981 Ecological and sociological importance of honey to the Mbuti net hunters, eastern Zaire. *African Study Monographs* 1:55-68.

Isaac, G.
1978 Casting the net wide: A review of archaeological evidence for early hominid land-use and ecological relations. In *Current Argument on Early Man*, edited by L.-K. Königsson, pp. 226-251. Pergamon Press, Oxford.

Isack, H. A., and H.-U. Reyer
1989 Honeyguides and honey gatherers: Interspecific communication in a symbiotic relationship. *Science* 243:1343-345.

Kajobe, R., and D. W. Roubik
2006 Honey-making bee colony abundance and predation by apes and humans in a Uganda forest reserve. *Biotropica* 38(2):210-218.

Kano, T.
1992 *The Last Ape: Pygmy Chimpanzee Behavior and Ecology*. Stanford University Press, Stanford, CA.

Lee, R. B.
1979 *The !Kung San: Men, Women and Work in a Foraging Society*. Cambridge University Press, Cambridge.

Lewis, J.
2002 Forest Hunter-Gatherers and Their World: A Study of the Mbendjele Yaka Pygmies of Congo-Brazzaville and Their Secular and Religious Activities and Representations. Ph.D. thesis, London School of Economics and Political Science.

Lonsdorf, E. V., S. R. Ross, S. A. Linick, M. S. Milstein, and T. N. Melbea
2009 An experimental, comparative investigation of tool use in chimpanzees and gorillas. *Animal Behaviour* 77:1119-1126.

Marlowe, F.

2007 Hunting and gathering: The human sexual division of foraging labor. *Cross-Cultural Research* 41:170-196.

Marlowe, F. W.

2010 *The Hadza: Hunter-Gatherers of Tanzania.* University of California Press, Los Angeles, CA.

McGrew, W. C.

1983 Animal foods in the diets of wild chimpanzees (Pan troglodytes): Why cross-cultural variation? *Journal of Ethology* 1:46-61.

1992 *Chimpanzee Material Culture: Implications for Human Evolution.* Cambridge University Press, Cambridge.

McGrew, W. C., L. F. Marchant, M. M. Beuerlein, D. Vrancken, B. Fruth, and G. Hohmann

2007 Prospects for bonobo insectivory: Lui Kotal, Democratic Republic of Congo. *International Journal of Primatology* 28:1237-1252.

Milton, K.

1987 Primate diets and gut morphology: Implications for hominid evolution. In *Food and Evolution: Towards a Theory of Human Food Habits,* edited by M. Harris and E. B. Ross, pp. 9-115. Temple University Press, Philadelphia, PA.

Murray, S. S., M. J. Schoeninger, H. T. Bunn, T. R. Pickering, and J. A. Marlett

2001 Nutritional composition of some wild plant foods and honey used by Hadza foragers of Tanzania. *Journal of Food Composition and Analysis* 14:3-13.

Queeny, E. M.

1952 The Wandorobo and the Honey-Guides. *Natural History* 61(9):392-396.

Rothschild, B. M., I. Hershkovitz, and C. Rothschild

1995 Origin of yaws in the Pleistocene. *Nature* 378:343-344.

Roubik, D. W.

1989 *Ecology and Natural History of Tropical Bees.* Cambridge University Press, Cambridge.

Sanz, C. M., and D. B. Morgan

2009 Flexible and persistent tool-using strategies in honey-gathering by wild chimpanzees. *International Journal of Primatology* 30:411-427.

Schaller, G. B.

1963 *The Mountain Gorilla: Ecology and Behavior.* University of Chicago Press, Chicago.

Short, L., and J. Horne

2001 *Toucans, Barbets and Honeyguides.* Oxford University Press, Oxford.

Skinner, M.

1991 Bee brood consumption: An alternative explanation for hypervitaminosis A in KNM-ER 1808 (*Homo erectus*) from Koobi Fora, Kenya. *Journal of Human Evolution* 20:493-503.

Skinner, M., K. E. Jones, and B. P. Dunn

1995 Undetectability of Vitamin-A in bee brood. *Apidologie* 26:407-414.

Sparrman, A.

1785-1786 *A Voyage to the Cape of Good Hope Towards the Antarctic Polar Circle Round the World and to the Country of the Hottentots and the Caffres from the year 1772-1776.* English translation V. S. Forbes, editor. Van Riebeeck Society, Cape Town, South Africa.

Spottiswoode, C. N., and J. F. R. Colebrook-Robjent

2007 Egg puncturing by the brood parasitic Greater Honeyguide and potential host counteradaptations. *Behavioral Ecology* 18(4):792-799.

Stanford, C. B.

2006 The behavioral ecology of sympatric African apes: Implications for understanding fossil hominoid ecology. *Primates* 47:91-101.

Takeda, J.

1976 An ecological study of the honey-collecting activities of the Tongwe, Western Tanzania, East Africa. *Kyoto University African Studies* 10:213-247.

de la Torre, I., R. Mora, and J. Martínez-Moreno

2008 The early Acheulean in Peninj (Lake Natron, Tanzania). *Journal of Anthropological Archaeology* 27:244-264.

Tutin, C. E. G., and M. Fernandez

1992 Insect-eating by sympatric lowland gorillas (*Gorilla gorilla gorilla*) and chimpanzees (*Pan t. troglodytes*) in the Lope Reserve, Gabon. *American Journal of Primatology* 28(1):29-40.

Visscher, P. K., R. S. Vetter, and G. E. Robinson

1995 Alarm pheromone perception in honey bees is decreased by smoke (*Hymenoptera: Apidae*). *Journal of Insect Behavior* 8:11-18.

Walker, A., M. R. Zimmerman, and R. E. F. Leakey

1982 A possible case of hypervitaminosis A in *Homo erectus. Nature* 296:248-250.

Watts, D. P.
2008 Tool use by chimpanzees at Ngogo, Kibale National Park, Uganda. *International Journal of Primatology* 29:83-94.

Whitfield, C. W., S. K. Behura, S. H. Berlocher, A. G. Clark, J. S. Johnston, W. S. Sheppard, D. R. Smith, A. V. Suarez, D. Weaver, and N. D. Tsutsui
2006 Thrice out of Africa: Ancient and recent expansions of the honey bee, *Apis mellifera*. *Science* 314:64-645.

Wobber, V., R. W. Wrangham, and B. Hare
2010 Bonobos exhibit delayed development of social behavior and cognition relative to chimpanzees. *Current Biology* 20:226-230.

Wood, B. M.
2006 Prestige or provisioning? A test of foraging goals among the Hadza. *Current Anthropology* 47(2):383-387.

Wrangham, R.
2009 *Catching Fire: How Cooking Made Us Human*. Basic Books, New York.

Wrangham, R. W., and R. Carmody
2010 Human adaptation to the control of fire. *Evolutionary Anthropology* 19:187-199.

Yamagiwa, J., and A. K. Basabose
2009 Fallback foods and dietary partitioning among *Pan* and *Gorilla*. *American Journal of Physical Anthropology* 140(4):739-750.

10

A Worm's Eye View
of Primate Behavior

Jeanne Sept

Introduction

When grappling with archaeological interpretations, Glynn Isaac was always one to advocate a "worm's eye-view" (Isaac 1971b), to think about behavior from the ground up. He encouraged students and colleagues to study different behavioral and natural processes that could contribute to the archaeological record in different ecological contexts, across space and through time. In his early work at Olorgesailie he used his deep knowledge of geology and natural history, in combination with taphonomic experiments, to interpret Acheulian assemblages. Later, finding creative ways to understand the complexities of site formation processes became a central theme of his co-leadership of the Koobi Fora Research Project as he sought to evaluate why tool-using and meat-eating had repeatedly converged at some Oldowan sites, marking a major turning point in human evolution (Isaac 1971a).

With his seminal papers on this theme, "Food sharing behavior of proto-humans" (Isaac 1978a, b), Glynn captured the scientific imagination of his students and colleagues in a way that has shaped years of research. Were these early sites home-bases? The concept of "home" has complex meanings in different cultures today (Bertram 2008; Jackson 1995; Manzo 2005; Moore 2000), but having a home-base is widespread, perhaps universal. How long has home been a special place, a location created through episodes of social interaction like food sharing, and layers of shared memory? If being home is

being human, then early evidence of home could be an important marker of human origins.

Glynn framed this issue as both an empirical and theoretical challenge for paleoanthropology. He urged archaeologists, in particular, to study the ways in which the earliest debris record was distinctively patterned. And then he led the charge to empirically test his home-base/food-sharing model while actively encouraging his students and others to explore alternative working hypotheses (Isaac 1981a). For example could sites have formed as overlapping palimpsests of debris from varied sources - spatial patterns resulting less from socio-cultural design than from ecological coincidence? How would the artifacts surviving such depositional episodes become patterned archaeologically as the distinctive features of site locations blurred with time? These debates are ongoing and beyond the scope of this chapter, but alternative interpretations have recently been summarized by several authors (Dominguez-Rodrigo et al. 2007; Plummer 2004).

Glynn also recognized that searching for the earliest homes posed theoretical challenges. What should we be looking for? Since we are so familiar with and impressed by the anthropological nuances of home in the modern world, it may be difficult for us to recognize subtle hints of home in antiquity and distinguish homologous precursors or analogous alternatives. The abiding risk, as we look deeper and deeper into the past for evidence, is that the familiar patterns we see in the prehistoric record may actually just be reflections of our own ideals that mask unfamiliar realities. As Glynn wrote (Isaac 1986:237):

> let us imagine looking down a deep well-shaft. Beyond the dimly lighted upper rim is darkness extending away from the watcher ... but in these depths is the gleam of light on water. If the well is not too deep, by straining our eyes we perhaps see a figure – a figure set in an unfamiliar context, but yet a familiar figure – familiar, because it is our own reflection.... Awareness is dawning that in part we have been using archaeology and the early evidence as a mirror by which to obtain more or less familiar images of ourselves.

In this context, he noted that his original "food-sharing hypothesis" should be recast as a "central-place-foraging hypothesis" (Isaac 1983a, b). Central place foragers are animals that use dens or nests as bases from which to search for food. Animals as diverse as birds, bats and beavers practice central place foraging to provision their young in protected breeding sites, but such havens are often abandoned as offspring mature (Daniel et al. 2008; Olsson et al. 2008; Raffel et al. 2009). Hediger distinguished the use of nests built by birds from mammalian dens, which he called "homes" (Hediger 1977), raising the question of whether a broad, ethological concept of home could be defined simply as the habitual use of a fixed point – a safe den or roost – and the extent to which such places could be distinguished from other types of habitual activity areas. Similarly, to avoid the de facto attribution of human qualities to early hominins, archaeological tests of the central-place-foraging hypothesis must try to tease apart possible evidence for sleeping/nesting and food-sharing behaviors from the spatial patterns of other sorts of activities such as incidental tool use and individualistic feeding.

To grapple with these challenges Glynn advocated that colleagues "cast the net wide," and use a combination of experimental approaches and actualistic studies to pull together information from many sources to help interpret archaeological data (Isaac 1980). He was an early and enthusiastic advocate of using comparative studies of primates to help interpret archaeological evidence, whether looking at tool-making (Isaac 1986), food-sharing and hunting behaviors (Isaac and Crader 1981), or assessments of ranging, nesting, and other activity patterns (Isaac 1980, 1981b) to interpret the Oldowan archaeological record within a behavioral-ecological framework. Such a comparative approach, whether pursued as field studies of primate material culture (McGrew 1992), "ethoarchaeology" (Sept 1992), "ethnoarchaeology" (Joulian 1995), or "primate archaeology" (Haslam et al. 2009) has slowly been gaining momentum in both primatological and archaeological research over the ensuing decades, but much remains to be done (McGrew and Foley 2009).

Here I present a brief overview of some recent primatological research which refreshes the comparative foundations for our evaluation of the home-base hypothesis. I focus on the behavioral and ecological processes which generate spatial distributions of nests, tools and feeding remains, and the circumstances under which behavioral markers of such activities might become preserved, leading to patterns of evidence that could potentially be recognized archaeologically – a worm's eye view of primate behavior.

Primate Perspectives: Spatial Patterns of Activities

Primate Beds, Behavior and Social place

As Glynn developed his arguments about the distinctive nature of the early stone age record, he made a concerted effort to find maps of primate ranging behavior that could be contrasted with the patterns typical of human foragers. Ethnoarchaeologists had begun to document how human camps served both as places of refuge and as spatial foci for food sharing and other debris-producing activities (e.g., Yellen 1977), but primatologists were not yet commonly reporting the spatial distribution of non-human primate activities. Ultimately Glynn illustrated travel patterns of baboons, chimpanzees and gorillas inferred from a compilation of descriptions of daily ranging patterns (Isaac 1981b, 1983a, 1984). He expected that comparative maps of such movement patterns would indicate the "major differences between species in their social organization and social relations" and, compared to hunter-gatherers, that any clustered "nodes" in primate ranging patterns "...would be far less conspicuous features and those which were represented would be found not to involve repeated overnight groupings of individuals, but only temporary aggregations at attractions such as groves of fruit trees in season. There would be no points in space that could be called 'settlements'" (Isaac 1981b:131).

Since Glynn's intellectual forays into the patterns of primate ranging, fieldwork has greatly expanded our understanding of how primates use, re-use and remember their landscapes. Primates are not typical social mammals, of course; only the smallest primates have nursery sites, and only a few species sleep in the same place on consecutive nights. Primates normally choose sleeping sites within their range, based on factors of safety and comfort, as well as ecology and social dynamics (Anderson 1998).

Playing it safe, primates will choose sleeping sites that afford protection from predators, whether in cryptic or remote spots or in comfortable locations which could enhance their vigilance against predators. The smallest primates, among the prosimians (Bearder et al. 2003; Kappeler 1998; Rasoloharijaona et al. 2008) and some new world monkeys (Di Bitetti et al. 2000; Hankerson et al. 2007), prefer to make nests in tree holes or in leafy tangles. Most monkeys do not build nests, but normally perch in trees, or sometimes in secluded or sheltered roosts; baboons, for example, sleep together either in trees, cliffs or caves (Hamilton III 1982; Schreier and Swedell 2008). Among the apes, gibbons sleep on natural branches high in the canopy (Fan and Jiang 2008; Reichard 1998), but the larger apes (orangutan, gorilla, bonobo and chimpanzee), all learn to construct arboreal nests as juveniles (Videan 2006). The great apes build fresh nests every night, avoiding parasites in the process, and frequently build day nests for resting, as well. The structure of these nests varies with function, age and sex (Fruth and Hohmann 1994a, 1994b; Koops et al. 2007). The largest primates, male gorillas, often face little predation risk and will build comfortable sleeping cushions on the forest floor, close to other members of their group, rather than in trees (Rothman et al. 2006; Tutin et al. 1995). Nesting in low branches or on the ground has also been observed in chimpanzee populations where natural predators are rare (Koops et al. 2007; Preutz et al. 2008), and chimpanzees may devote special efforts to select soft materials to line their nests in dry habitats (Stewart et al. 2007).

Recent research has examined how habitat variation can influence the placement and arrangement of sleeping sites. In particular, interesting patterns have been documented in the selection of locations for nightly nest construction among the great apes. Overall, nest distribution through a habitat can reflect the cumulative ranging behavior and habitat preferences of apes, as illustrated by maps of chimpanzee nests in the Tai forest (Kouakou et al. 2009) or comparisons of monthly nesting patterns by sympatric chimpanzees and gorillas (Stanford 2008). Yet finer-grained variations in the choice of nest locales are also emerging. For example, studies in the Bwindi Impenetrable Forest revealed that chimpanzees do not select nesting trees based on their availability alone; their choices may reflect a variety of other variables, such as proximity of feeding trees, tree structure and ambient comfort (Stanford and O'Malley 2008). Though chimpanzee nesting site locations in forest are often correlated with food abundance (Furuichi and Hashimoto 2004), more careful studies of nesting substrate will be needed before a comprehensive understanding of chimpanzee nesting location choice is possible. Fruth and Hohman document, for instance, how differences in ecological setting constrain the location, placement and form of bonobo nests, which they infer are due to such variables as predation risk and patchiness of feeding opportunities (Fruth and Hohmann 1994a).

A number of studies have also begun to explore how social status and interpersonal relationships influence choice of sleeping sites in many primate species (Anderson 1998). These range from the role of dominance in access to preferred sleeping sites among spider monkeys, to consort pairs of baboons sleeping together at night, to mountain gorillas nesting together as a social group. For example, at Mt. Nimba, in West Africa, with no natural predation pressure, over 6% of the chimpanzee night nests occur on the ground (Koops et al. 2007), and there are no obvious environmental patterns that the researchers found to explain this; since many of these nests were nests built by males in proximity to clusters of smaller (female and juvenile) nests in trees, it seems clear that social context is an important variable in nesting location in this chimpanzee population, and likely in others.

The complex interaction of social context with habitat constraints has a particularly significant influence on chimpanzee ranging and sleeping choices in dry habitats. Chimpanzees living in open woodlands or wooded savannas forage across a much larger territory and day range than chimpanzees in closed habitats (Moore 1992; Pruetz and Bertolani 2009). Sometimes their nests are isolated, and distributed widely, but at other times their travelling parties are large and they consequently nest together in clusters, often in single trees (Baldwin et al. 1981; Ogawa et al. 2007; Preutz 2006; Preutz et al. 2008). A comparative study of chimpanzee nests in two adjacent savanna habitats in Senegal (Preutz et al. 2008) reveals the important impact that predation pressure can play on nest location in such open habitats. The chimpanzees at the Mt. Assirik site face significant predation pressure, and position their nests higher in the canopy in the taller trees and in denser gallery forests, while at the neighboring Fongoli site chimpanzees have few natural predators and tend to build their nests in more easily accessible, comfortable situations either in shorter trees or on the ground, in both forest and open woodland. In such dry sites, nest comfort may be particularly important (Stewart et al. 2007).

We can thus chart the variable array of primate sleeping sites along two major dimensions (Figure 10.1). Viewed as artifacts, the "design" of sleeping accommodations ranges from casual to formal: from primates merely balancing on an available branch with no modification; to collecting soft nesting materials into a natural hollow; to weaving a resilient platform of branches and lining it with leafy bedding. The placement of sleeping sites also varies, with some primate species deliberately re-using a limited array of safe nesting locations, while others re-locate their nests every night. In combination these two

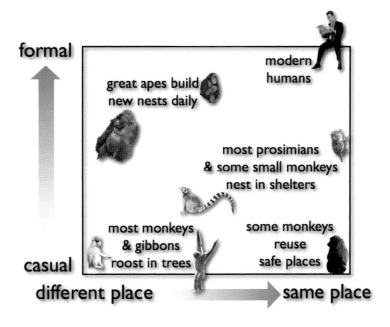

Figure 10.1 Resting and sleeping patterns of living primates. Living primates can be differentiated by their tendency to locate their sleeping sites in the same place or different places, and the extent to which they modify their sleeping sites, ranging from casual use (no deliberate modification) to formal design (e.g., deliberately structuring a nest or bed).

dimensions of primate sleeping situate modern human "nesting behaviors" at one end of a behavioral spectrum, since we typically sleep deliberately in the same locations with more formal bedding arrangements than other primates. Such an overview illustrates both the continuity and the distinctiveness of modern human nesting behaviors compared to other primates. For an archaeologist, however, this type of classification scheme lacks important elements of time and space that frame interpretations of the archaeological record. How distinctive would these patterns appear through taphonomic filters, from the ground, up?

Thinking about the home-base hypothesis in particular it is interesting to note that primate sleeping sites are sometimes re-used in significant frequencies, especially in settings when suitable nesting trees are limited. Among apes, the gibbons and orangutans maintain relatively small territories in forests with limited tree choice, and are reported to re-use trees to make fresh nests 20-30% of the time, especially in degraded habitats with a limited choice of preferred trees (Ancrenaz et al. 2004; Di Bitetti et al. 2000; Fruth and Hohmann 1994a). The African apes tend to have a larger choice of available nesting trees, and, on average, chimpanzees re-nest in trees 14% of the time, gorillas 4% and bonobos only 0.2% (Fruth and Hohmann 1994a). However, even among the bonobos, with relatively low proportions of tree re-use, repeated nesting can ultimately modify a tree through frequent manipulation of branches (Fruth and Hohmann 1994a, 1994c). This has also been reported for orangutans (Ancrenaz et al. 2004).

Such modifications create what archaeologists might call either a structural "feature" or an "ecofact," and demonstrate not only how repeated activities in the same locations can shape habitat, but also how short-term behaviors can indirectly influence the probability of future use of the site. Indeed, a detailed study by Fiona Stewart and colleagues of chimpanzee nesting in Tanzania (Stewart et al. in press) suggests that such successive use of nesting spots in individual trees enhances the structure of the trees, effectively creating pre-fab sleeping platforms and encouraging "nest fidelity." While almost all such organic evidence of activities will rot away, in unusual circumstances wood can be preserved and provide archaeological evidence of patterned landscape use, as illustrated by the type of taphonomic analyses done to identify the contributions of early hominin agents to ancient wood assemblages (Goren-Inbar et al. 2002).

Looking at the re-use of nesting sites, once again the nesting patterns of dry country chimpanzees stand out. The choices of nesting trees are constrained seasonally and geographically in savannas and open woodlands. So, despite the fact that they live at lower population densities and in larger home ranges than other chimpanzees (Ogawa et al. 2007), chimpanzees in these dry habitats more frequently re-use preferred nesting locations. For example, up to 90% of nesting sites and 27% of individual nesting trees were re-used by chimpanzees at Ugalla, Tanzania over several years (Hernandez-Aguilar 2009). During the dry season in these deciduous woodlands, chimpanzees in search of leafy refugia concentrated their nests in discontinuous patches of evergreen forest; but they also often re-used nesting sites during other seasons when a much wider range of suitable alternatives were available. The extent to which the re-use of resting/nesting groves by these chimpanzees is deliberate, socially mediated,

and/or a consequence of habitat constraints, is currently unknown. However, I think it is fair to assume that both comfort and familiarity are likely factors, as exemplified by the individuals and groups of Fongoli chimpanzees which apparently revisit rockshelters for cooling daytime stops during the hot, dry season (Pruetz 2007).

So, how might an understanding of such spatial variation in primate nesting help us address paleoanthropological questions? Before much of this nesting research on apes had been undertaken, I found an opportunity to train an archaeologist's eye on the nesting patterns and other activity patterns of a population of savanna chimpanzees. Seeking a new approach to testing the home-base hypothesis, I documented the patchiness of chimpanzee nesting along the Ishasha River (eastern Democratic Republic of the Congo) – a narrow ribbon of gallery forest cutting through semi-arid savanna grasslands, similar to paleoenvironmental settings which preserved a number of early Oldowan sites in east Africa. My study was designed to map samples of the nesting, tool-using and feeding behaviors of chimpanzees in a way that could facilitate comparisons with archaeological evidence (Sept 1992, 1998). I coined the term "ethoarchaeology" to represent the logical framework of this type of actualistic research because it is analogous to, though bioculturally distinct from, ethno-archaeological studies of the material remains of living peoples (David and Kramer 2002). The Ishasha study only ran for two field seasons before it was cut short by regional conflict. But even in that limited amount of time it was striking how repeated nesting and feeding activities by Ishasha chimpanzees produced patches of ephemeral debris along the riverbanks and channel floodplain that had the potential to mimic the density patterns of artifacts recovered at early archaeological sites in analogous habitats. Subsequently, the longer-term and more comprehensive ethoarchaeological studies of chimpanzee behavior in the miombo woodlands of Ugalla by Hernandez-Aguilar (2009) have also found dry country chimpanzees repeatedly clustering their nests in particular groves of trees, and at least some of these nesting sites stand out against a lower density background distribution of nests in the same microhabitats. It is interesting that while the Ugalla chimpanzee habitat includes some gallery forests along rivers incising broad floodplains, they only occasionally nest in these zones. Hernandez-Aguilar reports that the preferred nest locales occur along hill slopes, sometimes in evergreen ravines, and only rarely on plateaus; thus only a small fraction of the Ugalla chimpanzee nests are built in the types of sedimentary environments that could lead to burial and possible fossilization of remains, in contrast to the pattern at Ishasha, a significantly drier habitat where nesting opportunities are likely more constrained by vegetation structure.

Knowing that primates often rest and nest in the same places, how could studying such nesting sites help evaluate the Oldowan home-base/central-place foraging hypothesis among the many plausible alternatives, such as the recurring use of "favored places" like shade trees (Kroll 1994; Kroll and Isaac 1984; Schick 1987)? The more often a nesting site is revisited, the more likely it is that other types of more durable debris will accumulate at that spot. The next steps should involve an evaluation of how variation and redundancy in nesting localities fit into broader spatial patterns of primate socioecological activities, particularly those involving food and technology. And such studies should seek to use sampling scales and analytical approaches appropriate to comparative primatological and paleoanthropological research.

Primate Tools and Feeding Behaviors

Harking back to theories of "man the toolmaker," the first accounts of tool-making and using by primates in the wild fascinated paleoanthropologists, and Glynn was no exception (Isaac 1969, 1972). Since then, studies of animal tool use have grown exponentially (Shumaker et al. 2011), and an extensive ethogram of wild chimpanzee behaviors has been compiled as an audio-visual encyclopedia which includes intriguing nuances of tool-use (Nishida et al. 2010). These data are providing a rich collection of ethological and experimental research relevant both to understanding the cultural variation among primates and to inferring the adaptive significance of technology for early hominins (Toth and Schick 2009; Whiten et al. 2009). Indeed, researchers are increasingly analyzing the tool-use and resource exploitation strategies of wild primates using methods that will facilitate archaeological comparisons.

Primatologists now commonly record not just technological behaviors but also the morphology of tools used or modified by their study animals. Although the variety of detailed attributes that archaeologists often collect for ancient artifacts have rarely been recorded for primate tools, some measurements have allowed preliminary evaluation of ecological, functional and cultural variation between chimpanzee sites. The most detail has been described for stone hammers and anvils used in the wild by cebus monkeys (Visalberghi et al. 2007) and west African chimpanzees (Boesch and Boesch-Achermann 2000), though attributes of tools made from other materials have also been carefully documented, as in cross-continental comparisons of the termite fishing tools used by chimpanzees at Gombe and Mt. Assirik (McGrew 1992). Variation in honey foraging tools provides an interesting example; to pound-open or lever-open beehives chimpanzees choose sticks which are somewhat longer and thicker on average than tools used to dip and probe for the honey, though there is significant overlap in size between these two tool types, both within and between sites and depending on the species of bee being exploited (McClennan 2011; Sanz and Morgan 2009). Similarly, recent ethoarchaeological studies of chimpanzee hammer and anvil technology suggest that some tools may be selected based on a combination of morphological attributes which may not be represented effectively by simple functional classifications (Carvalho et al. 2009). It seems clear that developing common methodologies would help evaluate the ecological, cognitive and cultural underpinnings of assemblage variation, whether comparing the material culture of living peoples and animals or ancient artifacts (Gowlett 2009; Haslam et al. 2009; Stewart et al. in press).

Acknowledging the limits of a typological approach to understanding primate material culture, primatologists are also beginning to study the operational sequences, or *chaînes opératoires*, used by primates in making and using tools (Carvalho et al. 2008). Such ethoarchaeological research builds on descriptions of distinctive "cultural" gestures or actions (Joulian 1995; McGrew 1998), and the inferred cognitive complexity of manipulative tasks (Byrne 2007), but looks specifically at the stages of selection, modification, use, transport and discard patterns of artifacts. For example, chimpanzees in the Congo Basin have been observed to curate and re-use both puncturing sticks to break into termite nests and pounding sticks for opening honey hives (Sanz and Morgan 2007). Chimpanzees at Fongoli have been observed to "resharpen" the sticks they use to hunt bushbabies in several steps (Pruetz and Bertolani 2007). However, the most detailed work so far has focused on nut-cracking processes by cebus

monkeys and chimpanzees. Both wood and stone nut-cracking tools have been re-used at sites, creating tool assemblages with distinctive, often durable, and quite complex behavioral signatures, including: transported artifacts, pitted bedrock, battered and chipped hammers and anvils, wear traces and organic residues (Boesch and Boesch 1983, 1984a; Mercader et al. 2007; Ottoni and Izar 2008; Visalberghi et al. 2007). Influential early studies by the Boesches noted that Tai chimpanzees repeatedly carried hammers significant distances to crack nuts (Boesch and Boesch 1984a), and set the stage for detailed ethoarchaeological work at Tai (Joulian 1995), and at other sites. For example Carvalho has documented how chimpanzees repeatedly use and re-use composite tools to crack nuts, and has mapped the fine-grained movements and cumulative distribution patterns of artifacts at different nutting sites (Carvalho et al. 2009).

In such situations, we would expect the locations of profitable, long-term, fixed-point resources such as large nut trees or baobabs, termite nests or beehives to attract repeated episodes of extractive tool-use, just as groves of fruiting trees, cliffs and caves attract repeated episodes of primate nesting or repose. For example, because nut-bearing trees have a long lifespan and local ecological conditions can encourage groves of trees to regenerate, primates can revisit nutting sites over generations and the artifacts of these repeated behaviors can accumulate in significant concentrations. In particular, Mercader's excavations of nut-cracking sites in the Tai Forest have revealed a distinctive sedimentary record of fractured stone artifacts that likely derived from thousands of years of repeated nut-cracking by chimpanzees; the very large hammers and techniques used to produce many of the excavated stone pieces distinguish them from contemporary Later Stone Age assemblages

(Mercader et al. 2007, 2002). Another dramatic example of long-term nut-cracking sites can be seen in Brazil where cebus monkeys use bedrock platforms as anvils to crack open nuts from nearby trees, creating deeply pitted outcrops scattered with large hammer stones they have carried in (Visalberghi et al. 2007). Because the linear outcrops of suitable rocks form the edge of nut tree habitats, the intersection of food and rock has created a zone of repeated resource exploitation and relatively high archaeological visibility. A less-durable, but analogous example of this type of process was recorded when McBeath and McGrew collected assemblages of termite fishing tools from around Macrotermes mounds in different habitats, and noted that the highest densities of tools occurred at mounds in the zones where twigs from preferred Grewia bush species were abundant (McBeath and McGrew 1982). Even labor-intensive resources of more limited duration, such as tuber patches or reoccupied beehives, could form spatial foci for the accumulation of artifact and ecofact assemblages – as illustrated by the excavation holes and debris left behind by chimpanzees digging for roots at Ugalla and Semliki (R. Adriana Hernandez-Aguilar et al. 2007; McGrew et al. 2006), or digging for bee larvae and honey in Bulindi, Uganda (McClennan 2011). Though the feeding debris and artifacts from these activities will be ephemeral in most cases, these behavioral processes are habitual, and the predictable locations of such plant and insect foods mean that the use of places for food extraction and feeding will be super-imposed across generations, for dozens of years or more.

In contrast, when primates feed on vertebrates their meat-eating activities are assumed to rarely occur in the same place twice. Unfortunately, the spatial patterning and faunal traces of primate meat eating have only been documented specifically in a handful of studies. Rose (Rose and Marshall 1996; Rose 1997) observed that when cebus monkeys killed coatis or other small animals, the small carcasses were generally eaten on the spot, and those spots were distributed relatively randomly through the habitat, though it was not reported what bones were left in the scraps of carcasses the monkeys left behind. Chimpanzees hunt a variety of small mammals, especially monkeys, and tend to entirely consume their prey, bones and all. They occasionally use tools to acquire their prey or extract bits of meat and marrow (Boesch and Boesch 1989; Pruetz and Bertolani 2007; Sanz and Morgan 2007). These feeding patterns can modify bones in recognizable ways, as documented through experiments (Pickering and Wallis 1997), in the few studies of faunal remains left at chimpanzee meat-eating sites (Pickering and Wallis 1997; Plummer and Stanford 2000; Pobiner et al. 2007), or in fecal samples (Tappen and Wrangham 2000). However, while bone modification patterns across these compound assemblages have been detailed and collection spots schematically reported (e.g., Pobiner et al. 2007), no maps of the bone remains have been published which would allow an assessment of their distribution within habitats, or in relation to other spatial patterns of primate feeding activity, tool-use, ranging or nesting.

How likely is it that chimpanzee prey remains would accumulate in particular locations? Goodall reported that carcass remains were carried into a nest at Gombe (Goodall 1986), but that record is unique. Bone fragments that survive digestion might be the best bet, since chimpanzees will often defecate at their nesting sites or along frequently used trails. However, Tappen and Wrangham (2000) estimated that in the Kibale forest (Uganda) the annual bone fragment "rain" from chimpanzee dung would only be three fragments per hectare, probably dispersed relatively evenly throughout the community's home range through time. This contrasts markedly with the large assemblages of bones with distinctive bone damage which accumulate under the nests of crowned hawk-eagles, central-place foragers who also prey mostly on monkeys in the same forest (Pobiner et al. 2007; Trapani et al. 2006). Indeed, Pobiner et al. (2007) argue that because chimpanzees do not exhibit caching behavior, and because their prey is more likely to be scattered during consumption than transported to any particular location, their carcass feeding remains have a low probability of concentration. The 37 collections of bones which they analyzed at the Ngogo site were randomly dispersed over 20 km². On the other hand, Stanford argued that chimpanzee "predation core areas" at sites like Gombe are likely to be stable for decades because of the territoriality of both the hunters and the coincident locations of preferred plant foods and prey (Stanford 1996). This suggests that bone assemblages could accumulate during repeated episodes of predation within a narrow riparian forest zone, for example, though this has never been documented. In this context it is worth noting that our Ishasha studies found no evidence of chimpanzee carnivory in feces, dietary isotopes, or modified faunal remains, and the surface assemblages of "background density" animal bones scattered on the same transects showed no spatial correlations with the chimpanzee nests or other activity areas (Schoeninger et al. 1999; Sept 1994).

Primate meat-eating has intrigued paleoanthropologists because of the association of fossil bones and stones at the early Oldowan sites, and bone damage patterns suggesting that early hominins repeatedly used tools to acquire meat and marrow. Of course the decades since Glynn characterized the sharing of kills by Gombe chimpanzees as "tolerated scrounging" (Isaac 1978a) have seen an impressive documentation of hunting and strategic sharing of kills by monkeys and apes (Boesch and Boesch-Achermann 2000; Stanford 1996). Paleoanthropologists commonly note that the typical size (small) and taxon (primate, rodent, bird or young ungulate) of prey eaten by chimpanzees is different than the large mammal faunal remains typically associated with early stone tool assemblages, and this is probably one of the key behavioral shifts that accompanied the emergence of Oldowan ways of life. However, while several researchers including Glynn have considered the possibility of early hominins preying on microfauna (e.g., Isaac 1971), a combination of taphonomic biases and methodological specialization may have skewed our focus towards Oldowan large mammal meat-eating strategies. Renewed efforts to evaluate the exploitation of small animals by early hominins (Braun et al. 2010; Pobiner et al. 2008) or other predators (Arcos et al. 2010) will provide a welcome, expanded dataset to compare with the growing documentation of chimpanzee faunal assemblages. In turn, this will help us evaluate candidates for "pre-Oldowan" sites (Panger et al. 2002; Pobiner et al. 2007).

Primate Perspectives: A Worm's Agenda
Understandably, much primatological effort over the last few decades has been focused on documenting animal behaviors to record variation in endangered, isolated populations, often in the context of conservation efforts. The results have been extraordinary, and have significantly improved knowledge of our closest living relatives.

However, from an archaeologist's point of view, what might be missing from this emerging picture of primate behavior that would facilitate comparative research? As discussed earlier, a number of authors have commented on the need to develop rigorous and comparable methodologies for the analysis of artifacts and faunal remains, whether measuring standardized attributes or analyzing technological *chaînes opératoires*. But what other interdisciplinary challenges remain?

To take a worm's eye view of primates, we also must consider how artifacts, features and ecofacts are produced by primate behavior through space and time, and how these behavioral signatures, filtered by various taphonomic processes, have the potential to become directly or indirectly archaeologically visible. Worms, of course, would begin at a detailed spatial scale, and intriguing ethoarchaeological research has as well. For example the careful studies by Carvalho et al (2007) of localized tool use and transport help link variation in short-term process to site-specific depositional patterns. Stuart et al. (in press) trace local nest and tree attributes that provide behavioral signatures of a potentially greater time depth. These exemplary projects document fine-grained patterns of interest to both primatologists and archaeologists. New research is also needed to document variability in the density distributions and discard rates for a wide range of artifacts and feeding debris near landscape affordances such as the tools discarded around termite nests (e.g., McBeath and McGrew 1981), or knuckle-prints or other behavioral traces left behind at the caves at Fongoli (Preutz 2007). Ideally such research would follow Carvalho's lead, mapping patterns at comparable scales and recording both separate episodes of assemblage deposition (palimpsests of debris), and cumulative depositional patterns, whether the records were of durable artifacts or ephemeral traces.

Across a landscape to what extent are the specific locations of nesting, tool-use, meat-eating or other feeding behaviors likely to overlap through time? Visalberghi's (2007) mapping of pitted anvil sites, and Mercader's (2007) excavations of nutting sites in the Tai forest are two examples which emphasize the importance of documenting the longevity and potential taphonomic integrity of primate activity signatures in a broader landscape context, both within and between study areas. But how are the traces of activities distributed in relation to each other, or in relation to travel paths and core territories, social activity areas, or zones of predictable resources? Unfortunately such comprehensive patterns have rarely been systematically sampled and mapped even among modern human foragers, let alone primates. Current GPS technology and GIS software should greatly facilitate such work, but care must be taken to understand how such spatial data is sampled in relation to habitat variation as well as features from human land-use, such as trails or camps, that could impact the spatial redundancy of different types of behavior. Exciting new research by Carvalho is moving in this direction, as she is surveying, mapping and excavating a variety of chimpanzee activity sites in the forests of Guinea (Carvalho 2011a, b). Hernandez-Aguillar (2009) is also working towards this goal through systematic surveys that sample the spectrum of chimpanzee nest density in different habitats; by analyzing the geography of nesting in the dry Ugalla habitat, she has laid the groundwork for further spatial analyses of tool use and in the same region (R. Adriana Hernandez-Aguilar et al. 2007). This is particularly difficult to study in dry areas because the chimpanzees live in very large home ranges and are difficult to habituate. In particular she contrasted the large ranging area of the Ugalla chimpanzees with the tiny, unrepresentative fraction of chimpanzee behaviors that occur in sedimentary zones. And she describes how difficult it is to define a chimpanzee nesting site when in some cases nests are clustered in zones over a kilometer long. How are we to think about archaeological "sites" in analogous contexts (cf. Sept 1992, 1998)?

For years archaeologists have grappled with questions about how to define a "patch" or a "site" and what sampling strategies are most appropriate for different types of research questions in the early record (Kroll 1994). Glynn took an active interest in analyzing how the archaeological record was structured and how different kinds of sampling strategies could help evaluate alternative hypotheses - whether studying "sites," "mini-sites," or "scatters between the patches" (Isaac 1981b, 1984; Isaac et al. 1981), inspiring others to adapt landscape archaeology to the early stone age (see Behrensmeyer, Potts and Roche in this volume). Ethoarchaeological studies of primates can provide instructive analogues for different processes that could have influenced the accumulation and preservation of materials in different paleolandscape contexts, but we need to expand the sample of analogues to help evaluate equifinality inherent in the record (Gifford-Gonzalez 1991). Indeed, increasing the variety of species and habitats sampled will help avoid the "tyranny" of the ethological record, to paraphrase Wobst (1978), and avoid building interpretations unduly reliant on a limited range of familiar situations. And as we build the comparative dataset, we may gain a better sense of how the visibility of the ancient record is shaped not only by taphonomic processes but also by intersecting spatial and temporal rhythms of landscape ecology and culture (Figure 10.2).

Conclusions

The stage is set to anticipate what we can learn about site formation processes from living primate behavior. Perhaps the redundant spatial patterning sometimes evident in primate activities is not surprising, in retrospect, given the growing evidence that primates have sophisticated cognitive maps of their territories (Normand and Boesch 2009). But decades ago, when Isaac was developing his approach to analyzing the "stone age visiting cards" preserved in the archaeological record (Isaac 1980, 1981b), such clustered patterns of nests and other debris had not been documented systematically in non-human primates, and no one had done field studies to evaluate how the material residues of primate behavior in the wild could be compared to the early archaeological record. So whether focusing on studying the life histories of stone tools at individual sites, or mapping cultural behaviors in the context of taphonomic processes across a landscape, the progress is encouraging and should spur us on to do more. An ethoarchaeological approach to studying these relationships will bring us several steps closer to probing the ambiguity of these distinctive concentrations of stones and bones we call Oldowan sites.

It is not only the worms and the archaeologists who will learn from these inquiries. Bill McGrew has long advocated collaboration between primatologists and archaeologists (McGrew 1992, 2004; McGrew and Foley 2009) and has done much to encourage many of the promising new studies described here. He has demonstrated how expanding the methodological repertoire of primate studies to include attributes of material culture is valuable to primatologists interested in learning about details of cultural-ecological differences between groups, the time depth of behavioral traditions, and the cognitive complexity of our

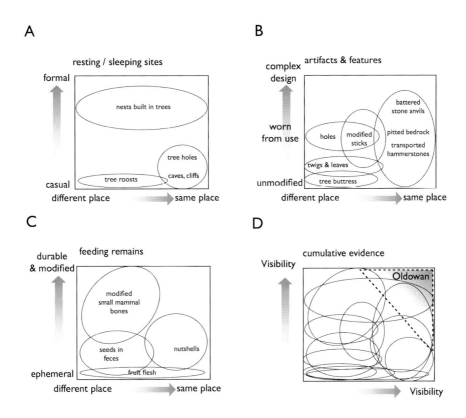

Figure 10.2 (a-d). Comparing the relative archaeological visibility of different primate activities: a) resting /sleeping situations range in spatial redundancy (situated in different places or same places) and formal design (from casual use to formal, deliberately structured design). Sleeping situations are more likely to be archaeologically recognizable if they are formally modified and reoccur in the sample place through time; b) artifacts and features created by primates range in spatial redundancy (situated in different places or same places) and formal design (from unmodified, to exhibiting wear from use, to formal and sometime complex design). Artifacts and artificial features are more likely to be archaeologically recognizable if they exhibit use-wear and/or deliberate shaping and reoccur and form assemblages in the same place through time; c) feeding remains created by primates range in spatial redundancy (situated in different places or same places) and durability after modification (from ephemeral food remains which will quickly become unrecognizable through digestion and decomposition, to food remains with a chance of preservation and evidence of having been fed upon, such as tooth marks or percussion damage). Food remains are more likely to be archaeologically recognizable if they are durable and behaviorally modified and if they reoccur and form assemblages in the same place through time; d) the patterns of evidence shown in Figures a-c are combined to represent the cumulative patterns of archaeological visibility of these behavioral signatures. These patterns are compared to an approximation of the visibility of the Oldowan record on the same scale. The Oldowan record has been recognized because it includes stone artifacts and large animal bones with distinctive surface damage which form assemblages which have been deposited in the same place through time. While some aspects of the Oldowan are distinctive and not represented in primate behavior today (such as flaked stone tools or butchered large mammal remains), there are other features (such as repeated use of space, use of stone hammers, deposition of small mammal remains) which do overlap with the behavior patterns of some living primates.

close living relatives. Similarly as researchers try to ferret out the underpinnings of niche separation between early hominin species, defining and evaluating the distinctive features of early Oldowan and pre-Oldowan records will be critical. Ethoarchaeological efforts promise to provide grist for the equifinality mill and create a rich comparative framework for paleoanthropological interpretations. I dare say Glynn would be pleased.

Acknowledgments

I was privileged to study with Glynn both as an undergraduate and graduate student at Berkeley, and as a visitor during his brief time at Harvard. His intellectual talents were matched by his whimsical enthusiasm and profound generosity. These gifts seem as fresh today as they were 25 years ago, and his inspiration continues.

References

Ancrenaz, M., R. Calaque, and I. Lackman-Ancrenaz
2004 Orangutan nesting behavior in disturbed forest of Sabah, Malaysia: Implications for nest census. *International Journal of Primatology* 25(5):983-1000.

Anderson, J. R.
1998 Sleep, sleeping sites, and sleep-related activities: Awakening to their significance. *American Journal of Primatology* 46:63-75.

Arcos, S., P. Sevilla, and Y. Fernández-Jalvo
2010 Preliminary small mammal taphonomy of FLK NW level 20 (Olduvai Gorge, Tanzania). *Quaternary Research* 74:405-410.

Baldwin, P. J., J. Sabater-Pi, W. C. McGrew, and C. E. G. Tuti
1981 Comparisons of nests made by different populations of chimpanzees (*Pan troglodytes*). *Primates* 22:474-486.

Bearder, S. K., L. Ambrose, C. Harcourst, P. Honess, A. Perkin, E. Pimley, S. Pullen, and N. Svoboda
2003 Species-typical patterns of infant contact, sleeping site use and social cohesion among nocturnal primates in Africa. *Folia Primatologica* 74:337-354.

Bertram, C.
2008 *Imagining the Turkish House: Collective Visions of Home.* University of Texas Press, Austin, TX.

Boesch, C., and H. Boesch
1983 Optimization of nut cracking with natural hammers in wild chimpanzees. *Behavior* 83:265-286.

1984a Mental map in wild chimpanzees: An analysis of hammer transports for nut cracking. *Primates* 25:160-170.

Boesch, C., and H. Boesch
1989 Hunting behaviors of wild chimpanzees in the Tai National Park. *American Journal of Physical Anthropology* 78:547-573.

Boesch, C. and H. Boesch-Achermann
2000 *The Chimpanzees of the Tai Forest: Behavioural Ecology and Evolution.* Oxford University Press, Oxford.

Braun, D. R., J. W. K. Harris, N. Levin, J. T. McCoy, A. I. R. Herries, M. K. Bamford, L. C. Biship, B. G. Richmond, and M. Kibunjia
2010 Early hominin diet included diverse terrestrial and aquatic animals 1.95 Ma in East Turkana, Kenya. *Proceedings of the National Academy of Sciences* 107(22):10002-10007.

Byrne, R.
2007 Culture in great apes: Using intricate complexity in feeding skills to trace the evolutionary origin of human technological process. *Philosophical Transactions of the Royal Society of London* 362B:577-585.

Carvalho, S.
2011a Diécké Forest, Guinea: Delving into chimpanzee behavior using stone tool surveys. In *The Chimpanzees of Bossou and Nimba*, edited by T. Matsuzawa, T. Humle and Y. Sugiyama, pp. 301-312. Springer, Tokyo.

2011b Extensive surveys of chimpanzee stone tools: From the telescope to the magnifying glass. In *The Chimpanzees of Bossou and Nimba*, edited by T. Matsuzawa, T. Humle and Y. Sugiyama, pp. 145-155. Springer, Tokyo.

Carvalho, S., E. Cunha, C. Sousa, and T. Matsuzawa
2008 *Chaînes opératoires* and resource-exploitation strategies in chimpanzee (*Pan troglodytes*) nut cracking. *Journal of Human Evolution* 55(1):148-163.

Carvalho, S., D. Biro, W. C. McGrew, and T. Matsuzawa
2009 Tool-composite reuse in wild chimpanzees (*Pan troglodytes*): Archaeologically invisible steps in the technological evolution of early hominins? *Animal Cognition* 12(Supplement 1):S103-S114.

David, N., and C. Kramer
2002 *Ethnoarchaeology in Action.* Cambridge University Press, Cambridge UK.

Daniel, S., C. Korine, and B. Pinshow
2008 Central-place foraging in nursing, arthropod-gleaning bats. *Canadian Journal of Zoology* 86(7):623-626.

Di Bitetti, M. S., E. M. L. Vidal, M. C. Baldovino, and V. Benesovsky

2000 Sleeping site preferences in tufted capuchin monkeys (*Cebus apella nigritus*). *American Journal of Primatology* 50:257-274.

Dominguez-Rodrigo, M., C. P. Egeland, and R. Barba

2007 The "home-base" debate. In *Deconstructing Olduvai: A Taphonomic Study of the Bed I Sites*, edited by M. Dominguez-Rodrigo, R. Barba and C. P. Egeland, pp. 1-10. Springer, Berlin.

Fan, P.-F., and X.-L. Jiang

2008 Sleeping sites, sleeping trees, and sleep-related behaviors of black crested gibbons (*Nomascus concolor jingdongensis*) at Mt. Wuliang, Central Yunnan, China. *American Journal of Primatology* 70:153-160.

Fruth, B., and G. Hohmann

1994a Ecological and behavioral aspects of nest building in wild bonobos (*Pan paniscus*). *Ethology* 94:113-126.

1994b Comparative analyses of nest-building behavior in bonobos and chimpanzees. In *Chimpanzee Cultures*, edited by R. W. Wrangham, W. C. McGrew, F. B. M. de Waal and P. G. Heltne, pp. 109-128. Harvard University Press, Cambridge, MA.

1994c Nests: Living artefacts of recent apes? *Current Anthropology* 35(3):310-311.

Furuichi, T., and C. Hashimoto

2004 Botanical and topographical factors influencing nesting-site selection by chimpanzees in Kalinzu Forest, Uganda. *International Journal of Primatology* 25(4):755-765.

Gifford-Gonzalez, D.

1991 Bones are not enough: Analogues, knowledge, and interpretive strategies in zooarchaeology. *Journal of Anthropological Archaeology* 10(3):215-254.

Goodall, J.

1986 *The Chimpanzees of Gombe*. Harvard University Press, Cambridge, MA.

Goren-Inbar, N., E. Werker, and C. S. Feibel

2002 *The Acheulian site of Gesher Benot Ya'aqov, Israel: The Wood Assemblage*. Oxbow Books, Oxford.

Gowlett, J. A. J.

2009 Artefacts of apes, humans, and others: Towards comparative assessment and analysis. *Journal of Human Evolution* 57(4):401-410.

Hamilton III, W. J.
1982 Baboon sleeping site preferences and relationships to primate grouping patterns. *American Journal of Primatology* 3:41-53.

Hankerson, S. J., S. P. Franklin, and J. M. Dietz
2007 Tree and forest characteristics influence sleeping site choice by Golden Lion Tamarins. *American Journal of Primatology* 69:976-988.

Haslam, M., A. Hernandez-Aguilar, V. Ling, S. Carvalho, I. de la Torre, A. DeStefano, A. Du, B. Hardy, J. Harris, L. Marchant, T. Matsuzawa, W. McGrew, J. Mercader, R. Mora, M. Petraglia, H. Roche, E. Visalberghi, and R. Warren
2009 Primate archaeology. *Nature* 460(7253):339-344.

Hediger, H.
1977 Nest and home. *Folia Primatologica* 28:170-187.

Hernandez-Aguilar, R. A.
2009 Chimpanzee nest distribution and site reuse in a dry habitat: Implications for early hominin ranging. *Journal of Human Evolution* 57(4):350-364.

Hernandez-Aguilar, R. A., J. Moore, and T. R. Pickering
2007 Savanna chimpanzees use tools to harvest the underground storage organs of plants. *Proceedings of the National Academy of Sciences* 104(49):19210-19213.

Isaac, G. L.
1969 Studies of early culture in East Africa. *World Archaeology* 1:1-28.

1971a The diet of early man: Aspects of archaeological evidence from lower and middle Pleistocene sites in East Africa. *World Archaeology* 2(3):278-299.

1971b Whither archaeology? *Antiquity* 45:123-129.

1972 Early phases of human behavior: Models in Lower Palaeolithic archaeology. In *Models in Archaeology*, edited by D. L. Clarke, pp. 167-199. Methuen, London.

1978a The archaeological evidence for the activities of early African hominids. In *Early Hominids of Africa*, edited by C. J. Jolly, pp. 219-254. Duckworth, London.

1978b The food-sharing behavior of protohuman hominids. *Scientific American* (238):311-325.

186   Casting the Net Wide

1980 Casting the net wide: A review of archaeological evidence for early hominid land-use and ecological relations. In *Current Argument on Early Man: Proceedings of a Nobel Symposium Organized by the Royal Swedish Academy of Sciences (May, 1978: Commemorating the 200th Anniversary of the Death of Carlous Linnaeus)*, edited by L. K. Konigsson, pp. 21-27. Oxford: Pergamon for the Swedish Academy of Sciences, Bjorkborns Herrgard, Karlskoga.

1981a Archaeological tests of alternative models of early hominid behavior: Excavation and experiments. *Philosophical Transactions of the Royal Society of London* 292 (Series B):177-188.

1981b Stone age visiting cards: Approaches to the study of early land-use patterns. In *Patterns in the Past*, edited by I. Hodder, G. L. Isaac and N. Hammond, pp. 37-103. Cambridge University Press, Cambridge.

1983a Aspects of human evolution, evolution from molecules to men. In *Essays on Evolution: A Darwin Centenary Volume*, edited by D. S. Bendall, pp. 509-543. Cambridge University Press, Cambridge.

1983b Bones in contention: Competing explanations for the juxtaposition of Early Pleistocene artifacts and faunal remains. In *Animals and Archaeology. I. Hunters and their Prey*, edited by J. Clutton-Brock, C. Grigson, vol. 3-19. British Archaeological Reports International series 163, Oxford.

1984 The archaeology of human origins: Studies of the lower Pleistocene in East Africa 1971-1981. In *Advances in World Archaeology*, edited by F. Wendorf, pp. 1-87. vol. 3. Academic Press, New York.

1986 Foundation stones: Early artifacts as indicators of activities and abilities. In *Stone Age Prehistory*, edited by G. N. Bailey and P. Callow, pp. 221-241. Cambridge University Press, Cambridge.

Isaac, G. L. and D. C. Crader
1981 To what extent were early hominids carnivorous? An archaeological perspective. In *Omnivorous Primates: Gathering and Hunting in Human Evolution*, edited by G. T. R. S. O. Harding, pp. 37-103. Columbia University Press, New York.

Isaac, G. L., J. W. K. Harris, and F. Marshall
1981 Small is informative: The application of the study of mini-sites and least-effort criteria in the interpretation of the early Pleistocene archaeological record at Koobi Fora, Kenya. In *Las Industrias mas Antiguas*, edited by G. L. Isaac and J. D. Clark, pp. 101-119. X Congreso, Union Internacional de Ciencias Prehistoricas y Protohistoricas, Mexico City.

Jackson, M.
1995 *At Home In the World*. Yale University Press, New Haven, CT.

Joulian, F.
1995 Mise en évidence de différences traditionnelles dans le cassage des noix chez les chimpanzés (*Pan troglodytes*) de la Côte d'Ivoire, implications paléoanthropologiques. *Journal des africanistes* 65(2)57-77.

Kappeler, P. M.

1998 Nests, tree holes, and the evolution of primate life histories. *American Journal of Primatology* 46:7-33.

Koops, K., T. Humle, E. H. M. Sterck, and T. Matsuzawa

2007 Ground-nesting by the chimpanzees of the Nimba Mountains, Guinea: Environmentally or socially determined? *American Journal of Primatology* 69:407-419.

Kouakou, C. Y., C. Boesch, and H. Kuehl

2009 Estimating chimpanzee populations size with nest counts: Validating methods in Tai National Park. *American Journal of Primatology* 71:1-11.

Kroll, E. M.

1994 Behavioral implications of Plio-Pleistocene archaeological site structure. *Journal of Human Evolution* 27:107-138.

Kroll, E. M., and G. L. Isaac

1984 Configurations of artifacts and bones at early Pleistocene sites in East Africa. In *Intrasite Spatial Analysis in Archaeology*, edited by H. J. Hietala, pp. 4-31. Cambridge University Press, Cambridge.

Manzo, L. C.

2005 For better or worse: Exploring multiple dimensions of place meaning. *Journal of Environmental Psychology* 25:67-86.

McBeath, N. M., and W. C. McGrew

1982 Tools used by wild chimpanzees to obtain termites at Mt. Assirik, Senegal: The influence of habitat. *Journal of Human Evolution* 11:65-72.

McClennan, M. R.

2011 Tool-use to obtain honey by chimpanzees at Bulundi: New record from Uganda. *Primates* 52(3):8.

McGrew, W. C.

1992 *Chimpanzee Material Culture. Implications for Human Evolution.* Cambridge University Press, Cambridge.

1998 Behavioral diversity in populations of free-ranging chimpanzees in Africa: Is it culture? *Human Evolution* 13(3-4, 1998):209-220.

2004 *The Cultured Chimpanzee: Reflections on Cultural Primatology.* Cambridge University Press, Cambridge, U.K.

McGrew, W., and R. A. Foley

2009 Palaeoanthropology meets primatology. *Journal of Human Evolution* 57(4):335-336.

McGrew, W. C., L. F. Marchant, and K. D. Hunt

2006 Etho-archaeology of manual laterality: Well-digging by wild chimpanzees. *Folia Primatologica* 78:240-244.

Mercader, J., H. Barton, J. Gillespie, J. Harris, S. Kuhn, R. Tyler, and C. Boesch

2007 4,300-year-old chimpanzee sites and the origins of percussive stone technology. *Proceedings of the National Academy of Sciences* 7 104:3043-3048.

Mercader, J., M. A. Panger, and C. Boesch

2002 Excavation of a Chimpanzee Stone Tool Site in the African Rainforest. *Science* 296:1452-1455.

Moore, J.

1992 "Savanna" chimpanzees. In *Topics in Primatology, Vol 1: Human Origins*, edited by T. Nishida and et al. pp. 99-118. University of Tokyo Press, Tokyo.

2000 Placing home in context. *Journal of Environmental Psychology* 20:207-217.

Nishida, T., K. Zamma, T. Matsusaka, A. Inaba, and W. C. McGrew

2010 *Chimpanzee Behavior in the Wild: An Audio-Visual Encyclopedia*. Springer, Tokyo.

Normand, E., and C. Boesch

2009 Sophisticated Euclidean maps in forest chimpanzees. *Animal Behavior* 77:1195-1201.

Ogawa, H., G. I. Idani, J. Moore, L. Pintea, and A. Hernandez-Aguilar

2007 Sleeping parties and nest distribution of chimpanzees in the savanna woodland, Ugalla, Tanzania. *International Journal of Primatology* 28:1397-1412.

Olsson, O., J. S. Brown, and K. L. Helf

2008 A guide to central place effects in foraging. *Theoretical Population Biology* 74:22-33.

Ottoni, E. B., and P. Izar

2008 Capuchin monkey tool use: Overview and implications. *Evolutionary Anthropology* 17:171-178.

Panger, M. A., A. S. Brooks, B. G. Richmond, and B. Wood

2002 Older than the Oldowan? Rethinking the emergence of hominin tool use. *Evolutionary Anthropology* 11:235-245.

Pickering, T. R., and J. Wallis

1997 Bone modifications resulting from captive chimpanzee mastication: Implications for the interpretation of Pliocene archaeological faunas. *Journal of Archaeological Science* 24:1115-1127.

Plummer, T.

2004 Flaked stones and old bones: Biological and cultural evolution at the dawn of technology. *Yearbook of Physical Anthropology* 47(118-164).

Plummer, T. W., and C. B. Stanford

2000 Analysis of a bone assemblage made by chimpanzees at Gombe National Park. *Journal of Human Evolution* 39:345-365.

Pobiner, B. L., J. DeSilva, W. J. Sanders, and J. C. Mitani

2007 Taphonomic analysis of skeletal remains from chimpanzee hunts at Ngogo, Kibale National Park, Uganda. *Journal of Human Evolution* 52:614-636.

Pobiner, B. L., M. J. Rogers, C. Monahan, and J. W. K. Harris

2008 New evidence for hominin carcass processing strategies at 1.5 Ma, Koobi Fora, Kenya. *Journal of Human Evolution* 55:103-130.

Preutz, J. D.

2006 Feeding ecology of savanna chimpanzees (*Pan troglodytes verus*) at Fongoli, Senegal. In *Feeding Ecology in Apes and Other Primates. Ecological, Physical and Behavioral Aspects*, edited by G. Hohmann, M. M. Robbins and C. Boesch, pp. 161-182. Cambridge Studies in Biological and Evolutionary Anthropology, K. B. Strier, general editor. Cambridge University Press, Cambridge.

Preutz, J. D., S. J. Fulton, L. F. Marchant, W. C. McGrew, M. Schiel, and M. Waller

2008 Arboreal nesting as anti-predator adaptation by savanna chimpanzees (*Pan troglodytes verus*) in southeastern Senegal. *American Journal of Primatology* 70:393-401.

Pruetz, J., and P. Bertolani

2009 Chimpanzee (*Pan troglodytes verus*) behavioral responses to stresses associated with living in a savannah-mosaic environment: Implications for hominin adaptations to open habitats. *PaleoAnthropology* 2009:252-262.

Pruetz, J. D.

2007 Evidence of cave use by savanna chimpanzees (*Pan troglodytes verus*) at Fongoli, Senegal: Implications for thermoregulatory behavior. *Primates* 48(4):316-319.

Pruetz, J. D., and P. Bertolani

2007 Savanna Chimpanzees, *Pan troglodytes verus*, Hunt with Tools. *Current Biology* 17(5):412-417.

Raffel, T. R., N. Smith, C. Cortright, and A. J. Gatz

2009 Central place foraging by beavers (*Castor canadensis*) in a complex lake habitat. *The American Midland Naturalist* 162(1):62-73.

Rasoloharijaona, S., B. Randrianambinina, and E. Zimmerman

2008 Sleeping site ecology in a rain-forest dwelling nocturnal lemur (*Lepilemur mustelinus*): Implications for sociality and conservation. *American Journal of Primatology* 70:247–253.

Reichard, U.

1998 Sleeping sites, sleeping places, and presleep behavior of gibbons (*Hylobates lar*). *American Journal of Primatology* 46:35–62.

Rose, L., and F. Marshall

1996 Meat eating, hominid sociality, and home bases revisited. *Current Anthropology* 37:307–338.

Rose, L. M.

1997 Vertebrate predation and food-sharing in *Cebus* and *Pan*. *International Journal of Primatology* 18(5):727–765.

Rothman, J. M., A. N. Pell, E. S. Dierenfeld, and C. M. McCann

2006 Plant choice in the construction of night nests by gorillas in the Bwindi Impenetrable National Park, Uganda. *American Journal of Primatology* 68:361–368.

Sanz, C., and D. Morgan

2007 Chimpanzee tool technology in the Goualougo Triangle, Republic of Congo. *Journal of Human Evolution* 52:420–433.

2009 Flexible and persistent tool-using strategies in honey-gathering by wild chimpanzees. *International Journal of Primatology* 30:411–427.

Schick, K. D.

1987 Modeling the formation of Early Stone Age artifact concentrations. *Journal of Human Evolution* 16(7/8):789–808.

Schoeninger, M. J., J. Moore, and J. M. Sept

1999 Subsistence strategies of two "savanna" chimpanzee populations: The stable isotope evidence. *American Journal of Primatology* (49):274–314.

Schreier, A. and L. Swedell

2008 Use of palm trees as a sleeping site for Hamadryas baboons (*Papio hamadryas hamadryas*) in Ethiopia. *American Journal of Primatology* 70:107–113.

Sept, J. M.

1992 Was there no place like home? A new perspective on early hominid sites from the mapping of chimpanzee nests. *Current Anthropology* 33(2):187–207.

1994 Bone distribution in a semi-arid chimpanzee habitat in eastern Zaire: Implications for the interpretation of faunal assemblages at early archaeological sites. *Journal of Archaeological Science* 21:217-235.

1998 Shadows on a changing landscape: Comparing refuging patterns of hominids and chimpanzees since the last common ancestor. *American Journal of Primatology* 46(1):85-101.

Shumaker, R. W., K. R. Walkup, and B. B. Beck
2011 *Animal Tool Behavior: The Use and Manufacture of Tools by Animals.* April 15, 2011 edition. The Johns Hopkins University Press.

Stanford, C. B.
1996 The hunting ecology of wild chimpanzees: Implications for the evolutionary ecology of Pliocene hominids. *American Anthropologist* 98(1):96-113.

2008 *Apes of the Impenetrable Forest. The Behavioral Ecology of Sympatric Chimpanzees and Gorillas. Primate Field Studies.* Pentice Hall, Upper Saddle River, NJ.

Stanford, C. B., and R. C. O'Malley
2008 Sleeping tree choice by Bwindi chimpanzees. *American Journal of Primatology* 70:642-649.

Stewart, F. A., A. K. Piel, and W. C. McGrew
in press Living archaeology: Artefacts of specific nest site fidelity in wild chimpanzees. *Journal of Human Evolution* 61(2011):8.

Stewart, F. A., J. D. Pruetz, and M. H. Hansell
2007 Do chimpanzees build comfortable nests? *American Journal of Primatology* 69:930-939.

Tappen, M., and R. Wrangham
2000 Recognizing hominoid-modified bones: The taphonomy of colobus bones partially digested by free-ranging chimpanzees in the Kibale forest, Uganda. *American Journal of Physical Anthropology* 118:217-234.

Toth, N., and K. Schick
2009 The Oldowan: The tool making of early hominins and chimpanzees compared. *Annual Review of Anthropology* 38(1):289-305.

Trapani, J., W. J. Sanders, J. C. Mitani, and A. Heard
2006 Precision and consistency of the taphonomic signature by crowned hawk-eagles (*Stephanoaetus coronatus*) in Kibale National Park, Uganda. *PALAIOS* 21(2):114-131.

Tutin, C. E. G., R. J. Parnell, L. J. T. White, and M. Fernandez
1995 Nest building by lowland gorillas in the Lope Reserve, Gabon: Environmental influences and implications for censusing. *International Journal of Primatology* 16(1):53-76.

Videan, E. N.
2006 Bed-building in captive chimpanzees (*Pan troglodytes*): The importance of early rearing. *American Journal of Primatology* 68:745-751.

Visalberghi, E., D. Fragaszy, E. Ottoni, P. Izar, M. G. de Oliveira, and F. R. D. Andrade
2007 Characteristics of hammer stones and anvils used by wild bearded capuchin monkeys (*Cebus libidinosus*) to crack open palm nuts. *American Journal of Physical Anthropology* 132(3):426-444.

Whiten, A., K. Schick, and N. Toth
2009 The evolution and cultural transmission of percussive technology: Integrating evidence from palaeoanthropology and primatology. *Journal of Human Evolution* 57(4):420-435.

Wobst, H. M.
1978 The archaeol-ethnology of hunter-gatherers or the tyranny of the ethnographic record in archaeology. *American Antiquity* 43(2):303-309.

Yellen, J. E.
1977 *Archaeological Approaches to the Present. Models for Reconstructing the Past.* Academic Press, New York.

11

A GEOGRAPHIC OVERVIEW OF
NEANDERTHAL-MODERN HUMAN ENCOUNTERS

Ofer Bar-Yosef

Opening Remarks

Recent publications indicate that interbreeding between Neanderthals and non-African modern humans took place in the past and that the flow of the former genes to the more recent population is estimated at 1-4% (Green et al. 2010). The lack of such evidence among African populations restricts the zones of biological encounters to Eurasia. Similar proposals concerning an interbreeding between the two populations were offered in the past on the basis of morphological characteristics of local Neanderthals at the edge of their geographic distribution such as the Levant and the Iberian peninsula (Arensburg and Belfer-Cohen 1998, and references therein; Trinkaus 2005; Zilhao 2006). Hence, the challenges to current archaeological research are: a) demonstrating 'where' the two populations overlapped allowing for the interchange of genes; and, b) identifying the archaeological markers for such interactions.

In order to achieve these goals in a short essay we need to examine the patterns of behaviors common among dispersing hunter-gatherers, the geographic distribution of Neanderthals during and after the MOIS4 prior to the arrival of the last wave of modern humans (ca. 55/50-40 Ka BP, known as the Initial Upper Paleolithic), and then speculate how we may detect in the archaeological records along the dispersal routes of modern humans the results of their social interactions with the Neanderthals.

It is important for understanding the dynamics of prehistoric demography to remember that the Neanderthals, prior to the arrival of modern humans, were technically and economically successful population, and dispersed from western Europe eastward into regions that were either populated (even sparsely) by archaic humans (such as the Skhul-Qafzeh group) or were uninhabited. The notion that Neanderthals were also colonizers and not solely a population that succumbed to the invading modern humans should be stressed. Perhaps it is only one of many examples we tend to ignore, that during the Pleistocene prehistoric groups could be and were colonizers responsible for the extinction of the others.

Genetic evidence indicates that around 50-35 Ka modern humans spread from Africa. We may therefore expect that both archaic and modern populations behaved in a similar way as they moved into others territories, and that their behavior resembled what is known from current studies of foragers. In this context I believe, we should abandon the image of the "noble savage", more a reflection of political ideas of the 17th century than concrete evidence from the field (e.g., Otterbein 1997).

Ethnographic studies of hunter-gathers societies in both the Old and New World indicate that while expanding from their original homeland into an area already occupied by other local foragers migrants face three alternative modes of interactions: a) ignore the presence of local foragers while dispersing around them, camp in suitable locales, gather and hunt in the immediate territory without considering the needs of their 'neighbors'; b) view the locals as competitors or enemies and engage in confrontational violence – winners or losers may reflect in part the efficiency of hunting devices as weapons; and, c) employ different ways and means for mutual interactions that may benefit both populations, or may cause the spread of diseases brought by the foreigners into the "new territory". The nature of these interactions may change through time. I also intentionally do not employ in this context the term "acculturation" that is correctly seen as indicating "asymmetrical relationships of dominating and dominated" (d'Errico et al. 1998:S3). "Interactions" carries a more balanced meaning when two populations meet each other and it is our duty as archaeologists to figure out the nature of the relationships.

Taking these possible patterns into account we should recognize that the study of prehistory from 50,000 to 35,000 years ago is still plagued with two well-known difficulties: first, verifying the dates for the arrival of modern humans in a given region, or the demise of the Neanderthals, or the time span of their contemporaneity; second, attributing material cultural elements, i.e., the lithic industries as well as bone, ivory and antler objects, allowing us to characterize a given population by identifying an "archaeological culture".

The degree of chronological resolution of radiocarbon dates suffers from ambiguities due to uncertain information concerning the depositional context of the dated samples, and the use of one or another calibration curves. Therefore, this paper will employ uncalibrated BP dates as reported in the literature, and will make two bold statements concerning these issues in order to present a comprehensive hypothesis.

Accepting the radiocarbon dates as "correct" will be done by assuming that the Neanderthal population spread from west to east within Eurasia (thus older dates in the west and younger in the east) and that modern humans spread from Africa in two directions – westward into Europe (thus younger dates at the Atlantic front) and eastward into central and eastern Asia (where contextual ambiguities prevail; I assume the Australian early dates as 45±5 Ka are correct).

Along their dispersal avenues, modern humans as foragers may have behaved in variable ways, i.e., conducting amiable relationships with the Neanderthal groups (Zilhao 2006), ignoring them and sometimes killing them as they had better weaponry (Shea and Sisk 2010). We should also consider the role of imported diseases.

Identifying prehistoric cultures will be based on objects that can serve as "cultural markers". During the last three decades a growing awareness among archaeologists resulted in cautiously rejecting the idea that form and function of lithic artifacts are directly related. Accepting that core reduction strategies are part and parcel of a group's technological tradition – often with the shaping of particular tools such as foliates – facilitates the identification of past populations. These general observations emerged from recognizing that in every society of foragers teaching and learning were determined by the traditional knowledge involved in choosing raw material and the techniques employed for obtaining blanks and shaping them into desired tools (Pelegrin 1985, 1990, 1995). The goals of teaching were the 'know how' to make tools from hard rocks and organic matters that efficiently serve in food acquisition techniques critical for keeping the biological survival of the group. By watching and mimicking adolescents artisans and adults younger members of the band could achieve these technical goals (Karlin 1991) but the learning process would be best served with oral explanations (e.g., Bar-Yosef and Van Peer 2009).

The importance of the teaching process is well exemplified in the Levantine records in western Asia. The regional sequence of prehistoric lithic industries demonstrates that the same general knapping methods (sometimes referred to as techniques) were used during a portion of the Middle Paleolithic sequence that lasted from ca. 250 Ka through 47/45 Ka. The Levantine Middle Paleolithic is subdivided today into three major entities, labeled after the three main stratigraphic units in Tabun cave as "Tabun -B, C, and D type" (also called today Lower, Middle and Late Mousterian). Here I take only as an example the lithic industries from MIS 5 (Last Interglacial) through MIS 4 and 3 (essentially "Tabun C and B types"). Exploiting the same good quality flint sources and in spite of evidence for regional climatic fluctuations (Bar-Matthews et al. 2003) major typo-technological changes were recognized during the detailed lithic analysis (Bar-Yosef 1989, 2000; Meignen 1990, 1995; Meignen and Bar-Yosef 1992; Hovers 2006, 2009; Shea 2008). In particular this was noted when the Late Mousterian, produced by Neanderthals, was studied in detail.

During the period of this entity (ca. 75–45 BP) a dominant method known as unidirectional convergent strategy, with some flexibility around the standard, was recorded, for instance, in Kebara and Amud caves (Meignen and Bar-Yosef 1992; Meignen 1995; Goren-Inbar and Belfer-Cohen 1998; Hovers 1998). We therefore should not ignore the number of generations during which the same rigid teaching led to the production of the same lithic assemblages with minimal variability. Differences among individual knappers within the group and during training of the young children probably caused the observable minimal techno-typological variability. In addition to stressing that the

raw material sources did not had any affect on the kind of lithic industries, suffice to note that at least from the Levantine Late Acheulian (ca. 450 Ka BP) through the Upper Paleolithic in Mount Carmel major changes in the techniques for fabricating stone blanks and shaped tools types occurred while the same good quality flint was available within 2-5 km of every site (Jelinek 1991). Resource stability characterized this small area and it is also expressed in the availability of the same animals and plant foods documented for a long period (e.g., Stiner et al. 2009). Thus, cultural changes were not related to raw material availability or quality but most probably characterized different human groups that practiced different technical traditions. Whether they all belonged to the same population is hardly a justified question in view of the accepted differences between the earlier Skhul-Qafzeh group and the local Neanderthals (even if they were already genetically mixed (Tillier et al. 2008) during the "take over" of the Levant by the Neanderthals). This example hints at a potential explanation in that many of these cultural shifts could be due to repeated occupations of the same landscape by new and different populations of foragers. Thus, among each group the role of cultural teaching and rigid social rules of how to make the required stone tools were probably stronger than the desire for innovation.

In sum, particular knapping techniques and shaping tools or a particularly designed object may serve as 'cultural markers' for tracing 'people'. Indeed, adopting this approach allows us to potentially recognize geographically wide-ranging archaeological phenomena. Similarly to modern demographic studies, and although 'the devil is in the details', I will ignore certain 'details' that may be considered as important ones by other scholars for the benefit of the overall picture as exposed in the following pages.

Several ambiguities in the interpretations of the archaeological data sets emanating from poorly-dated assemblages or layers will be mentioned, but they do not change the overarching continental picture of prehistoric cultures at the boundary of the so-called Middle and Upper Paleolithic.

Before presenting the hypothesis concerning claims for cultural contacts between two different populations we need to briefly examine the question, "how does one recognize archaeologically the signs of cultural interactions?" This issue is extremely controversial. For example, it was raised during the debate concerning whether the bone, antler, ivory objects and teeth pendants found in the Châtelperronian layers (X-VIII) at Grotte du Renne, together with numerous Mousterian tools and the typical Châtelperronian blade industry as well as 16 Neanderthals teeth (Bailey and Hublin 2006), were the results of Neanderthals original activities or due to cultural interaction with incoming modern humans (e.g., d'Errico et al. 1998 and comments therein). The dominant opinion attributed the making of body decorations to Neanderthals, while other examples of their capacity for making body decoration were recently documented (Zilhao et al. 2010). Several archaeologists felt that the production of the pendants, bone and antler objects may possibly have resulted from exchange between the two groups. In the case of this site it seems that the deposits that contained Neanderthal teeth and Mousterian and Châtelperronian stone tools and organic objects were naturally mixed, due to penecontemporaneous and post-depositional processes (Bar-Yosef and Bordes 2010; Higham et al. 2010). However, this does not exclude the possibility that the groups interacted, as will be mentioned below. For example, we suggested (Bar-Yosef 2006; Bar-Yosef and Bordes 2010)

that the St. Cesaire Neanderthal burial (Vandermeersch 1993) reflects emotional attachment to a location that was perhaps originally the home of a group of Neanderthals. In this scenario, the area was taken over by modern humans, the new folk, but as often sites were occupied seasonally the Neanderthals who used to be the cave dwellers returned to bury one of their dead. Bringing the bones of a dead individual to be entered in an ancestral home is a known pattern of behavior among foragers (e.g., Pardoe 1988, and references therein). In addition this case supports the contention that both Neanderthals and modern humans, the makers of the Châtelperronian industry (Bar-Yosef 2006) were present contemporaneously in the region.

The Neanderthal Lands

To review the interactions between the two populations we need to briefly summarize the continental-wide archaeological information concerning the lithic industries of local Neanderthals. The best records are available in Europe and the eastern Mediterranean, including the Levant. Given the numerous sites and regional reports, a description concerning the nature of Middle Paleolithic occupations, their contents, and the variable labels given to the identifiable units (cultures), is readily simplified.

Neanderthals in temperate Europe responded to climatic calamities by shifting their territories, but they undoubtedly were successful survivors. During the cold period of MIS4 (ca. 75-57 Ka) Neanderthals in the north European plain moved into refugia in southwest and southeast Europe resulting in the depopulation of a large region and causing a population "bottle neck" (e.g., Jöris 2006). In addition, they moved also into Anatolia, the Levant and the Zagros mountains (e.g., Bar-Yosef 1988). Their occupations in the area north of the Caucasus took place at an earlier time.

Further east they succeeded in reaching beyond the Caspian Sea to Teshik-Tash and other caves in Uzbekistan (Glantz et al. 2008), and to the Altai mountains (Derevianko and Markin 1995). That their migrations were not all conducted at the same time is demonstrated by the differences between the Middle Paleolithic assemblage of the upper layer of the Mousterian in Denisova cave that falls squarely within the eastern Micoquian with a few bifacial pieces and Okladnikov cave with its mostly short flakes heavily retouched as scrapers (Derevianko et al. 2003; Derevianko and Shunkov 2005).

It is possible that certain groups continued eastward into northwest China and were the makers of a Levallois type industry at Shuidonggou Locality 1, (Ningxia 2003). We may expect that crossing the landmass of central Asia Neanderthals would have shown the variable behavior of hunter-gatherers who migrate into new environments.

Looking westward into Europe, we note that more than one knapping technique was employed during the early part of the Upper Pleistocene as recognized through the studies of several hundreds of assemblages across Europe, all made by the Neanderthals. At about 50–40 ka a great variability among the European Mousterian entities was recognized e.g., (Delagnes and Meignen 2006; Delagnes et al. 2007; Jöris 2004). The largest cultural province is in temperate Europe, north of the Pyrenees, the Alps, the Carpathian Mountains and the Caucasus. Entities with bifacially shaped handaxes (such as the Mousterian of Acheulian Tradition) are largely represented in western Europe, while the makers of the keilmesser (bifacially shaped knives) are present in central and eastern Europe; these industries are mostly incorporated under the term eastern Micoquian (Burduckeiwicz 2000; Kozlowski 2000).

The lack of these bifacially shaped pieces characterizes the Neanderthal lithic assemblages in the Mediterranean lands from the Iberian and Italian peninsulas through Anatolia, the Zagros Mountains and the Levant. Instead the artisans employed one of the Levallois methods for manufacturing variable, intensively retouched side scrapers and points in the Taurus, the Zagros, the southern foothills of the Caucasus but not in the Levant (e.g., Iberia Cabrera-Valdes and Bernaldo de Quiros 1992; Mussi 2001; Panagopoulo et al. 2002/2004; Baumler and Speth 1993; Dibble 1993; Doronichev and Golovanova 2003; Adler 2002). In this 'big picture' some geographic exceptions occur, such as the non-biface industries of Quina-type, Denticulate-type Mousterian, and Ferrassie-type Mousterian known from southwest France, and the Levallois industries in Crimea where they were contemporary with the eastern Micoquian (Chabai and Monigal 1999)

In the Levant two Middle Paleolithic entities were recognized during the Upper Pleistocene and labeled as "Tabun C-type and B-type". Deposits with the "Tabun C-type" (Middle Levantine Mousterian; Bar-Yosef 1998; Shea 2008; Hovers 2009) contain an industry mostly made by one of the Levallois methods as well as burials of modern humans often referred to as the 'Skhul-Qafzeh' group. The overlying deposits contained assemblages labeled as "Tabun-B-type" (Late Levantine Mousterian) always associated with Neanderthal remains. These assemblages characterized by numerous triangular Levallois points made by reducing convergent Levallois cores, (Meignen and Bar-Yosef 1991, 1992; Meignen 1995) were fabricated by local Neanderthals and demonstrate a great similarity from Dederiyeh cave in the north to Tor Faraj in the south, regardless of their different ecological niches (e.g., Bar-Yosef

1989; Meignen and Bar-Yosef 1992; Shea 2003; Hovers 2009; Henry 2003). Other objects in the same contexts were made by employing the method the recurrent Levallois (Boëda 1995) for the production of flakes and blades (Meignen 1988, 1991).

Burials of these Neanderthals were uncovered in Kebara Dederiyeh, Amud, and the woman from Tabun that I prefer to attribute to "Tabun B-type industry (Bar-Yosef and Callendar 1999). It should be noted that at the same time, the proliferation of isolated bones and teeth of humans resembles the same situation as in European sites. As a reminder, in contrast Early Mousterian in the Levant ("Tabun-D type") in excavated caves yield human relics (mostly isolated teeth) rarely, and these do not yet allow population identification. The same is true when we consider most Early Upper Paleolithic contexts across the western Eurasia.

Dispersals of Modern Humans Represented by the Upper Paleolithic Industries

The proposed dispersal routes from Africa into Eurasia of modern humans are both northern and southern (Figures 11.1–11.3). The northern path into Europe led human groups through the "Danube Corridor" into the western lands (e.g., Conard 2006). The southern path, not well recorded, crossed the European Mediterranean region from the southern Balkans, into the Italian peninsula (as the Adriatic sea was mostly a dry land) reaching southern France (Figure 11.2). Other groups, after crossing Anatolia, continued into the area of Bulgaria and eastward to the Russian plains. Another group of later migrants picked the more difficult path through the mountains of the Zagros and eastern Anatolia, reaching the Caucasus and then proceeding though the mountain ranges moved after a short time into

the other areas of the Russian plains. Their presence in Mezmaiskaya on the northern slopes and in Ortvale Klde and Dzudzuana on the southern hills occurred within one or two millennia.

We should note that the migrations into new territories involved different groups as indicated by dental research (Bailey and Hublin 2010). Their intrinsic variability differs considerably from the ubiquitous similarity of all studied Neanderthals' teeth (Bailey and Hublin 2006, 2010). The main way into Europe from the Levant was through the "Danube corridor" and this was apparently the choice of several different groups who left the remains of their lithic industries such as the Bachokirian and Bohuniciuan.

The Bacholirian is one of the earliest industries in the Balkans and that bears particular characteristics of an Initial Upper Paleolithic industry (Kozlowski 1982, 2000; Tsanova 2006; Tsanova and Bordes 2003). The Bohuncian produced a different IUP assemblages, technically similar to of Boker Tachtit in the Israeli Negev (Svoboda 2003; Skrdla 2003). One can mention the slightly later Châtelperronian as another example (Pelegrin 1985, 1995).

This scenario is merely a tentative reconstruction of routes based on the dating of the Initial Upper Paleolithic assemblages as marked in Figure 11.2. The rate of movements is reflected in the time difference between the presence of modern humans in the staging area of the Levant, and their final destinations in western Europe. It took approximately 3,000–5,000 years to reach the Atlantic front. But their route from east Africa into eastern Asia, as suggested by Forster (2004), passing through the Bab-el-Mandeb straights, through Oman, the Hormuz straights, and then following the coast of the Indian Ocean to southeast Asia, and Sahul (New Guinea and Australia, took approximately the same time, arriving at ca. 45±5 ka BP

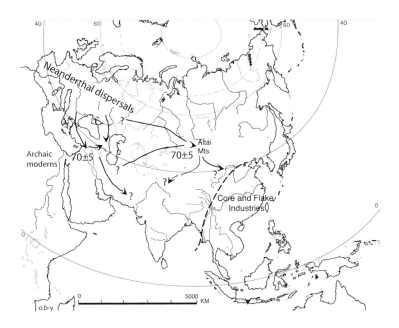

Figure 11.1 The dispersal of the Neanderthals from their European homeland.
Question marks designate uncertain interpretations as explained in the text.

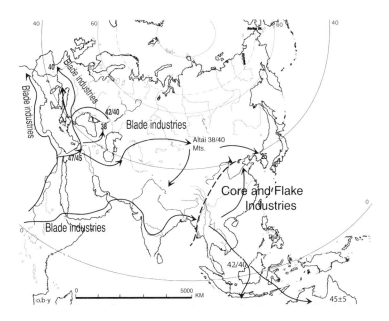

Figure 11.2 The pathways of modern human groups moving into Eurasia,
splitting the Neanderthal territories with uncalibrated dates for main sites.

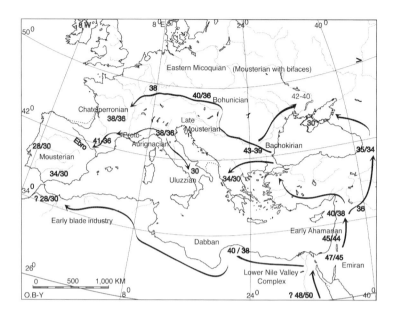

*Figure 11.3 The results of the colonization of modern humans in Europe
and the two fronts of interactions with retreating Neanderthal populations.*

(Summerhays et al. 2010). However, we note that in certain areas there were clear delays. From the Levant where their arrival was ca. 47/45 Ka BP to the Zagros and the Caucasus it took some 10,000 years for moderns to arrive. Possibly the numerous mountain ranges that had to be crossed were responsible for this retardation in establishing firm occupations.

As Papuans carry the same Neanderthal contribution as Chinese and French, the route taken by early moderns should have been through places where Neanderthal population was present. Where? While basic information is still missing there are several possibilities. One is that Zagros Neanderthals represent a group that also reached the coast of the Indian Ocean; if this was the route out of Africa, they met there. Another option is somewhere south of the Himalayas, along the Ganges Valley, or on the northern side of the mountains in Afghanistan. Did

Neanderthals from Uzbekistan reach Afghanistan? Without a few modern excavations in this region we do not know.

The HUGO Pan-Asian SNP Consortium (2009) provides the most comprehensive genetic data sets from modern populations in southeast and east Asia and its interpretation in terms of human migrations. The main finding of this research is that the southern migration across Asia was the major route of human dispersals resulting in the colonization of southeast and east Asia, Australia and the islands. There is a general agreement that another route of modern humans dispersal was through northern Asia and is represented archaeologically by microblade (microlithic) assemblages. This set of industries, characterized by several techniques for obtaining bladelets, is dated at least to the LGM (Bae 2010).

The hunter-gatherers who produced this microlithic industry exploited also cores and flakes detached from local raw materials such as quartz and quartzite, and made pottery during the Terminal Pleistocene and early Holocene. I wonder whether the ancestors of the makers of the microblade assemblages were in contact with the Neanderthals in the Altai area. The suggested origin of this industry as the Upper Paleolithic blade assemblages in this region (Figure 11.2; Kuzmin 2007) suggests that this may be the case.

At this point we need to turn to the debatable issue of the materialist expression of short- and long-term interactions.

The Impacts of the Encounters between Neanderthals and Modern Humans

The aim of this paper is to speculate where and when Neanderthals and modern humans either met and interbred for sometime or had no contact at all. As it is currently accepted that the Levantine and the European areas were the first where both populations met, I begin with a few comments on the Levant.

The IUP of the Levant is hard to interpret because the early phase of the new technology emerged from the Levallois and the Nubian methods of the Nile Valley and therefore it is a true "cultural transition" as understood in prehistoric research. Thus the technical and typological attributes are still close to the ancestral lithic industry. In spite of the proposal that the nature of encounters between the local Neanderthals and the incoming modern humans was competitive exclusion, as suggested for the earlier time when the Neanderthals were the invaders, we are not aware of clear indications of the kind of interactions. However, we do not need to stress that reliance only on the evidence from artifacts is the main obstacle to a fuller picture.

Indeed, one of the three kinds of alternative or complementary behaviors of invading foragers, causing extinction, requires sound evidence for

violence, and this is hardly available until the terminal Pleistocene when several sites, dated to Mesolithic Europe, produced the skeletal evidence for massacres. However, in the context of interactions we should therefore consider the earlier expansion of the Neanderthals into western Asia. Competitive exclusion, for example, was the suggested model for the success of the Neanderthals in the Levant. This could have ended in the same way by mass killings but there is no evidence for such an event or events (Shea and Bar-Yosef 2005; Shea 2008). In addition, as this is one of the first well-documented expansions of Neanderthals some 70,000 years ago, we should wonder what were the effects of the encounters. We do not have any clear material evidence for possible exchanges for that time period. One option to test in detail is the degree of similarity between the lithic industries of "Tabun C type" (Middle Mousterian and "Tabun B-type" (Late Mousterian) and whether the Levallois technique could have been evidence for interbreeding aside from the morphological attributes (Arensburg and Belfer-Cohen 1998). Whether intimate interactions between the two populations took place in Anatolia is yet unknown due to the paucity of Middle Paleolithic published reports from the latter region. Briefly, the typical heavily retouched Mousterian points and scrapers, often referred to as a "Charentian-type", are known from Karain cave but typical, much less retouched Levallois assemblages are also present (Yalçinkaya 1995; Slimak et al. 2008).

This brings us back to the expansion of modern humans into Europe around 50/45 Ka. Most scholars accept that during that time the Neanderthal population became extinct. Although it is not easy to demonstrate the various claims suggested in the literature for the demise of the Neanderthals I believe that Figure 11.4 offers a geographic view that indicates, although

Figure 11.4 The flow of modern humans into Eurasia based on the distribution of Initial Upper Paleolithic blade industries.

statically, what was probably an evolving process during a few thousands of years. Under the pressure of incoming modern humans, Neanderthal retreated from major parts of their traditional territories. Without detailing the spatial changes, the model predicts that splitting the Neanderthal territories across the European plains, some portions of the Atlantic lands, and (rarely) the European Mediterranean lands, deprived them of important food resources such as the preferred game and plants, and in addition disrupted their mating networks. Under these conditions we may expect that the total fertility rate of particular groups declined, and even a small decrease in the number of successful births by females would lead to a decrease in population size, and disappearance within several centuries. The extinction of the Neanderthals did not take place at the same time everywhere across their lands, as documented by late survival in Iberia, southeast Europe, Crimea and the Caucasus.

When incoming people interact with locals we may detect evidence in the archaeological assemblages. In central Europe, it was already suggested that the Szeletian (Figure 11.4) was probably made by Neanderthals who adopted the technique of detaching blades from prismatic cores (Svoboda 2005). In addition, the dates of the IUP assemblages (such as the Bohunician) and those of the Szeletian partially overlap in central Europe. The same could be said about the so-called Danubian Szeletian that is located in the path of modern humans moving around the western side of the Black Sea (compare Figure 11.3 with Figure 11.4). A similar case is the Jerzmanovician entity in Poland that is rich in foliates and dates to the same period, namely 36-28 Ka BP while the Krakow-Zweirzyniec with its proliferation of arched back blades which could indicate the presence of modern humans (Kozlowski 2000).

It is hardly surprising that further east there is additional evidence for interactions between these two populations in the area of Kostenki, the middle Don River area, and Crimea (Chabai 2003, 2007; Marks and Chabai 2006;

Anikovich et al. 2007). The observed variability of lithic industries, in spite the lack of human remains, suggested to researchers that several of these entities represented Neanderthals in this resource rich peninsula, such as the Mousterian of western Crimea and the eastern Micoquian. The radiocarbon dates of both entities demonstrate a high degree of contemporaneity (36-28 Ka according to Chabai 2003) with the Streletskaya (ca. 36-27 Ka) and the Spitsynska (ca. 36-32 Ka) "cultures". The Streletskaya entity contains bifacial points resembling typical arrowheads, foliates, discoidal cores and "flat faced" opposed platform cores, and resembles the "Eastern Szeletian" in Buran Kaya III (Marks and Monigal 2000), and thus could be interpreted as Neanderthal lithic technology influenced by interactions with modern humans. In addition it has a wide geographic distribution that incorporates Crimea, the middle and lower Don valley, and also the central and northern Urals (Chabai 2007, and references therein). Given the northern dispersals of the makers of the Streletskaya industry, I suggest interpreting this as reflecting the geographic retreat of the Neanderthals.

The Gorodtsovskaya 'culture' (ca. 30/28-26 Ka) is seen as fully made by modern humans (with a rich bone and ivory objects). A similar interpretation is suggested for the Spitsynskaya entity due to its dominant blade industry and the bone and ivory elements. Hence, the archaeological data from southeastern Europe supports the notion of non-violent encounters between the two populations.

In a recent review of the Kostenki area, Anikovich and associates (2007) propose a model of acculturation, the result of interactions and possible interbreeding between modern humans and Neanderthals. They stress that in Kostenki, and in general on the Russian plain,

there is no real Middle Paleolithic. They describe what they refer to as "symbiotic industries", interpreted as a result of meetings in the areas where both the populations were newcomers.

Closing Remarks

The goal of this paper was to offer a few comments on where Neanderthals and modern humans interacted and interbred. I therefore started with the difficulties in getting clear and fully accepted answers from the archaeological records. Laying out a few assumptions about my proposal and how to use the archaeological information I have tried to demonstrate that the Neanderthals were an expanding and colonizing population, which speaks for their social organization and technological capacities. I did not discuss - and had no intention to discuss - issues of cognition or symbolic behavior. Suffice it to say that those who can master the Levallois technique had somewhat more knowledge and experience in obtaining particular blanks than the makers of a simple blade industry. Neanderthals widened their territories by expanding into western, central and possibly eastern Asia. I avoided the issue of who were the locals that the Neanderthals displaced.

The clearest evidence for cultural markers resulting from interactions, meaning cultural influence between the two populations, is provided by the central and eastern European records of Early Upper Paleolithic times (ca. 42-30 Ka uncalibrated BP). Most of the described cases rely on the interpretation of the Szeletian, Jerzmanovician and the Streletskayan entities.

The archaeological information from the so-called Initial Upper Paleolithic in western Europe does not allow us to suggest clear cases of contact that left visible archaeological evidence except for a few cases in the Iberian peninsula (Zilhao 2006). The genetic as well as the paleoanthropological evidence should keep us examining the archaeological records of the so-called MP–UP transition, with an open eye on different interpretations. Not less important would be additional field projects using modern techniques for assuring that we understand the nature of the occupations through the study of site formation processes as well as extensive dating programs.

Acknowledgments

I am grateful to David Pilbeam and Jeanne Sept for inviting me to contribute a paper to the Glynn Isaac memorial volume. I had the pleasure of spending together Glynn's last year at Harvard when I was on sabbatical leave from the Hebrew University. I thank L. Meignen (CEPAM, CNRS) for many useful comments, Zinovi Matskevich for assistance with the Russian literature and David Pilbeam for editing the text. Needless to stress that all shortcomings are mine.

References

Adler, D. S.
2002 Late Middle Palaeolithic Patterns of Lithic Reduction, Mobility, and Land Use in the Southern Caucasus. Ph.D. Dissertation, Department of Anthropology, Harvard University.

Anikovich, M. V., N. K. Anisiutkin, L. B. Vishniatsky
2007 *Uzlovye problemy perekhoda k verkhnemu paleolitu v Evrazii* (Key Problems of the Upper Palaeolithic Transition in Eurasia). Nestor-istoria, St. Petersburg.

Arensburg, B., and A. Belfer-Cohen
1998 *Sapiens* and Neandertals: Rethinking the Levantine Middle Paleolithic Hominids. In *Neandertals and Modern Humans in Western Asia*, edited by T. Akazawa, K. Aoki and O. Bar-Yosef, pp. 311-322. Plenum Press, New York.

Bae, K.
2010 Origins and patterns of Upper Paleolithic in the Korean peninsula and movement of modern humans in East Asia. *Quaternary International* 211(1-2):103-112.

Bailey, S. E., and J-J. Hublin
2006 Dental remains from Grotte du Renne at Arcy-sur-cure (Yonne). *Journal of Human Evolution* 50:485-508.

Bailey, S. E., and J-J. Hublin (Editors)
2010 *Dental Perspectives on Human Evolution: State of the Art Research in Dental Paleoanthropology (Vertebrate Paleobiology and Paleoanthropology)*. Springer-Verlag.

Bar-Matthews, M., A. Ayalon, M. Gilmour, A. Matthews, and C. J. Hawkesworth
2003 Sea-and oxygen isotopic relationships from planktonic foraminifera and speleothems in the Eastern Mediterranean region and their implication for paleorainfal during interglacial intervals. *Cheochimica et Cosmochimica Acta* 67(17):3181-3199.

Bar-Yosef, O.
1988 The date of Southwest Asian Neanderthals. In *L'Homme de Néandertal, Vol. 3*, edited by M. Otte, pp. 31-38. ERAUL, Liège.

1989 Geochronology of the Levantine Middle Palaeolithic. In *The Human Revolution: Behavioral and Biological Persectives on the Origins of Modern Humans*, edited by P Mellars and C. Stringer, pp. 589-610. Edinburgh University Press, Edinburgh.

1998 Chronology of the Middle Paleolithic of the Levant. In *Neandertals and Modern Humans in Western Asia*, edited by T. Akazawa, K. Aoki and O. Bar-Yosef, pp. 39-56. Plenum Press, New York.

2000 The Middle and Early Upper Paleolithic in Southwest Asia and neighboring regions, In *Geography of Neandertals and Modern Humans in Europe and the Greater Mediterranean*, edited by Ofer Bar-Yosef and David Pilbeam, pp. 107-156. Peabody Museum, Cambridge, MA.

2006 Neanderthal and modern humans: A different interpretation. In *Neanderthals and Modern Humans Meet*, edited by N. Conard. Tübingen Publications in Prehistory, Kerns Verlag, Tübingen .

Bar-Yosef, O., and J.-G. Bordes
2010 Who were the makers of the Châtelperronian culture? *Journal of Human Evolution* 59:586-593.

Bar-Yosef, O., and J. Callander
1999 The woman from Tabun: Garrod's doubts in historical perspective. *Journal of Human Evolution* 37:879-885.

Bar-Yosef, O., and P. Van Peer
2009 The *chaîne opératoire* approach in Middle Paleolithic archaeology. *Current Anthropology* 50(1):103-131.

Baumler, M. F., and J. D. Speth
1993 A Middle Paleolithic Assemblage from Kunji Cave, Iran. In *The Paleolithic Prehistory of the Zagros-Taurus*, edited by D. L. Olszewski and H. L. Dibble, pp. 1-74. University Museum Monograph 78. University of Pennsylvania, Philadelphia.

Boëda, E.
1995 Levallois: A volumetric construction, methods, technique. In *The Definition and Interpretation of Levallois Technology*, edited by H. Dibble and O. Bar-Yosef, pp. 41-69. Monographs in World Archaeology 23. Prehistoric Press, Madison.

Burduckeiwicz, J. M.
2000 The backed biface assemblages of east central Europe. In *Toward Modern Humans: Yabrudian and Micoquian, 400-500 K years ago*, edited by A. Ronen and M. Weinstein-Evron, pp. 155-166.

Cabrera Valdes, V., and F. Bernaldo de Quirós
1992 Approaches to the Middle Paleolithic in Northern Spain. In *The Middle Paleolithic: Adaptation, Behavior, and Variability*, edited by H. L. Dibble and P. Mellars, pp. 97-112. University Museum Symposium Series, Vol. IV (University Museum Monograph 78). The University Museum of Archaeology and Anthropology, University of Pennsylvania, Philadelphia.

Chabai, V. P.
2003 The chronological and industrial variability of the Middle to Upper Paleolithic transition in eastern Europe. In *The Chronology of the Aurignacian and of the Transitional Technocomplexes: Dating, Stratigraphies, Cultural Implications*, edited by J. Zilhão and F. D'Errico, pp. 71-88. Instituto Português de Arqueollogia, LISBOA.

2007 The Middle Paleolithic and Upper Paleolithic in the northern Black Sea region. In *The Black Sea Flood Question*, edited by V. Yanko-Homback et al., pp. 279-296. Springer.

Chabai, V. P., and K. Monigal
1999 *The Middle Paleolithic of Western Crimea*, vol. 2. ERAUL 87, Liège

Conard, N. J.
2006 Changing views of the relationship between Neanderthals and modern humans. In *When Neanderthals and Modern Humans Met*, edited by N. J. Conard, pp. 5-20. Tübingen Publications, Kerns Verlag, Rottenburg.

Delagnes, A., J. Jaubert, and L. Meignen
2007 Les technocomplexes du Paleolithique moyen en Europe occidentale dans leur cadre diachronique et geographique. In *Les Neandertaliens. Biologie et Cultures*, pp. 213-229. Editions du CTHS, Paris.

Delagnes, A., and L. Meignen
2006 Diversity of lithic production systems during the Middle Paleolithic in France: Are there any chronological trends ? In *Transitions Before the Transition: Evolution and Stability in the Middle Paleolithic and Middle Stone Age*, edited by E. Hovers and S. Kuhn, pp. 85-108. Springer, New York, Boston, Dordrecht, London, Moscow.

d'Errico, F., J. Zilhão, M. Julien, D. Baffier, and J. Pelegrin
1998 Neanderthal acculturation in Western Europe? *Current Anthropology* 39(Supplement):S1-S44.

Derevianko, A. P., and S. V. Markin
1995 The Mousterian of the Altai in the context of the Middle Paleolithic culture of Eurasia. In *The Definition and Interpretation of Levallois Technology*, edited by H. Dibble and O. Bar-Yosef, pp. 473-484. Prehistory Press, Madison.

Derevianko, A. P., and M. V. Shunkov
2002 Middle Paleolithic Industries with foliate bifaces in Gorny Altai. *Archaeology Ethnology and Anthropology of Eurasia* 1(9):16-42.

Derevianko, A. P., and M. V. Shunkov, A. Agadjanian, et al.
2003 *Natural Environments and Humans in the Paleolithic of the Altai Mountains.* IAET SO RAN, Novosibirsk.

Dibble, H., and S. Holdaway
1993 The Middle Paleolithic industries of Warwasi. In *The Paleolithic Prehistory of the Zagros-Taurus*, edited by D. L. Olszewski and H. L. Dibble, pp. 75-100. University Museum Monograph 78. University of Pennsylvania, Philadelphia.

Doronichev, V. B., and L. V. Golovanova
2003 Bifacial tools in the Lower and Middle Paleolithic of the Caucasus and their contexts. In *Multiple Approaches to the Study of Bifacial Technologies*, edited by H. Dibble and M. Soressi, pp. 77-107. University of Pennsylvania Museum of Archaeology and Anthropology, Philadelphia.

Forster, P. I.
2004 Ice ages and the mitochondrial DNA chronology of human dispersals: A review. *Philosophical Transactions of the Royal Society, London* B 359:255-264.

Glantz, M., B. Viola, P. Wrinn, T. Chikisheva, A. Derevianko, A. Kriovoshapkin, U. Islamov, R. Suleimanov, and T. Ritzman
2008 New hominin remains from Uzbekistan. *Journal of Human Evolution* 55(2008):223-237.

Goren-Inbar, N., and A. Belfer-Cohen
1998 Technological Abilities of the Levantine Mousterians. In *Neandertals and Modern Humans in Western Asia*, edited by T. Akazawa, K. Aoki, and O. Bar-Yosef, pp. 205-221. Plenum Press, New York and London.

Green, R. E., et al.
2010 A draft sequence of the Neandertal genome. *Science* 328:710-722.

Henry, D. O. (Editor)
2003 *Neanderthals in the Levant. Behavioral organization and the Beginnings of Human Modernity.* Continuum, London.

Higham, T., R. Jacobi, M. Julien, F. David, L. Basell, R. Wood, W. Davies, C. B. Ramsey
Chronology of the Grotte du Renne (France) and implications for the context of ornaments and human remains within the Chatelperronian. *PNAS* 107(47):20234-20239.

Hovers, E.

1998 The lithic assemblages of Amud Cave: Implications for understanding the rnd of the Mousterian in the Levant. In *Neandertals and Modern Humans in Western Asia*, edited by T. Akazawa, K. Aoki, and O. Bar-Yosef, pp. 143-163. Plenum Press, New York.

2006 Neandertals and modern humans in the Middle Palaeolithic of the Levant: What kind of interaction? In *When Neanderthals and Modern Humans Met*, edited by N. J. Conard, pp. 65-85. Tubingen Publication in Prehistory. Kerns Verlag, Tubingen.

2009 *The Lithic Assemblages of Qafzeh Cave*. Oxford University Press.

HUGO Pan-Asian SNP Consortium

2009 The HUGO Pan-Asian SNP Consortium (2009). Mapping human genetic diversity in Asia. *Science* 326:1541-1545.

Jelinek, A. J.

1991 Observations on reduction patterns and raw materials in some Middle Paleolithic industries in the Perigord. In *Raw Material Economies Among Prehistoric Hunter-Gatherers*, edited by A. Montet-White and S. Holen, pp. 7-32. Publications in Anthropology, 19. University of Kansas, Lawrence.

Jöris, O.

2006 Bifacially backed knifes (*Keilmesser*) in the Central European Middle Palaeolithic. In *Axe Age - Acheulian Toolmaking from Quarry to Discard. Approaches to Anthropological Archaeology*, edited by N. Goren-Inbar and G. Sharon, pp. 287-310. Equinox, London.

Karlin, C.

1991 Connaissances et savoir-faire: Comment analyser un processus technique en Préhistoire: Introduction. In *Tecnologia y cadenas operativas liticas. Bellaterra: Departament d'Historia de les Societats Pre-capitales i d'Antropologia Social.* (Treballs d'Arqueologia, 1), edited by R. Mora, X. Terradas, A. Parpal, and C. Plana, pp. 99-124.

Keeley, L. K.

1996 *War Before Civilization*. New York, Oxford University Press.

Kozlowski, J. E. (Editor)

1982 *Excavation in the Bacho Kiro Cave (Bulgaria): Final Report*. Pansowowe Wydawnictowo Naukowe, Warsaw.

Kozlowski, J. K.

2000 The problem of cultural continuity between the Middle and the Upper Paleolithic in central and eastern Europe. In *The Geography of Neandertals and Modern Humans in Europe and the Greater Mediterranean*, edited by O. Bar Yosef and D. Pilbeam, pp. 77-105. Peabody Museum, Harvard University, Cambridge, MA.

Kuzmin, Ya. V.

2007 Geoarchaeological aspects of the origin and spread of microblade technology in northern and central Asia. In *Origin and Spread of Microblade Technology in Northern Asia and North American*, edited by Y. V. Kuzmin, S. G. Keates, and C. Shen, pp. 115-124. Archaeology Press, Simon Fraser University, Burnaby.

Marks, A. E., and V. P. Chabai

2006 Stasis and change during the Crimean Middle Paleolithic. In *Transitions before the Transition: Evolution and Stability in the Middle Paleolithic and Middle Stone Age*, edited by E. Hovers and S. L. Kuhn, pp. 121-136. Springer, New York.

Marks, A. E., and K. Monigal

2000 The Middle to Upper Palaeolithic interface at Buran-Kaya-III, eastern Crimea. In *Neanderthals and Modern Humans - Discussing the Transition: Central and Eastern Europe from 50.000-30.000 B.P.*, edited by J. Orschiedt, and G.-C. Weniger, pp. 212-226. Neanderthal Museum, Mettman.

Meignen, L.

1990 Le Paléolithique moyen du Levant: Synthèse. *Paléorient* 14:168-173.

1995 Levallois lithic production systems in the Middle Paleolithic of the Near East: The case of the unidirectional method. In *The Definition and Interpretation of Levallois Technology*, edited by H. Dibble and O. Bar-Yosef, pp. 361-381. Prehistory Press, Madison, WI.

1998 Hayonim Cave lithic assemblages in the context of the Near Eastern Middle Paleolithic: A preliminary report. In *Neandertals and Modern Humans in Western Asia*, edited by T. Akazawa, K. Aoki and O. Bar-Yosef, pp. 165-180. Plenum Press, New York.

Meignen, L., and O. Bar-Yosef

1988, Variabilité technologique au Proche Orient: L'exemple de Kebara. In *L'Homme de Neandertal*, vol. 4, edited by M. Otte, p. 81-95, La Technique. ERAUL, Liege.

1991 Les outillages lithiques moustériens de Kebara (fouilles 1982-1985): Premiers résultats. In *Le Squelette oustérien de Kebara 2*, edited by O. Bar-Yosef and B. Vandermeersch, pp. 49-75. Editions du CNRS. (Cahiers de Paléoanthropologie), Paris.

1992 Middle Palaeolithic variability in Kebara Cave (Mount Carmel, Israel). In *The Evolution and Dispersal of Modern Humans in Asia*, edited by T. Akazawa, K. Aoki, and T. Kimura, pp. 129-148, Hokusen-Sha, Tokyo.

Mussi, M.

2001 *Earliest Italy: An Overview of the Italian Paleolithic and Mesolithic*. Kluwer, Plenum, New York.

Ningxia (Ningxia Kaogu Yanjiusuo)
2003 *Shuidonggou: The Report of 1980 Excavation.* Science Press, Beijing (in Chinese).

Otterbein, K.
1997 The origins of war. *Critical Review* II(2):251-277.

Panagopoulou, E., P. Karkanas, G. Tsartsidou, E. Kotjabopoulou, K. Harvati, and M. Ntinou
2002-2004 Late Pleistocene archaeological and fossil human evidence from Lakonis Cave, Southern Greece. *Journal of Field Archaeology* 29(3 and 4):323-349.

Pardoe, C.
1988 The cemetery as symbol. The distribution of prehistoric aboriginal burial grounds in southeastern Australia. *Archaeology in Oceania* 23-1(April 1988), pp. 1-16.

Pelegrin, J.
1985 Réflexions sur le comportement technique. In *La SignificatiCulturelle des Industries Lithiques*, edited by M. Otte, pp. 72-91. B.A.R. (Studia praehistorica belgica 4. BAR. International Series 239), Oxford.

1995 *Technologie lithique: Le Chatelperronien de Roc-de-Combe (Lot) et de La Côte (Dordogne).* CNRS Editions (Cahiers du Quaternaire, n°20), Paris.

1990 Prehistoric lithic technology: Some aspects of research. *Archaeological Review of Cambridge* 9:116-125.

Shea, J.
2003 Neandertals, competition and the origin of modern human behavior. *Evolutionary Anthropology* 12(173-187).

2008 Transitions or turnovers? Climatically-forced extinctions of *Homo sapiens* and *Neanderthals* in the East Mediterranean Levant. *Quaternary Science Reviews.*

2010 *Neanderthals* and Early *Homo sapiens* in the Levant. In *South-Eastern Mediterranean Peoples Between 130,000 and 10,000 years ago*, edited by E. A. A. Garcea, pp. 126-143. Oxbow Books, Oxford.

Shea, J., and O. Bar-Yosef
2005 Who were the Skhul/Qafzeh people? An archaeological perspective on Eurasia's earliest modern humans. *Journal of the Israel Prehistoric Society* 35:449-466.

Shea, J., and M. L. Sisk
2010 Complex projectile technology and *Homo sapiens* dispersal into Western Eurasia. *Paleoanthropology Society* 10(36):100-122.

Skrdla, P.

2003 Bohunician technology: A reffiting approach. In *Stranska Skala: Origins of the Upper Paleolithic in the Brno Basin, Moravia, Czech Republic,* edited by J. Svoboda and O. Bar-Yosef, pp. 119-151. Peabody Museum of Archaeology and Ethnography, Harvard University, Cambridge.

Slimak, L., S. L. Kuhn, H. Roche, D. Mouralis, H. Buitenhuis, N. Balkan-AtlI, D. Binder, C. Kuzucuoglu, and H. Guillou

2008 Kaletepe Deresi 3 (Turkey): Archaeological evidence for early human settlement in Central Anatolia. *Journal of Human Evolution* 54(1):99-111.

Stiner, M., R. Barkai, and A. Gopher

2009 Cooperative hunting and meat sharing 400-200 kya at Qesem Cave, Israel. *Proceedings of the National Academy of Sciences USA* 106:13207-132012.

Summerhays, G. R., M. Leavesley, A. Fairbairn, H. Mandui, J. Field, A. Ford, and R. Fullagar

2010 Human adaptation and plant use in highland New Guinea 49,000 to 44,000 years ago. *Science* 330:78-81.

Svoboda, J. A.

2003 The Bohuncian and the Aurignacian. In *The Chronology of the Aurignacian and of the Transitional Technocomplexes: Dating, stratigraphies, cultural implications.* Proceedings of the Symposium 6.1 of the XIVth Congress of the UISPP. Trabalhos de Arquelogia 33, pp. 123-131. Instituto Portugues de Arquelogia, Lisbon.

2005 The Neandertal extinction in eastern Central Europe. *Quaternary International: The Journal of the International Union for Quaternary Research* 137:69-76.

Svoboda, J., and P. Skrdla

1995 Bohunician technology. In *The Definition and Interpretation of Levallois Technology,* edited by H. Dibble and O. Bar-Yosef, pp. 432-438. Prehistory, Madison.

Tillier, A. M., Arensburg, B. and J. Bruzeck

2008 Identité biologique des artisans moustériens de Kébara (Mont Carmel, Israël). Réflexion sur le concept de Néanderthalien au Levant méditerranéen. *Bulletins et Mémoires de la Société d'Anthropologie de Paris,* n.s. 20, 1-2:33-58.

Trinkaus, E.

2006 Modern human versus *Neandertal* evolutionary distinctiveness. *Current Anthroplogy* 47(4):597-620.

Tsanova, T.

2006 Les débuts du Paléolithique supérieur dans les Balkans. Réflexion à partir des études taphonomiques et techno-économiques des ensembles lithiques des sites de Bacho-Kiro (couche 11), Temnata (couche VI et couche 4) et Kozarnika (niveau VII). Université de Bordeaux I, unpublished Ph.D. Thesis.

Tsanova, T., and J-G. Bordes
2003 Contribution au débat sur l'origine de l'Aurignacien: Principaux résultats d'une étude technologique de l'industrie lithique de la couche 11 de Bacho Kiro. In *The Humanized Mineral World: Towards Social and Symbolic Evaluation of Prehistoric Technologies in South Eastern Europe. Proceedings of the ESF Workshop, Sofia. Liège*, edited by T. Tsonev and M. Kokelj, pp. 41-50. ERAUL.

Vandermeersch, B.
1993 Was the St. Césaire discovery a burial? In *Context of a Late Neandertal: Implications of Multidisciplinary Research for the Transition to Upper Paleolithic Adaptations at Saint-Césaire, Charente-Maritime, France*, edited by F. Lévêque, A. M. Backer and M. Guilbaud, pp. 129-131. Prehistory Press, Madison.

Yalçinkaya, I.
1995 Thoughts on Levallois technique in Anatolia. In *The Definition and Interpretation of Levallois Technology*, edited by H. Dibble and O. Bar-Yosef, pp. 399-412. Prehistory Press, Madison.

Zilhão, J.
2006 Neandertals and moderns mixed, and it matters. *Evolutionary Anthropology* 15(5):183-195.

Zilhão, J., et al.
2010 Symbolic use of marine shells and mineral pigments by Iberian Neandertals. *Proceedings of the National Academy of Sciences* 107(3):1023-1028.

12

Sorting Out the Muddle in the Middle East: Glynn Isaac's Method of 'Multiple Working Hypotheses' Applied to Theories of Human Evolution in the Late Pleistocene Levant

John J. Shea

Introduction

Glynn Isaac's scholarly legacy includes his many important empirical contributions to human origins research and to African prehistory (e.g., Isaac 1977, 1997). His theoretical and methodological contributions were no less profound; yet, surprisingly, they have received less attention by historians of archaeology. The method of "multiple working hypotheses" (Isaac 1983) was a cornerstone of his approach to prehistory. Briefly, Isaac argued that if one cannot falsify rival explanations for the same set of observations, one should retain them and work to winnow down their number by gathering new evidence. This approach contrasts with the more common judicial/adversarial practice in which individual scientists assert a particular hypothesis and defend it against all challengers. Powerful incentives encourage this latter approach, and they inevitably lead its practitioners to treat favored hypotheses with less critical scrutiny than competing explanations.

Isaac himself was not immune to these incentives or to their effects. He first discussed multiple working hypotheses in the context of a debate about human versus non-human factors in Early Pleistocene site formation processes. In "Bones in Contention: Competing Explanations for the Juxtaposition of Early Pleistocene

Artifacts and Faunal Remains" Isaac (1983:5) proposed several contrasting scenarios/hypotheses about how concentrations of stone tools and broken bones of large terrestrial vertebrates became associated with one another in African Early Pleistocene contexts. These included a hydraulic jumble, a common amenity (e.g., a shade tree) used independently by carnivores and hominins, bones accumulated by carnivores and later used by hominins, hominin scavenging of natural death sites, and hominin central place foraging. However, as Binford (1985) noted, few of Isaac's models presented an alternative to his core hypothesis that early hominins routinely shared food with one another (Isaac 1978). Potts' (1984) "strategic caching" model proposed such alternative for Early Paleolithic site formation processes.

This paper explores the method of multiple working hypotheses using recent debate about Neanderthal and early *Homo sapiens* in the east Mediterranean Levant as a case study. Since the discovery of human fossils in the caves of Mount Carmel and the Galilee in the 1920s–1930s, paleoanthropologists have speculated about the evolutionary relationships among these hominins. I summarize the main theories about these relationships currently advocated by researchers working on the Levant and consider how they would be affected by new kinds of paleoanthropological evidence. This "thought experiment" suggests that theories invoking long-term hominin evolutionary continuity are impossible to falsify. I examine why such unfalsifiable explanations are so popular in human origins research and propose how to integrate them into a method of multiple working hypotheses.[1]

Human Evolution in the Late Pleistocene of the East Mediterranean Levant

The east Mediterranean Levant encompasses the modern states of Lebanon, Syria, Israel, Jordan, and parts of adjacent countries, such as the Egyptian Sinai. Pleistocene human settlement in this relatively small region was concentrated in the Mediterranean (oak-terebinth) woodland and its ecotone with the Irano-Turanian steppe. This paper concerns itself mainly with that part of the Late Pleistocene between 128-18 ka. There are so few hominin fossils from the Levantine Middle Pleistocene contexts (728-128 ka) that little can be said with certainty about evolutionary continuity or discontinuity there (Dennell 2003; Vandermeersch 1989). There are many human fossils from contexts younger than 18 ka, and evolutionary continuity between these humans and recent populations in the Levant is usually not disputed by prehistorians (Smith 1995).

Levantine Paleolithic research began in the early years of the 20th Century. The first major human fossil discoveries occurred in the 1920s-1930s; and, the most famous of these were from Tabun and Skhul caves on Mount Carmel (then Palestine, now Israel). Discovered during limestone quarrying operations, the caves of the Wadi el-Mughara (Valley of the Caves) were excavated between 1928-1934 by an American School of Prehistoric Research expedition led by Garrod (Garrod and Bate 1937; McCown and Keith 1939).

The fossils from Tabun include a female Neanderthal (Tabun C1) buried with a (unrecovered) neonate, a mandible fragment of uncertain affinities (Tabun C2), and numerous dental remains attributed to Neanderthals. The *Homo sapiens* fossils from Skhul included several in anatomical articulation (Skhul 1, 4, and 5) thought likely to have been burials and numerous fragmentary remains. The human fossils from Tabun and Skhul were initially described as separate Neanderthal and *Homo sapiens* samples, but the final monograph grouped them together into "Mount Carmel Man" (Keith 1937). Their shared association with Middle Paleolithic stone tools and broadly similar sets of large mammal fossils influenced this decision to combine them (Garrod and Bate 1937). McCown and Keith may also have been influenced by the waning of typological approaches to human/hominin variation that was part of the "New Synthesis" in evolutionary biology (Bowler 1986).

During the 1950s-1980s, the Mount Carmel samples were divided on biostratigraphic grounds and combined with fossils from other southwest Asian sites (Howell 1959). Southwest Asian Late Pleistocene hominin fossils were rearranged into three chronologically-successive samples. These included an earlier Neanderthal group (Tabun C1, Amud 1, and Kebara 1 and 2), an "early modern" Middle Paleolithic *Homo sapiens* group (Skhul and the Middle Paleolithic Qafzeh fossils) and an Upper Paleolithic *Homo sapiens* group (Ksar Akil, and the Upper Paleolithic Qafzeh fossils; Trinkaus 1984).

During the mid-1980s, the Levantine Late Pleistocene fossil record underwent many changes owing to advances in geochronology (Bar-Yosef 1989; Shea 2003a). The length of the Middle Paleolithic Period more than doubled (from 90-40 ka to 245-45 ka). Additional fossils were discovered (most notably at Amud, Kebara,

Table 12.1 Geochronology, hominin paleontology, and archaeology of the Late Pleistocene Levant

MIS	Dates ka	Hominin Fossils	Archaeology	Selected Dated Contexts
2	18-32	*Homo sapiens*	Later Upper Paleolithic (Levantine Aurignacian, Atlitian)	Hayonim D Ohalo 2
late 3	32-45/47	*Homo sapiens*	Early Upper Paleolithic (Emiran, early Ahmarian)	Ksar Akil 4-25 Qafzeh E Üçagizli B-H
early 3-4	45-75	Neanderthals	Later Middle Paleolithic	Kebara VI-XII Amud B1-B4
5	75-128	*Homo sapiens* and Neanderthals	Interglacial Middle Paleolithic	Skhul B Qafzeh 17-24 Tabun Upper I
6-7	128-250	Neanderthals (Tabun C1) and *Homo sp. indet.* (Tabun C2)	Early Middle Paleolithic	Tabun Lower I
7-8	250-350	*Homo sp. indet.* (Zuttiyeh 1)	Acheulo-Yabrudian	Tabun IX-XI Hayonim E-F Zuttiyeh

and Dederiyeh), but Neanderthal and *Homo sapiens* fossils do not occur together in the same stratigraphic deposits. Rather, occurrences of Neanderthal and *Homo sapiens* fossils alternate with one another along with major shifts in regional climate (Shea 2008; see Table 12.1).

Theories about Neanderthal versus *Homo sapiens* Evolutionary Relationships

Models for evolutionary relationships between Levantine Neanderthals and early *Homo sapiens* are divisible into continuity and discontinuity theories. Continuity theories see Neanderthal and *Homo sapiens* fossils samples as representing a single population that is ancestral to later Levantine human populations. Discontinuity theories see these fossils as representing at least two distinct populations of which only one (if any) is ancestral to later Levantine humans. Considerable variation exists within these broad

categories of continuity and discontinuity theories, variation that needs to be understood before one can examine their underlying structural differences.

Continuity Theories

The earliest synthesis (McCown and Keith 1939) described Levantine Middle Paleolithic hominins as a population "in the throes of an evolutionary transition," but this choice of phrasing left much interpretive leeway. McCown's and Keith's other publications suggest that what they meant was a geographic transition between west Eurasian Neanderthals and more "modern" Asia humans, rather than a chronological transition (Howell 1958).

The hypothesis of a chronological transition was largely a product of the 1950s and 1960s. In reviewing the Levantine paleoanthropological record, Howell (1959) grouped the fossils from

Skhul with those from Qafzeh Units XVII–XXIV and the Tabun female Neanderthal with other Neanderthal fossils from Amud and Shanidar. This "transitional" variant of continuity theories gained considerable support through the 1960s–1980s. Archaeologists and physical anthropologists proposed a variety of models explaining the transformation of Neanderthal ancestors to *Homo sapiens* descendants in terms of a biocultural "transition" (Binford 1968; Clark and Lindly 1989; Jelinek 1982; Marks 1983; Smith et al. 1989; Trinkaus 1984; Wolpoff 1989). There was by no means unanimity on this issue (Vandermeersch and Bar-Yosef 1988), but the overwhelming majority of paleoanthropologists who wrote about the Levantine record in the 1960s–1980s argued for evolutionary continuity between Neanderthals and *Homo sapiens*.

In a way, the geochronological restructuring of the Levantine record in the 1990s led to a revival of evolutionary models strikingly similar to those of McCown and Keith some sixty years earlier. Simmons (1994) proposed that the Levant was a contact zone within which separate Neanderthal and African-derived *Homo sapiens* populations interbred. Kramer and colleagues (2001) proposed a similar model, although theirs argues that there was never a significant interruption of gene flow between Eurasia and Africa.

Continuity theories have proven remarkably adaptive in dealing with radiometric dates that reversed the long assumed ancestor-descendant relationship between Levantine Neanderthals and early *Homo sapiens* (Arensburg and Belfer-Cohen 1998). For example, Hovers' (2006) most recent iteration of a continuity theory argues that there was a wide range of variability in Neanderthal-early *Homo sapiens* evolutionary relationships but that there was also an overall thread of continuity across the Middle and Upper Paleolithic periods.

Discontinuity Theories

Discontinuity theories are largely products of the last 20 years of Levantine prehistoric research. Rak (1993) was the first to propose that Levantine Neanderthals and *Homo sapiens* were different species whose comings and goings in the Levantine fossil record reflected ecological vicarism - population dislocations driven by paleoenvironmental change. This hypothesis was supported by studies showing that Neanderthals were better able to conserve body heat under cold conditions than early *Homo sapiens*, who retained the heat-shedding body shape of their presumed tropical ancestors (Churchill 2006; Holliday 2000).

If Neanderthals and *Homo sapiens* were different species, the similarities apparent in their archaeological records suggest they would have competed for the same sets of resources. Such competition can result in niche partitioning, competitive exclusion, or some combination of both coevolutionary relationships (Pianka 1988).[2] Henry (1995) has proposed a niche partitioning model, suggesting Neanderthal settlement and subsistence focused on the Mediterranean woodlands and early *Homo sapiens* adaptations focused on the steppe. I (2003a, 2003b) contended that changes in the hominin fossil record reflected successive competitive displacements of one hominin species by the other.

Competition is not the only plausible mechanism for extinction and turnover in the fossil record. Small populations of large mammals are at a much greater risk of extinction than larger populations of smaller mammals (Cardillo et al. 2005). Noting a close correspondence between change in the hominin fossil record and major paleoclimatic shifts, I have argued (2008, 2009) that the Levantine paleoanthropological record is congruent with an hypothesis of multiple climatically-forced extinctions and turnovers among hominin populations.

Use of the Archaeological Record

Proponents of continuity and discontinuity theories both claim support from the Levantine archaeological record, but quasi-paradigmatic biases lead them to differ in what they infer from this evidence (Clark 2002). Proponents of continuity models frequently cite the similar stone tool technology, faunal remains, and settlement preferences seen in the Levantine Middle Paleolithic record as evidence of close social and biological links between Neanderthals and *Homo sapiens* (e.g., Clark 1992; Hovers 2009; Wolpoff 1989). This practice follows a longstanding continental European practice of treating Paleolithic industries as proxies for social divisions among prehistoric humans. Proponents of discontinuity models interpret these similarities as evidence of adaptive convergence and any inferred behavioral differences as evidence of divergent adaptive strategies. This interpretation follows a more "adaptationist" approach popular in Anglo-American research traditions that sees Paleolithic variability mainly as reflections of abilities and activities and less as expressions of "culture" (e.g., Lieberman and Shea 1994; Shea 2006).

The Effects of New Evidence in Six Scenarios

Paleoanthropologists often speculate about what kind of evidence they would need in order to answer major questions in their field. "More evidence" is a perennial favorite, but the history of science shows that merely gathering more of the same kind of data rarely solves the big research questions (Kuhn 1962). Rather, substantive progress usually results from insights gleaned from truly new kinds of evidence and the new ways of thinking such evidence engenders. Here I explore how such admittedly conjectural new

evidence from geochronology, archaeology, and paleo-genetics could affect the falsifiability of the principal continuity and discontinuity theories about human evolution in the Late Pleistocene Levant.

Chronostratigraphy

What chronostratigraphic discoveries might influence current theories? As matters stand today, there is no stratigraphic evidence for overlapping Neanderthal and *Homo sapiens* occupations in the Levant. Results of the various trapped-electron dating methods can be read to indicate close or even coincident occupations of different sites by these hominins, but all such dates have enormous standard errors. What could change this situation? Let us imagine two scenarios.

In Scenario 1, caves from several disparate parts of the Levant each contain multiple fossils of Neanderthals and *Homo sapiens* in the same stratigraphic level and enclosed in sediments deposited rapidly under low-energy conditions over the course of several thousand years. Having multiple sites from different regions within the Levant would be crucial, because much of the known fossil evidence comes from caves along the Mediterranean Coast. The overwhelming majority of prehistorians would regard this as clear and convincing evidence for prolonged sympatry between Neanderthals and *Homo sapiens*.

Such evidence poses no obvious difficulties for continuity theories. Indeed, arguments for culture contact, interbreeding and gene flow have long been challenged by inadequate geochronological and stratigraphic evidence for Neanderthals and *Homo sapiens* living in the same place at the same time. Discontinuity models, in contrast, would be in trouble. Prolonged Neanderthal and *Homo sapiens* sympatry contradicts the vicarism and extinction/turnover models. It is also inconsistent with the "competitive exclusion principle" from which the niche partitioning and competition hypotheses are derived. One could probably find cases of such sympatry among closely-related competitor species in the world today (e.g., wolves and coyotes in North America) but even the strongest exponents of discontinuity hypotheses (myself included) would consider such evidence a fair falsification of that hypothesis.

A contrasting Scenario 2 continues current trends in the Levantine evidence, which show no overlapping Neanderthal and *Homo sapiens* occupations. Such evidence is consistent with all of the various discontinuity hypotheses. Yet, it can also be accommodated by continuity theories by appeal to sample error for a single, highly-polymorphic, Levantine hominin population. Unless one is prepared to accept *a priori* the reality of species-level distinctions among Levantine hominin fossils, something continuity proponents could argue amounts to assuming the consequent, additional evidence showing non-overlapping Neanderthal and *Homo sapiens* occupations would not falsify continuity hypotheses.

Archaeology

Great strides have been made in reconstructing Neanderthal and early *Homo sapiens* behavior in the Levant. The longstanding perception that both are associated with the "same" archaeological residues has given way to a more nuanced appreciation that this record also includes evidence for behavioral differences among Neanderthals and both Middle and Upper Paleolithic *Homo sapiens* (Lieberman and Shea 1994, Shea 2007a). To consider the effects of new discoveries about Levantine hominin behavior, we have to imagine two different scenarios.

In Scenario 3, two archaeological assemblages are recovered from different sites, one associated with Neanderthal fossils and the other

with *Homo sapiens* fossils. Both sample a relatively prolonged period of time and broadly similar ecological conditions. Archaeological analysis reveals essentially identical behavior in all dimensions of archaeological variability – in raw material economy, tool manufacturing techniques and uses, dietary choices, pyrotechnology, site structure, seasonality, symbolic behavior, and mortuary practices. Such a finding would obviously be good news for proponents of continuity models. It would show no obvious behavioral barriers to the kinds of interaction and assimilation continuity models infer. This evidence would contradict the niche partitioning theory. Proponents of vicarism, competition, and climatically-forced extinctions might be inclined to dismiss such archaeological similarities as convergent behavioral evolution. I could accept this counterargument if the similarities in question were of a general nature (e.g., use of flint versus other materials, hunting the same prey species, living in caves; Shea 2006). If, as imagined in this scenario, the similarities involved identical choices among functionally-equivalent tool designs, nutritionally-equivalent foods, and symbolic artifacts, I think such evidence would be strongly inconsistent with all discontinuity theories.

The circumstances of Scenario 4 are essentially the same as for Scenario 3, except that all aspects of the archaeological evidence show significant behavioral differences – differences greater than those among ethnographically distinct humans today. Such evidence would be consistent with all of the discontinuity hypotheses. It would not be necessarily inconsistent with continuity models. The kinds of behavioral differences in technology, land-use, and subsistence that have been inferred between Levantine Neanderthals and early *Homo sapiens* in the Levant are comparable in scale to differences among recent human societies. They are trivial compared to the behavioral differences between living humans and panins (chimpanzees, bonobos, and gorillas), and yet, humans have been able to develop effective methods for communicating with captive African apes. Such communication has not led to hominin-panin interbreeding, but one disposed to defend continuity models could argue that documented behavioral differences between Neanderthals and *Homo sapiens* are unlikely to have posed significant barriers to communication and interaction and (if possible) interbreeding, gene flow and evolutionary continuity.

Ancient DNA (aDNA)

The successful recovery of aDNA from Neanderthal fossils is one of the most exciting recent developments in human origins research (Krings et al. 1997). aDNA has not yet been recovered from Levantine hominin fossils; one can speculate about how the recovery of such aDNA would affect continuity and discontinuity models.

In Scenario 5, aDNA recovered from Levantine Neanderthals and early *Homo sapiens* is more similar to one another than either Neanderthal or early *Homo sapiens* sub-samples are to any group of living humans or to the known sample of Neanderthal aDNA. This would be great news for proponents of continuity models, for it would remove one of the major obstacles to hypotheses about gene flow between Neanderthals and *Homo sapiens* – molecular evidence that such gene flow actually occurred. This evidence would not conclusively falsify discontinuity theories, for recent human groups move in response to climate change, compete with one another, partition niches, and become "extinct" (at least at a local level), but it would require these theories to be reformulated as intra-specific rather than inter-specific evolutionary processes.[3]

In Scenario 6, Levantine Neanderthals' and early *Homo sapiens*' DNA is discovered to be significantly different, not only from each other but also from the DNA of any two groups of living humans and other Neanderthal DNA. Such a finding would favor discontinuity models, nearly all of which assume there was interbreeding between Levantine Neanderthal and *Homo sapiens*. However, this evidence would not necessarily contradict continuity models. Different genera of African baboons (*Theropithecus* and *Papio*) readily interbreed with one another and yet remain distinct species (Jolly 2001). If the number of hominin fossils preserving recoverable aDNA remained small relative to the number of fossils known, one could also invoke sample size/sample error to explain the differences.[4]

Overview

Based on the preceding conjectures, I think discontinuity models for evolutionary relationships between Neanderthals and early *Homo sapiens* in the east Mediterranean Levant could be proven wrong (Table 12.2), but that continuity theories could withstand challenges from all permutations of the best possible paleoanthropological evidence. This finding raises serious questions about how to deal with continuity theories in a method of multiple working hypotheses. For observations and models, the issue is clear. If they cannot be proven wrong, they are out of the running. Where theories are concerned, where one is dealing with arguments about what observations should be made and what they mean, the question is more complex.

Dealing with Unfalsifiable Continuity Theories

How do we deal with theories if they are impossible to falsify? A strictly positivist Popperian approach would be to remove them from consideration in scientific discourse. This is obviously the correct course of action when the

Table 12.2 Effect of different scenarios on continuity and discontinuity theories about hominin evolution in the Late Pleistocene Levant

Scenario	Description	Continuity Models	Discontinuity Models
1	Stratigraphic and chronological evidence for prolonged chronological overlap between Neanderthals and *Homo sapiens*	Consistent	Inconsistent
2	Stratigraphic and chronological evidence against prolonged chronological overlap between Neanderthals and *Homo sapiens*	Arguably consistent	Consistent
3	Archaeological evidence indicates same behavior by Neanderthals and *Homo sapiens*	Consistent	Inconsistent
4	Archaeological evidence indicates different behavior by Neanderthals and *Homo sapiens*	Arguably consistent	Consistent
5	Fossil DNA shows close genetic relationship between Neanderthals and *Homo sapiens*	Consistent	Inconsistent
6	Fossil DNA shows distant genetic relationship between Neanderthals and *Homo sapiens*	Arguably consistent	Consistent

explanations invoke supernatural causes or other abrogations of uniformitarianism and parsimony. The problem with doing this in an historical science like paleoanthropology is that continuity theories might actually be right. It is not impossible that there was long-term evolutionary continuity among the east Mediterranean Levant's Late Pleistocene hominin populations. It is just difficult to disprove such a continuity argument in the same way one can falsify discontinuity theories. In this respect, continuity theories are more like historians' evaluations of documentary evidence concerning the Fall of the Roman Empire, the causes of the American Revolution, or other complex events.

Continuity theories are popular, not just among paleoanthropologists working in the Levant, but virtually the world-over. Few regions lack for vigorous proponents of long-term evolutionary continuity from the earliest signs of hominin occupation to the present day. Before considering what to do about continuity

arguments, it is worth considering why such arguments attract so much support among hominin paleontologists and archaeologists. I argue this affection for continuity arguments reflects careerism, regionalism, and the lingering effects of narrative explanations in human origins research.

Careerism, Regionalism and Narrative Explanations

"Careerism" refers to professional strategic choices made for individual self-interest rather than for the good of some larger collaborative enterprise. The unusual "ecosystem" of human origins research creates careerist incentives that explain continuity theories' persistent popularity (White 2000). In paleoanthropology, a small number of producers (field scientists, lab directors) supply high-value resources (fossils, archaeological sites, and data derived from them) to a vastly larger number of potential consumers. Access to those resources can be controlled, sometimes by the producers themselves, but also by other actors (e.g., government bureaucrats, museum officials). In practical terms, this means that a consumer aspiring to professional advancement in human origins research faces fewer obstacles to claiming "ownership" of a theory and promoting/defending it than they do to gaining exclusive access to paleontological or archaeological evidence.

Claiming a theory is not risk-free. If the theory in question is based on a falsifiable hypothesis, the effort expended in its cause becomes a loss when it ends up being falsified, as most hypotheses eventually do. Under these circumstances, there are powerful selective pressures favoring arguments about human origins that appear falsifiable, but that are actually immune to refutation. This, I contend, is an underlying appeal of continuity theories. Advocating them means one will never have to incur the intellectual "retooling" costs of being proven wrong. One has to remain agnostic over whether proponents of continuity arguments consciously choose this course of action out of narrow careerist motives; but it is beyond dispute that the "ecology" of human origins research favors and rewards such a choice.

"Regionalism," as used here, refers to paleoanthropologists' habit of specializing in the evidence from one particular region. Regionalism is most pronounced among archaeologists (Otte and Keeley 1990). Regionalism is less pronounced among physical anthropologists, whose sparse fossil evidence is often broadly dispersed over time and space. The analogous problem for hominin paleontology arises from specialization on particular fossil taxa or particular fossil specimens and it reflects the lasting influence of narrative frameworks in human origins research.

As Landau (1991) showed, both early and recent explanations of human evolution were often framed in terms of simple, linear "anthropogenic" narrative, one in which a single hominin protagonist is transformed by a single cause (or "donor") into an ancestral human. The reality of human evolution is vastly more complex and ambiguous. There are multiple possible protagonists, the chronology is uncertain, and there are multiple possible causes for nearly every major event. Reconstructing the course of human evolution is very much like trying to reconstruct the complex, nonlinear plot of Tarantino's film, *Pulp Fiction*, but from a partial print. We all know this, and yet time and again we allow the fossils to be touted as the missing link without reference to the improbabilities involved in such an inference. "Science by press release" arguments asserting evolutionary continuity between a particular fossil and living humans carry considerable rewards in popular interest and funding with little risk of being proven wrong. In contrast, there are few professional incentives for claiming one's hard-won fossil evidence represents an evolutionary "dead end".[5]

A Reality Check and a Better Way Forward

Evolution has winners and losers. It is simply not possible that all regions and all hominin fossils were equally important for human evolution. We all individually suspect our chosen region and its fossils are among the former, but the odds are against it. Fossil preservation does not necessarily favor ancestral individuals. Even if we could devise a method for probabilistically sampling the fossil record of a particular hominin taxon, the chance that such a sample would include an individual whose descendants walk among us today is vanishingly small. Most paleoanthropologists accept that even if individual fossils and their associated archaeological remains were not demonstrably created by ancestral individuals, those fossils and archaeological evidence are representative of ancestral individuals in one way or another. It is important to remember that this is an assumption, not a fact, and we have to be alert to potential sources of error in this enterprise.

In attempting to reconstruct hominin evolution from the paleoanthropological record, we face the classic statistical problem of inferring population parameters from sample statistics. That one can only make such inferences from probabilistic samples is a bedrock principle of statistical inference (Sokal and Rohlf 1995). And yet, the process of paleoanthropological discovery is decidedly non-probabilistic. Sites located near major cities in peaceful countries with temperate climates (e.g., Abri Pataud, in France) are vastly more likely to be investigated and well-documented than their opposites (e.g., Darra-i-Kurr, in Afghanistan). There have been cases where the search for human fossils and archaeological sites was guided by prior theory, as in Dubois' explorations in Java (Shipman 2002). Far more often, though, paleoanthropological discovery is more haphazard. It results from an injudiciously-placed limestone quarry, as in the case of the Mount Carmel Caves (Garrod and Bate 1937), or from someone looking out an airplane window at just the right moment, as in Richard Leakey's discoveries in east Turkana (Leakey and Lewin 1978).

A non-probabilistically-sampled record inevitably allows for well-principled debate about the ancestral status and evolutionary significance of particular fossils. In such debates, most paleoanthropologists also recognize that each hominin taxon and regional archaeological sample is worthy of study in its own right, not just for the light it sheds on the origins of our species or those of us living in one region or another. But, it is also a fact that careerism, regionalism, and narrative explanatory frameworks affect the way in which we integrate our observations about

particular fossils and archaeological assemblages into broader explanations of human evolution. Right now, it is very easy to predict individual researchers' positions about continuity in human evolution; such continuity usually involves their particular region of expertise and the hominin fossils they are studying at the moment.

There has to be a better way to frame predictions about hominin evolutionary continuity. It would be much better for continuity arguments to be based on first principles derived from evolutionary and ecological theory, as well as from relevant insights from ethnology (Foley 1987). Hominins are relatively large mammals with a generalized omnivorous diet that evolved in tropical to warm-temperate woodlands under relatively stable climatic conditions. These shared features of our primate heritage yield reasonable predictions about factors likely to promote or to constrain long-term evolutionary continuity in a given region (Table 12.3). Large mammals fare poorly in small isolated habitats. Larger regions, particularly ones networked to other large regions, are more optimal habitats because they offer room for population growth and opportunities for dispersal. Human and nonhuman primate diets are dominated by plant foods of the kind that are most readily available in equatorial and warm temperate woodlands. Plant and animal foods are readily available in other regions, of course, but the habitats most like those in which ancestral hominins evolved require the fewest energetic costs associated with exosomatic processing (i.e., tool use). Humans need to drink water regularly, and thus proximity to either polar or subtropical deserts decreases the probability of long-term human evolutionary continuity. Lastly, because we reproduce slowly, wide variations in foraging returns caused by habitat variability also represent significant obstacles to evolutionary continuity.

Table 12.3 Conditions likely to influence long term hominin evolutionary continuity

	Continuity Probable	Continuity Improbable
Geographic Extent	Large	Small
Connectivity	Networked (e.g., Corridors)	Isolated (e.g., Islands)
Climate	Equatorial, Warm Temperate	Boreal, Cold Temperate
Aridity	Humid (e.g., Woodlands)	Arid (e.g., Deserts)
Climate Variability	Stable	Unstable
Hominin Population Size	Small	Large
Hominin Ecological Niche	Wide	Narrow

Some regions plainly offer more favorable conditions for long-term human and hominin evolutionary continuity than others do. Long-term evolutionary continuity is less likely among hominins living on the north European plain or the British Isles than among hominins living in equatorial Africa or mainland southeast Asia. The paleoanthropological records for these regions, as currently understood, seem to bear out these predictions (Barham and Mitchell 2008; Dennell 2009; Gamble 1986). Each of them features numerous instances of the cessation of archaeological site formation and presumed abandonment of entire regions during periods of lower temperatures and depressed environmental productivity. Such evidence is starkly clear in regions that have been the focus of prolonged and extensive archaeological survey and paleoanthropological research, such as the United Kingdom (Stringer 2006).

Viewing the Late Pleistocene paleoanthropological evidence in the light of these considerations results in a mixed verdict. The Levant is a small region whose Late Pleistocene climate was mostly cold and arid with wide short-term fluctuations in temperature and humidity (Bar-Matthews et al. 2000). These are all factors that would seem to work against continuity. On the other hand, the paleontological record for the Levant shows that it was also a biogeographic corridor linking Africa, Asia, and Europe for much

of the Late Pleistocene (Kurten 1965; Tchernov 1997). Between 128-18 ka periods of extreme aridity were relatively brief and humid conditions of more prolonged duration. These factors favor regional continuity. Inferences about continuity and discontinuity in Levantine Late Pleistocene human evolution diverge sharply depending on how one weighs these different factors (e.g., compare Hovers 2009; Shea 2009).

Conclusion

Glynn Isaac was an enthusiastic proponent of Clarke's (1973) call for archaeologists to develop a "critical self-consciousness" - an awareness of how the ways in which we think about scientific research limits our understanding of the past. Paleoanthropology, the larger enterprise of which prehistoric archaeology is a part, could profit from such critical self-consciousness as well. As scientists, we strive for objectivity, but countless subjective factors enter into our professional judgements. There are strong careerist incentives for researchers to identify themselves with either popular positions or more iconoclastic arguments. Cultural differences, language barriers, and nationalism express themselves in choices of fieldwork sites, in citation patterns, in invitations to meetings, and in other ways. Careerism, regionalism and the structural constraints on narrative explanation all influence models of human origins. Paleoanthropological research will make significant strides towards a real critical self-consciousness when paleoanthropologists recognize these factors and take them into account in our deliberations.

The method of multiple working hypotheses is Glynn Isaac's most important theoretical and methodological legacy for paleoanthropology. It provides a means by which to sort those explanations that could be proven wrong with evidence from those that could not be proven wrong no matter what kind of evidence is discovered. Having sorted these explanations, we now need to consider what to do with them. If only for clarity's sake, we need to start formally distinguishing among them. Following standard scientific practice, explanations for change in prehistory that can be proven wrong with evidence should be called "hypotheses". Explanations that cannot be proven wrong with evidence should be called "arguments". No one should dismiss continuity arguments because

they are not falsifiable in strict positivist terms; but we need to be more explicit about why we retain such arguments in scientific debate. Do they present a coherent accounting of the facts in evidence? Is it that they simply cannot be proven wrong as rival hypotheses can? Or, do they satisfy the other, non-scientific agendas in human origins research?

Acknowledgements

I thank Jeanne Sept and David Pilbeam for inviting me to contribute to this volume and for their editorial advice. I am deeply indebted to Glynn Isaac for taking me on as a student and encouraging me to study Paleolithic archaeology. I also thank Ofer Bar-Yosef for taking up where Glynn left off, for introducing me to Levantine prehistory, and for his support and guidance over the years. Katheryn Twiss, John Fleagle, and Frederick Grine provided constructive comments on earlier drafts of this paper.

Endnotes

1   Throughout this paper, I use the term "paleoanthropology" as a catch-all synonym for human origins research encompassing both prehistoric archaeology and hominin paleontology/physical anthropology. Most of the comments below may seem directed at paleontologists, but they apply to archaeologists as well.

2   Extremely low population densities might lead to a third outcome, a kind of meta-stable equilibrium with no overt competition (Wang et al. 2002). Levantine Late Pleistocene humans were probably few in number (Shea 2007b), their dependence on high-quality food sources (large animal carcasses) and the geographic circumscription of the Mediterranean woodlands by ocean, desert, and mountains were factors that almost certainly would have led to intense competitive encounters if the two hominins were present in the Levant at the same time. Similar factors demonstrably cause intense competition among living carnivores (Van Valkenburgh 2001).

3   If Levantine Neanderthal and early *Homo sapiens* DNA was more similar to that of living southwest Asian humans than to other living humans this would settle the argument in favor of continuity as a longer-term model for Levantine Late Pleistocene human evolution. This kind of phylopatry has not been demonstrated for aDNA for Pleistocene-age *Homo sapiens* fossils, and it seems inappropriate to include it in these scenarios.

4   If evolutionary development analyses were able to show that differences between Levantine Neanderthal and early *Homo sapiens* DNA posed a reproductive barrier, perhaps yielding sterile hybrids, then such differences would be a more convincing argument against continuity. Again, studies of DNA recovered from extinct species has not yet yielded this kind of information, and thus one cannot easily or credibly include it in a discussion of possible future research findings.

5   Other than, of course, the satisfaction of knowing that one is likely correct.

References

Arensburg, B., and A. Belfer-Cohen

1998 *Sapiens* and Neandertals: Rethinking the Levantine Middle Paleolithic hominids. In *Neandertals and Modern Humans in Western Asia*, edited by T. Akazawa, K. Aoki and O. Bar-Yosef, pp. 311-322. Plenum, New York.

Bar-Matthews, M., A. Ayalon, A. Kaufman, and G. J. Wasserburg

2000 The Eastern Mediterranean paleoclimate as a reflection of regional events: Soreq Cave, Israel. *Earth and Planetary Science Letters* 166:85-95.

Bar-Yosef, O.

1989 Geochronology of the Levantine Middle Paleolithic. In *The Human Revolution*, edited by P. A. Mellars and C. B. Stringer, pp. 589-610. Edinburgh University Press, Edinburgh.

Barham, L., and P. Mitchell

2008 *The First Africans: African Archaeology from the Earliest Toolmakers to Most Recent Foragers*. Cambridge University Press, New York.

Binford, L. R.

1985 Human ancestors: Changing views of their behavior. *Journal of Anthropological Archaeology* 4:292-327.

Binford, S. R.

1968 Early Upper Pleistocene adaptations in the Levant. *American Anthropologist* 70(4):707-717.

Bowler, P. J.

1986 *Theories of Human Evolution: A Century of Debate, 1844-1944*. Johns Hopkins University Press, Baltimore.

Cardillo, M., G. M. Mace, K. E. Jones, J. Bielby, O. R. P. Bininda-Emonds, W. Sechrest, C. D. L. Orme, and A. Purvis

2005 Multiple causes of high extinction risk in large mammal species. *Science* 309(5738):1239-1241.

Churchill, S. E.

2006 Bioenergetic perspectives on Neanderthal thermoregulatory and activity budgets. In *Neanderthals Revisited: New Approaches and Perspectives*, edited by K. Harvarti and T. Harrison, pp. 113-134. Springer, Dordrecht.

Clark, G. A.

1992 Continuity or Replacement? Putting Modern Humans in an Evolutionary Context. In *The Middle Paleolithic: Change, Adaptation, and Variability*, edited by H. L. Dibble and P. A. Mellars, pp. 183-206. University of Pennsylvania Museum Press, Philadelphia, PA.

2002 Neandertal archaeology - implications for our origins. *American Anthropologist* 104(1):50-67.

Clark, G. A., and J. Lindly

1989 Modern human origins in the Levant and western Asia: The fossil and archeological evidence. *American Anthropologist* 91:962-985.

Clarke, D. L.

1973 Archaeology: The loss of innocence. *Antiquity* 46:237-239.

Dennell, R.

2003 Dispersal and colonisation, long and short chronology: How continuous is the Early Pleistocene record for hominids outside East Africa. *Journal of Human Evolution* 45:421-440.

2009 *The Palaeolithic Settlement of Asia.* Cambridge, UK, Cambridge University Press.

Foley, R.

1987 *Another Unique Species: Patterns in Human Evolutionary Ecology.* Longman Group, Harlow, UK.

Gamble, C.

1986 *The Palaeolithic Settlement of Europe.* Cambridge University Press, Cambridge.

Garrod, D. A. E., and D. M. A. Bate (Editors)

1937 *The Stone Age of Mount Carmel, Vol. 1: Excavations in the Wady el-Mughara.* Clarendon Press, Oxford.

Henry, D. O.

1995 Late Levantine Mousterian patterns of adaptation and cognition. In *Prehistoric Cultural Ecology and Evolution: Insights from Southern Jordan*, edited by D. O. Henry, pp. 107-132. Plenum, New York.

Holliday, T. W.

2000 Evolution at the crossroads: Modern human emergence in western Asia. *American Anthropologist* 102(1):54-68.

Hovers, E.

2006 Neanderthals and Modern Humans in the Middle Paleolithic of the Levant: What Kind of Interaction? In *When Neanderthals and Modern Humans Met*, edited by N. Conard, pp. 65-85. Kerns Verlag, Tübingen.

2009 *The Lithic Assemblages of Qafzeh Cave.* Oxford University Press, Oxford, UK.

Howell, F. C.

1958 Upper Pleistocene men of the southwest Asian Mousterian. In *Hundert Jahre Neanderthaler*, edited by G. H. R. von Koenigswald, pp. 185-198. Kemik en zoon, Utrecht.

1959 Upper Pleistocene stratigraphy and early man in the Levant. *Proceedings of the American Philosophical Society* 103:1-65.

Isaac, G. L.
1977 *Olorgesailie: Archaeological Studies of a Middle Pleistocene Lake Basin in Kenya.* The University of Chicago Press, Chicago.

1978 Food sharing and human evolution: Archaeological evidence from the Plio-Pleistocene of East Africa. *Journal of Anthropological Research* 34:311-325.

1983 Bones in contention: Competing explanations for the juxtaposition of Early Pleistocene artifacts and faunal remains. In *Animals and Archaeology: Hunters and Their Prey,* edited by J. Clutton-Brock and G. Grigson, pp. 3-19. British Archaeological Reports International Series 163, Oxford.

Isaac, G. L., and B. Isaac (Editors)
1997 *Koobi Fora Research Project Series, Volume 5: Plio-Pleistocene Archaeology.* Clarendon, Oxford, UK.

Jelinek, A. J.
1982 The Middle Paleolithic in the Southern Levant with comments on the appearance of modern *Homo sapiens.* In *The Transition from Lower to Middle Paleolithic and the Origins of Modern Man,* edited by A. Ronen, pp. 57-104. British Archaeological Reports International Series 151, Oxford.

Jolly, C. J.
2001 A proper study for mankind: Analogies from Papionin monkeys and their implications for human evolution. *Yearbook of Physical Anthropology* 44:177-204.

Keith, A.
1937 Mount Carmel Man: His bearing on the ancestry of modern races. In *Early Man,* edited by G. G. MacCurdy, pp. 41-52. J. B. Lippincot Co., New York.

Kramer, A., T. L. Crummett, and M. H. Wolpoff
2001 Out of Africa and into the Levant: Replacement or admixture in Western Asia. *Quaternary International* 75(1):51-63.

Krings, M., A. Stone, R. W. Schmitz, H. Krainitzki, M. Stoneking, and S. Pääbo
1997 Neandertal DNA Sequences and the origin of modern humans. *Cell* 90:19-30.

Kuhn, T. S.
1962 *The Structure of Scientific Revolutions.* University of Chicago Press, Chicago, IL.

Kurten, B.

1965 The carnivora of the Palestine caves. *Acta Zoologica Fennica* 107:1-74.

Landau, M. L.

1991 *Narratives of Human Evolution.* Yale University Press, New Haven, CT.

Leakey, R. E., and R. Lewin

1978 *People of the Lake: Mankind and Its Beginnings.* Avon, New York.

Lieberman, D. E., and J. J. Shea

1994 Behavioral differences between archaic and modern humans in the Levantine Mousterian. *American Anthropologist* 96:300-332.

Marks, A. E.

1983 The Middle to Upper Paleolithic transition in the Levant. In *Advances in World Archaeology,* edited by F. Wendorf and A. E. Close, pp. 51-98. vol. 2. Academic Press, New York.

McCown, T. D., and A. Keith

1939 *The Stone Age of Mt. Carmel, Volume 2: The Fossil Human Remains from the Levalloiso-Mousterian.* Clarendon Press, Oxford.

Otte, M., and L. H. Keeley

1990 The Impact of Regionalisation on the Palaeolithic Studies. *Current Anthropology* 31(5):577-582.

Pianka, E. R.

1988 *Evolutionary Ecology, 4th Edition.* Harper and Row, New York.

Potts, R.

1984 Home bases and early hominids. *American Scientist* 72:338-347.

Rak, Y.

1993 Morphological variation in *Homo neanderthalensis* and *Homo sapiens* in the Levant: A biogeographic model. In *Species, Species Concepts, and Primate Evolution,* edited by W. H. Kimbel and L. B. Martin, pp. 523-536. Plenum, New York.

Shea, J. J.

2003a The Middle Paleolithic of the east Mediterranean Levant. *Journal of World Prehistory* 17(4):313-394.

2003b Neandertals, competition, and the origin of modern human behavior in the Levant. *Evolutionary Anthropology* 12(4):173-187.

2006 The Middle Paleolithic of the Levant: Recursion and convergence. In *Transitions Before the Transition: Evolution and Stability in the Middle Paleolithic and Middle Stone Age*, edited by E. Hovers and S. L. Kuhn, pp. 189-212. Plenum/Kluwer, New York.

2007a Behavioral differences between Middle and Upper Paleolithic *Homo sapiens* in the East Mediterranean Levant: The roles of intraspecific competition and dispersal from Africa. *Journal of Anthropological Research* 64(4):449-488.

2007b The boulevard of broken dreams: Evolutionary discontinuity in the Late Pleistocene Levant. In *Rethinking the Human Revolution*, edited by P. Mellars, K. Boyle, O. Bar-Yosef, and C. Stringer, pp. 219-232. McDonald Institute for Archaeological Research Monographs, Cambridge, UK.

2008 Transitions or turnovers? Climatically-forced extinctions of *Homo sapiens* and Neanderthals in the East Mediterranean Levant. *Quaternary Science Reviews* 27(23-24):2253-2270.

2009 Bridging the gap: Explaining the Middle-Upper Paleolithic transition in the Levant. In *Transitions in Prehistory: Essays in Honor of Ofer Bar-Yosef*, edited by J. J. Shea and D. E. Lieberman, pp. 77-105. Oxbow, Oxford, UK.

Shipman, P.
2002 *The Man Who Found the Missing Link: Eugéne Dubois and His Lifelong Quest to Prove Darwin Right*. Harvard University Press, Cambridge, MA.

Simmons, T.
1994 Archaic and modern *Homo sapiens* in the contact zones: Evolutionary schematics and model predictions. In *Origins of Anatomically Modern Humans*, edited by M. D. Nitecki and D. V. Nitecki, pp. 201-225. Plenum, New York, NY.

Smith, F. H., A. B. Falsetti, and S. Donnell
1989 Modern human origins. *Yearbook of Physical Anthropology* 32:35-68.

Smith, P.
1995 People of the Holy Land from prehistory to the recent past. In *Archaeology of Society in the Holy Land*, edited by T. E. Levy, pp. 58-74. Facts on File, New York.

Sokal, R. R., and F. J. Rohlf
1995 *Biometry, Third Edition*. W. H. Freeman and Company, New York.

Stringer, C.
2006 *Homo Britannicus: The Incredible Story of Human Life in Britain*. Penguin, London.

Tchernov, E.

1997 Are Late Pleistocene environmental factors, faunal changes, and cultural transformations causally connected? The case of the southern Levant. *Paléorient* 23(2):209-228.

Trinkaus, E.

1984 Western Asia. In *The Origins of Modern Humans*, edited by F. H. Smith and F. Spencer, pp. 251-293. Alan R. Liss, New-York.

Van Valkenburgh, B.

2001 The dog-eat-dog world of carnivores: A review of past and present carnivore community dynamics. In *Meat Eating and Human Evolution*, edited by C. B. Stanford and H. T. Bunn, pp. 101-121. Oxford University Press, New York.

Vandermeersch, B.

1989 The evolution of modern humans: Recent evidence from Southwest Asia. In *The Human Revolution*, edited by P. Mellars and C. B. Stringer, pp. 155-164. Princeton University Press, Princeton, NJ.

Vandermeersch, B., and O. Bar-Yosef

1988 Evolution biologique et culturelle des populations du Levant au Paléolithique Moyen. Les données récentes de Kebara et Qafzeh. *Paléorient* 14(2):115-117.

Wang, Z.-L., F.-Z. Wang, S. Chen, and M.-Y. Zhu

2002 Competition and coexistence in regional habitats. *American Naturalist* 159(5):498-508.

White, T. D.

2000 A view on the science: Physical anthropology at the millennium. *American Journal of Physical Anthropology* 113:287-292.

Wolpoff, M. H.

1989 The place of Neanderthals in human evolution. In *The Emergence of Modern Humans*, edited by E. Trinkaus, pp. 97-141. Cambridge University Press, New York.

13

SCATTERS BETWEEN THE SITES – FARM SHELTERS, HUNTING AND GATHERING AND THE FARMING CYCLE IN WEST AFRICA: LESSONS FOR ARCHAEOLOGICAL DISTRIBUTIONS

Merrick Posnansky

I shared a history with Glynn Isaac in that I lived at Olorgesailie for 19 months from 1956-1958 some three years before Glynn came to Kenya and began work at Olorgesailie himself. After six months, during which we restored the museums on the sites and the visitors' center, my interests shifted away from the Stone Age to the Iron Age sites in Kenya that became the principal focus for my later time in east and west Africa. Nevertheless, I often walked across the Olorgesailie landscape looking for sites that I envisaged as concentrations of stone tools and the debris from the making of tools as well as fossilized bones on old land-surfaces sandwiched between lacustrine deposits that were normally diatomaceous silts. These were our envisaged occupation sites. What I in fact usually discovered were smaller scatters of stone tools or flakes which did not appear to be part of occupation sites and which were unassociated with faunal remains. Had they just been dropped there? In a lecture by Glynn in the States I learned of his own reflections, that he discussed in his book (Isaac 1977:80-84), in which he mused about the larger quantity of material away from the concentrations that he regarded as camp sites sometimes termed as "central" sites. One of his conclusions was that Olorgesailie folk moved away from the lake flats and camps near to water, where mosquito activity was too vexatious, to grassland ridges where they could have a more restful time.

It was Glynn's attention to the meaning of artifact distribution away from major occupation sites that ultimately led to my own work on the historical archaeology of former farming populations in the Begho area of Ghana. The problem was how to react to scatters of artifacts between the established sites.

The emphasis that Glynn gave to looking at the scatters between sites was a necessary corrective. Archaeologists had become fixated on camp sites, rock shelters, houses, palaces or cemeteries. Areas between sites and even around sites had been neglected. This applied not only to sites from the Stone Age but particularly to later periods. When not neglected, such sites were not open to interpretation but left as meaningless archaeological noise between sites.

Archaeologists looked at houses, but not at yards or activity areas around houses. As soon as a new approach was adopted, in African-American sites on American plantations in the Caribbean, the nature of African American life was opened up and diasporan archaeology has never been quite the same. GPS is allowing archaeologists to plot scatters and thus if we can understand them further we have an important tool for the interpreting the finds and understand the lifestyles of the people.

Begho was a medieval to early modern town in Ghana some 270 km northwest of the capital of Accra. We excavated there from 1970-1979 in

eight different sessions. It dates between AD
1150 and perhaps 1850, though it is difficult to
date the chronological parameters more precisely. At its heyday there were at least six distinct
quarters or occupation areas surrounding a central market zone. The quarters were associated
with people of different ethnicities, the largest
quarters were those of the Bron who were
indigenous to the area. The Kramo or merchants
occupied perhaps two of the quarters. Kramo
refers to people of the book, and thus Muslims,
who were of the same linguistic grouping as the
Mande in Mali. The final quarter was that of the
metalworkers or Dwinfuor. Estimating the number of house mounds suggests there had been
possibly up to 1,500 courtyard houses containing a population that was in excess of 10,000 in
Begho's heyday around the seventeenth century.
Archaeologically we could say a great deal about
the technology and trade of the people who
dyed, spun and wove cotton cloth; fashioned
ornaments and functional items from local
ivory; wrought objects from locally smelt iron;
cast brass objects from scrap metal; made blue
glass beads; traded ceramics with surrounding
villages and marketed gold and local forest
products like kola nuts, shea butter, animal skins
and dried and smoked game meat. In return
they received rugs, brassware, salt, beads, high-
er-grade ceramics and religious items. We have
learned from oral history, as well as documents
in Arabic, about their social structure and cultural practices.

We attempted to estimate the size of the
town first by mapping the extent of the distinct
quarters and then by conducting radiating surface surveys from the centers of the quarters as
far as the scatter of potsherds and other finds
persisted. We were also able to indicate where
the settlement sites ended by noting subtle vegetation changes. The present successor village of

Hani is ringed by trash heaps, toilet areas and an
abundance of such tree varieties as lemons and
limes used in washing and giving the body a
more acceptable smell and plants useful in
sedating upset stomachs and curing some of
worst effect of diarrhea. These are grown intentionally around the fringes of settlements.
Though the elephant grass returns once a settlement is abandoned, the fringe "forest" persists
for several generations. Using these methods we
were able to indicate that at its peak Begho may
have comprised up to 6 km^2.

The Archaeologists and Farming Practices

Though we know from archaeology about what
the people of Begho ate, the proportions of wild to
domesticated taxa and also how they prepared
their food, we know less about how their food was
acquired. Archaeologists in the past have excavated at set seasons, often in the dry periods when
few agricultural activities are taking place, or possibly in the harvest season that in west Africa coin-
cides with academic long vacations, when farmers
have less time to spend with interviewers. The
problem ensures that archaeologists observe the
agricultural cycle but only for short periods. They
do not experience the agricultural routine
through the seasons noting the different seasonal
activities of the farmers. In 1980 and 1983 we
endeavored to accompany the farmers on all their
tasks and find exactly what they did and when it
happened. We also made short visits at different
seasons. We discovered that even with good
informants out of sight was out of mind and they
forgot some of their most routine activities if there
was nothing to jog the memory. Various key tools
were stored on the farm and not in their main res-
idence, many tools had been fashioned by the
farmers themselves on the farm whittling away at
available wood. One particular tool was the wood-
en dabba used in planting yams into the soft earth

of the yam mounds and taking them out when fully grown (Figure 13.1). Such wooden tools are preferred to iron machetes, known in Ghana as cutlasses, as they do not damage the tubers which would quickly lead to rotting. These were overlooked when the farmers described their agricultural implements since they were left near the fields where they were used. Rather than observing highlights we were privileged to record what took place in off periods of activity. We were particularly interested in farmers as hunters and gatherers and the farmers' use of field shelters.

Farmers in Hani were away from the village, the equivalent of the campsite for hunter-gatherers, when they foraged for food from non domesticated sources and also when they farmed. In much of Africa farmers are to some extent migratory, many in west Africa practice slash and burn agriculture, also known as brandwirtschaft. Farmers and their families start by clearing a manageable area of scrub (secondary) forest that can cover as much as 2 hectares (ha; up to 5 acres) though is normally smaller. They cultivate their plot until they've exhausted the fertility of the soil. Normally in Hani they plant yams (dioscorea) the first year and also possibly in the second year, together sometimes with cocoyams (taro) and maize whilst groundnuts and cassava are cultivated in the third year. That particular area of their farm is then abandoned for as much as 17 or 18 years, depending on the pressure on the land by competing farmers. Out of 25 farmers only one, an old man, used his farm shelter for a fourth year and I only learned of one instance in which a shelter had been used for five years. This way the long term fertility of the soil is assured. The process is constantly being repeated. In fact they rotate around a set piece of land assigned to them by the elders. There are no fixed barriers but the farmers are acutely aware of the land to which they are

Figure 13.1 Hani farmer planting seed yam in a yam mound using a wooden dabba. Photo by Merrick Posnansky.

assigned. Land is held communally and as a farmer grows old and his family moves away his farm, the size of his allotment is also reduced. The women cultivate such plants as tomatoes, peppers, garden eggs, beans, okra and onions on plots nearer the village, though a few women, such as young widows, may also have large farms. In the late 1990s the population of the village was around 2,200 and their farm shelters were scattered over an area of around 40 km^2.

Hunting and Collecting

Farmers leave their immediate settlement area either for slash and burn agriculture or for hunting and gathering. Each activity leaves its mark

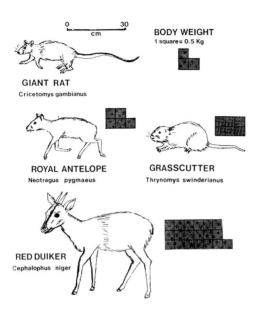

Figure 13.2 Small animals hunted and trapped by Hani farmers.

but the evidence for hunting and gathering is less identifiable. In periodic questionnaires administered over 28 years (Posnansky 2004) all the farmers claimed they were active hunters. Hunting took them to a catchment more than 10 km from their homes (over 120 km² as opposed to the 0.15 km² of their nucleated village). In our 1980 survey of 25 households farmers had shot or caught 2,193 animals, mostly grass cutters (*Thrynomys swinerderianus*), also known as cane rats and agouti in the Francophone countries, and giant forest rats (*Cricetomys gambianus*). Small antelope including duikers (*Cephalophus niger* and *Neotragus pymaeus*) are also targeted (Figure 13.2). These are normally dried and smoked and strips of meat are used in stews. Some of the dried meat is traded and helps provide carbohydrate foods and fruits such as oranges in return when yams are scarce in the village. Seventy-six % of the

householders have guns, referred to historically as "Dane" guns, locally manufactured from bicycle frame tubing, unrifled and cap detonated. More than 50% of the households catch birds, including Guinea fowl, and each farmer averages at least 50 birds a year. Sixty-five % of the farmers have dogs that dig out rodents and chase down the small antelope. In this way the protein content of their diet is higher than initially assumed. All the households collect borassus palm nuts that they boil with corn meal and sugar for over 24 hours to make porridge. All the farmers also collect white, fatty oil palm grubs (akokonu) that fill a need for fat at the end of the dry season and into the rainy season before any crops can be harvested. At the beginning of the rainy season whole families (88% of respondents) go into the forest to collect giant forest snails (*Archachatina* sp.) that even when taken from the shell and dried can yield up to 500 grams of rich protein. All the villagers collect Shea butter nuts (*Butyrospermun parkii*) that can be used both for processing into cooking oil or for facial and body ointment. A further five varieties of oleaginous nuts are collected as well as up to 500 roots, leaves, fruits, barks etc that are used for food, medicinal and even building purposes. Honey collecting is a more discriminating pursuit and only 50% of the farmers have obtained honey and most is collected by an even smaller number of woodsmen with special skills. Many other natural commodities like silk cotton from the kapok tree along with ropes and vines and bark cloth from a variety of fig tree enrich their diets and material culture, though bark cloth making is an art known to only two or three families.

Most unfortunately the hunters and gathers leave little imprint on the areas they exploit. They enlarge the hollows of trees looking for honey, tap some trees for medicinal sap like the nyante used for stomach complaints, but nevertheless their presence can be observed particularly through floral distributions. Trees like the shea butter nut, though wild, are protected by them and thus indicate the forager's presence as does the borassus palm whose fermented nuts drive elephants crazy, but are an essential corollary of settlement over a wide area of Africa. Hunting is often ritualized. At Hani, a nyifie festival signals the start of the hunting season when grassland is burnt to flush out the grass cutters for easier hunting. At the site where chickens are sacrificed for the nyifie festival to begin, there is a small pile of stones, a shrine that would go normally unnoticed. Other shrines are located in thickets, away from the ravages of sheep and goats, sometimes with broken pots on top of the stones. All have explanations that enrich their folklore. Such artifactual evidence is only accessible through oral history but their existence provides clues to former as well as present foraging activities. The burning, as well as the slash and burn agriculture, ensures that the cleared areas around settlements remain relatively open. Such clearings are markers of past agricultural activity and can be picked out on aerial photographs. Arrowheads and guns within the village suggest hunting activities but the ceramic corpus of the excavated sites contains sherds with pierced holes that are almost certainly colanders used for rendering honey from hive material using fires. Such pots are heavier and bigger than normal colanders that are shallower and have less trace of the material being drained.

The two activities, slash and burn agriculture and hunting and gathering, however, use the same landscape. Trapping does not leave a physical mark on the landscape; however, the associated farm shelters do, which means that the farm shelters provide some indication of the territorial range of the agricultural hunter-gatherer (Figure 13.3). We also found that during the severe

African drought of 1982-1984 the farming community turned much more to hunting and gathering to supplement their meager holdings of food. Farmers normally worked close to their shelters with 70% of the time spent less than 1 km away. Much food preparation, particularly drying of cassava strips took place around the farm shelter. Yams, weighing as much as 10 kg, are stored near the shelters in stacks with a good airflow and grass covers to ensure maximum dryness. This seems to be preferred over storage in the village where vermin infestation is high. Storage at the shelter also had the advantage of keeping one's tradable food assets secret. Eighty-five % of all farmers stored food on their farms of which 34% was maize. The principal threats to the piles of yams are grass cutters and rats both of which are preferred species hunted by the farmers. The yam piles are lures for both these creatures so the farmers sets traps around his yam piles (Figure 13.4). In some cases up to 25 fall traps, with rafts of weighted stones triggered by slender sticks that are easily dislodged, guard the piles. In addition there are wire traps more useful against birds and occasional small antelope. In one instance a large black mamba was caught in a wire snare and eventually eaten. The farmer also has his gun, both as a deterrent and in hopeful anticipation that some creature will be incautious enough to approach the piles while he is in the vicinity. The traps are quite effective and it is not unusual for the farmer to obtain four or five creatures in a day. It is easy to underestimate the value of trapping and hunting but economically a single smoked grass cutter was worth up to 20 times the urban minimum wage in the 1980s when there was a real food crisis. Game meat still is preferred in the town to beef and goat and can be kept for long periods without refrigeration.

The knowledge of hunters is often as important as the testimonies of chiefs but is all too often

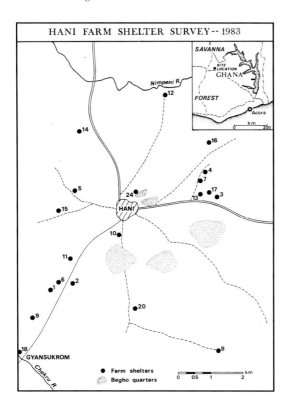

Figure 13.3 Map of farm shelters studied around Hani (from Posnansky 2004).

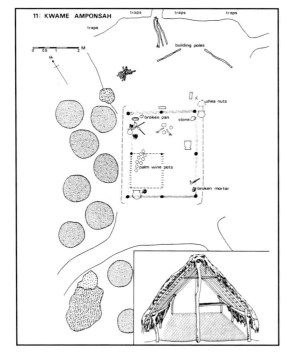

Figure 13.4 Plan of Kwame Amponsah's farm shelter. Note the line of traps at the northeast side. The shea nuts and palm wine pots indicate a farmer who forages. A rubbish dump and cassava peelings at southwest fringes.

neglected. Hunters often pick out village sites in advance of settlement, indicate boundaries and are an essential link to past practices of food acquisition and a pharmacopoeia that becomes important in times of environmental crisis. They play important roles in funeral ceremonies firing off their muskets and though often looking unkempt they have a rich lore of stories probably passed on between the generations on hunting forays that appeal to other villagers.

Farm Shelters

From 1970 until 1979 while excavating Begho we had studied the ethnoarchaeology of Hani (Posnansky 2004). In 1980 I turned my full attention to the ethnoarchaeology. A primary tool was a questionnaire on farming and resource management administered to approximately one tenth of the active farmers chosen using random sampling normally every other year. In 1980 and 1983 I particularly studied farm shelters and hunting and gathering. Some 25 farms were studied intensively and at least one of our team, all but myself from the University of Ghana, accompanied the farmers in their daily tasks. All the farmers erected farm shelters on their farms. A repeated proverb reiterates that a farmer who cannot build a pata on his house will never have a proper house (*onipa a otumi usi wo nafuomno otuminsi dan da*). Plans were made of each shelter and the objects used there itemized. Chickens were noted at several sites and over 40% of the farmers regularly kept chickens with one having as many as 19, though I was told of a case in which a farmer tried to keep as many as 100. Farmers often brought out their dogs that also kept away predators. The farms were situated from 1-6 km away from their homes. Few farmers walked more than 10 km a day suggesting a catchment area radial extent of around 5-6 km from the village. It was a given that the farmers

stayed out all day. As farm shelters were often close to one another, a certain amount of socializing took place with an emphasis on exchanging information about farming activities. Farmers normally rested during the heat of the day and often took the opportunity to pass on knowledge to the young sometimes playing the board game, *oware*, that has a mathematical basis and is an invaluable tool for teaching counting skills. Oware was played using holes scooped out of an area of smoothed earth. In some few instances children also learnt musical skills on the xylophone or short side blown trumpets (aben). Simple meals were prepared, and shared, on the farm so hearths consisting of three stones were constructed. In some cases up to three hearths existed at a shelter. Earthenware pots were used for storing liquids, as were calabashes. Tools were tucked into the eaves of the huts and the occasional hurricane lamps suggested that the farmers stayed over night on some occasions. The migrant Dagarti farm workers often used the shelters for three or four days at a time during harvesting.

In an inventory of items in one shelter that was fairly typical, we enumerated one hoe, two cutlasses, an axe blade, the inner rim of a tire with the extracted wire used for traps, a bag of salt, tomato seeds, tobacco leaves, old clothes, paper, tins and a harvest festival envelope. This same farmer also had pineapple tops for planting. The farmers husked corn and processed food like cassava at the farm shelter cutting the tubers into slithers that were dried on the ground or on sheets of abandoned aluminum roofing. The piles of cassava peel and gratings provided a fertile environment for growing mushrooms during the early rains.

The shelters, known as patas in the Bron language, are always built on the same basis. First an "A" frame of sturdy logs is cut from nearby trees. They are always rectangular in shape and roughly

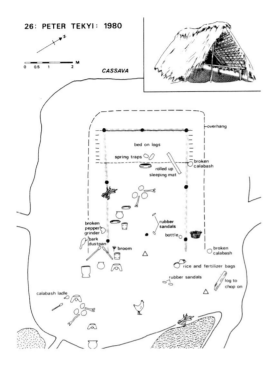

Figure 13.5 Peter Teky's farm shelter. The bed and large number of personal items indicate a successful farmer. Note chickens kept there and hearths within the shelter on south side.

in a north-south orientation with the north side often partially walled by grass. The orientation avoided the sun from the west and the rains from the east and appeared to have had no symbolic meaning. Their average dimensions were 4.7 m long, 3.4 m broad and 2.4 m high (Figures 13.5-6). A few were longer. They were all thatched with local cut grass or borassus palm fronds and built in several days. All have at least one hearth within the shelter composed of three boulders between which branches and logs can be progressively pressed into the central fire. Some have one or more outside hearths. All have an assortment of logs and rough cut wooden stools, some made at the site, on which they sit.

All the farmers have wooden pestles and mortars to pound yam into fufu and a pottery grinder and small hourglass shaped wooden grinders for grinding peppers, tomatoes and onions for the sauce they use to accompany the fufu or other pounded carbohydrate. Normally there are up to four or five whole, or occasionally partially broken, globular pots used in cooking as well as the occasional battered aluminum saucepan or cast aluminum or iron pot. For those who kept chickens on the farm nesting baskets were kept inside the shelters sometimes above the bed. Scattered around are old rubber sandals, broken baskets, iron from old metal traps, calabash spoons and containers though the interior and the few meters around the hut are kept cleared and intermittently swept clean with a whiskbroom made from stiff grasses. There was normally a trash heap a few meters from the shelter at the south end, in some case as much as 70 cm high. Only 11% of the shelters had two or more trash heaps.

Whilst working at Notse in Togo, an Ewe speaking area in a rather different environmental setting with a lower precipitation, similar shelters were also noted and planned. Others have been recorded elsewhere, an indication of the widespread nature of rotational slash and burn agriculture. In the more open environment of southern Togo foraging activities were of a lesser importance.

A year after we mapped the Hani shelters we returned to those shelters that had been abandoned after three years of use. None were still standing. On abandonment of the farms, nearby farmers took away useful building poles and the farmers themselves took away complete pots and pestles. All the sites were overgrown and without my plans I could not easily have recognized their locations but farmers told me that they could recognize old pata sites for up to 30 years. Sixty-five % recognized the old patas from the hearth stones,

Figure 13.6 Photograph of a farm shelter indicating informality and large number of calabash containers.
Photograph by Merrick Posnansky.

47% from the trash heaps and 47% by seeing the scatter of earthenware sherds. Knowing the tell-tale signs I was also able to find some of the shelters after 15 years. With clearance, however, it would be possible for archaeologists to identify these sites. Early identification came from the degraded trash heaps but this was a feature that became less clear as the heaps subsided over the years. Scattered over the area were pieces of broken pot, the hearth-stones, fragments of metal such as the stubs of hoes or parts of cutlass blades too small to be taken back for recycling by resourceful smiths and pieces of plastic, though these latter would not indicate pre-mid 20th century sites. There were scarcely any bones. Scavenging animals had either taken them away or they had degraded in the acidic soils. On the farms the farmers had largely cooked vegetal matter. Meat and dried fish were saved for eating at home. Careful examination of the flora was also insightful as a fringe vegetation of useful species was found around the site where farmers defecated and threw their organic trash that included mango pits as well as citrus and papaya seeds. In walking through the secondary forest I had often been surprised to see the occasional fruit trees and wondered about their origin. Now I knew they indicated the use of the catchment resources of former or existing settlements.

Conclusion

What the study clearly indicates is that if we want to understand human settlement in the Iron Age and Historic periods we cannot rely solely on mapping major settlement sites. In areas of river drawdown, as in the Niger delta around Jenne or around Lake Tchad, where the fertility of the soils was enriched with fresh silts annually, agriculture could consistently take place. Elsewhere in west Africa soils are relatively thin and become exhausted rapidly. In such areas agriculture was on a shifting pattern. New shelters were erected on the area being farmed every three or four years. Pots were broken, hearths burned leaving tell tale evidence of ash and discolored hearth stones. A very small scatter of artifacts is all that we can now find but the scatters over the landscape indicate the catchment area of the related archaeological site. The wider the spread and the greater number of scatters imply a larger settlement from which the farmers came; a small catchment area suggests a smaller settlement population. If we can plot the scatters, accurately date them to show synchronous occupation, we have a powerful tool for indicating demographic changes. These scatters, ephemeral as they are, combined with evidence about foraging activities give a good idea about resource strategies. What I set out to demonstrate is that scatters between sites are not random but their interpretation depends as much on ethno-archaeology as it does on straightforward archaeology. In order to understand what is happening in a rural environment it is just as important to know about all aspects of the societies we study including the most ephemeral of shrines, that can consist of a seemingly meaningless jumble of stones and pot sherds and provide further largely untapped information about activities away from the central village site. What this study indicates is that seemingly ephemeral distributions of artifacts between sites are often as informative about human behavior in the past as detailed studies of central sites.

Acknowledgments

This study was assisted financially by a grant in 1983 from the L. S. B. Leakey Foundation and a brief description of the work, without figures, appeared in *Anthroquest* 30:11-12. The UCLA Academic Senate Research Committee provided other assistance and help in kind was afforded by the Department of Archaeology of the University of Ghana. The author is grateful for the cooperation and enthusiastic support of the late Hanihene Kofi Ampofo II and the farmers of Hani. The late Tim Seymour drew the plans of the shelters illustrated.

References

Isaac, Glynn L.

1977 *Olorgesailie Archeological Studies of a Middle Pleistocene Lake Basin in Kenya.* University of Chicago Press, Chicago and London.

Posnansky, Merrick

2004 Processes of change - A longitudinal ethno-archaeological study of a Ghanaian village. *Hani* 1970-98, *African Archaeological Review* 21(1):31-47.

14

In Pursuit of the Past: A Consideration of Emergent Issues in African Archaeology in the Last Fifty Years

Francis B. Musonda

Introduction

In 1973, I embarked on a career in archaeology that has become my lifelong pursuit of a fascinating study of ancient humans and cultures under the mentorship of some of the best archaeologists in the world. This pursuit began at the University of Ghana under the direction of Merrick Posnansky, Andrew B. Smith, James Anquandah, Leonard Crossland, Jeane Dombrowski and Signe Nygaard. With strong background training in the natural sciences from the University of Zambia, I built a strong foundation in a program that introduced me to a broad range of approaches and interests in Stone Age and Iron Age studies, the origins of agriculture, subsistence and resource use, early hominid social organization, and the development of cities and urbanism. In 1979, I enrolled in a program at Berkeley that was developed by J. Desmond Clark and Glynn Isaac (Clark 1986), which introduced me to a new and broader range of research interests.

About a decade earlier, at the start of my training, Lewis Binford and his colleagues had advocated some very fundamental changes in the way archaeology was being conducted (Binford 1962; Binford and Binford 1968). Their attention was directed at two phases of development in African archaeology conveniently subdivided by Clark (1990:189) into: 1) the pioneer phase (1850s to around 1950); and, 2) the formative phase (1930 to 1960), which was associated with the recovery of archaeological data, and the dating and interpretation of data. They were unhappy with the methods that were being used and the limitations of descriptive archaeology and saw explanation as a better approach to the analysis of evidence. In the decades that followed, there were new advances in the way archaeology was being done and this period, which is associated with New Archaeology, contrasted significantly with the earlier one characterized by the culture-historical approach (Trigger 1990, 1996:235). It was during the 1970s and beyond that those archaeologists who accepted the new approaches began to make attempts to apply new ideas developed in the 1960s to actual archaeological situations. The best example perhaps comes from Koobi Fora, east of Lake Turkana, where Glynn Isaac developed new theoretical concepts in the study of early hominids and their ancient environments. I was part of the Koobi Fora Research Project in 1979. Emphasis was placed on the interdependence and complexity of a number of variables that were responsible for cultural and human evolution and he guided students to analyse the context in which the occurrences took place.

The results of his efforts are the basis of this paper in which I examine some of the most challenging approaches and methods that have emerged in African archaeology during the last fifty years and the advances made in the way archaeology is done today. The paper highlights those approaches developed during Glynn Isaac's career that I feel have helped in the interpretation of Africa's past from the perspective of research undertaken in eastern Africa at Koobi

Fora, East of Lake Turkana. These developments undoubtedly resulted in new interpretations of past human behavior.

Formative Phase

The difficulties faced by archaeologists working in Africa make it exceedingly difficult to attempt a synthesis of developments made across the continent (Musonda 1990). The overall archaeological picture is blurred because some regions are devoid of recent work while other regions such as eastern, southern, and western Africa have made exceptional progress and this concern has also found expression elsewhere (Ucko 1993:xxxi). It would even be ambitious to argue for growth and development of African archaeology considering the problems those doing archaeology on the African continent face. Ucko has alluded to these problems which raise a number of pertinent questions on the nature of Africa's contribution to the world stage. Accordingly, whereas east Africa is a leader in early hominid and urban civilization studies, South Africa is leading in the study of stone tool assemblages and early hominids while west Africa is well versed in early civilizations and North Africa in other forms of civilizations. Regions in between deal with a myriad of studies ranging from Stone Age to Iron Age and pre-colonial studies. However, considering the size and diversity of the African continent, there has undoubtedly been a myriad of developments and issues that have taken place and emerged that are not captured in this paper as the discussion is biased in favor of Eastern Africa where Glynn Isaac spent considerable time doing archaeological work.

Development of systematic study of African prehistory began in the mid-20th century when several European and American Scholars took up teaching and research positions in some African countries (Robertshaw 1990). This marked the beginning of prehistoric studies on the continent and the first archaeologists were inspired by developments that took place in their countries of origin and often employed models that had been developed there (Clark 1986, 1990; Shaw 1990; Shinnie 1990). It was during this phase that paradigms of African archaeology were dominated by theories of diffusion, which were later challenged by nationalists resulting in the revision of archaeological knowledge (Holl 1990:301). Until recently, little progress was made to diminish colonialist mentality surrounding the racist controversies which surrounded interpretations of Africa's past such as Great Zimbabwe, which according to early debate was part of prehistoric white colonization in South Africa (Ndoro 1997; Garlake 1973; Caton Thompson 1931). Today, the debate is firmly centered on achievements made by Africans resulting in new interpretations and demonstrating that Great Zimbabwe was of Bantu origin (Mufuka 1983; Garlake 1973, 1983, 1984). Research in the recent phases of African prehistory had not attracted Europeans because they assumed nothing significant had taken place in Africa prior to the arrival of Europeans (Trigger 1996:275; Trevor-Roper 1966:9). However, as interest in the study of the African past grew in the 1960s, it began to show significant advances and there was emphasis on technological and typological characteristics of Stone Age assemblages which made archaeology substantially different from the way it was studied in the earlier decades. There was a tremendous concern with typology, technology and terminology of artifacts and a systematic effort to standardize terms in an effort to enhance reporting and communication (Bishop and Clark 1967). Terms such as "Iron Age" and "Stone Age" and "Neolithic" were developed within a

European context but their continued applicability to the material evidence of Africa's past whose interpretation and study has continued to diversify and taken on a new meaning has been questioned (Sinclair 1988).

However, toward the end of the 1950s, few changes had begun to take place in stone tool classification since the adoption of the stone tool nomenclature proposed by Goodwin and Van Riet Lowe in the late 1920s (Goodwin and Van Riet Lowe 1929). The European nomenclature of Paleolithic, Mesolithic and Epipaleolithic which was being used alongside the African system was discarded except as a source of a few designations for tool-manufacturing techniques (Bishop and Clark 1967; Posnansky 1982:348). This paved the way for the introduction of African terminologies that best described African stone tool assemblages (Clark 1974; Leakey 1971; Isaac 1977). These typological changes in stone artifacts were reinforced by the principle established by Sahlins and Service (1960) that both cultural and biological evolutions move in two directions at the same time. This idea helped transform interpretation of changes in African prehistory that occurred during the process of cultural development from one stage to another. It was this realization that evolution should be considered to be responsible for both progression and diversity of characteristic adaptation that contributed to the fascinating interpretation of variability in stone tool industries at several sites on the African continent (Binford 1972b:26; Kleindienst 1961a, 1961b; Leakey 1971; Isaac 1977).

The changes taking place in classification were accompanied by one major development that drastically changed the culture historical approach to interpretation of the African past. The rejection and demolition of the pluvial-interpluvial climatic framework by geologists (Flint 1959; Cooke 1958) that was generally believed and used by archaeologists to correlate African ancient climates with glacial and interglacial periods in Europe resulted in Archaeologists re-orienting their thinking and interpretation of the past. The new approach to the study of African past was supported by ^{14}C dating that eventually revolutionized interpretation in African prehistory.

Emerging Issues

Following developments that occurred on the African continent concurrent with European powers taking control of the continent and the arrival of researchers, new pursuits began to emerge in the study of the activities of early hominids. Glynn Isaac, Mary Leakey, and others working in east Africa pioneered the development of what is termed, "Paleolithic living floor archaeology". This specialized branch of archaeological excavation was conducted at sites such as Olorgesailie (Isaac 1977) and Olduvai Gorge (Leakey 1971). These sites exhibited some of the most spectacular lithic assemblages on a hominid floor of which the African continent can boast. This new approach to doing archaeology offered challenges to scholars interested in the study of the past. The African continent began to receive unprecedented attention from international scientists leading to important discoveries and development of new methods and techniques. African archaeology began to benefit from outside influences as a result of these important discoveries. Holl (1990:304) has argued that the African continent has benefited from a plurality of research problems, methods and techniques arising from involvement in research by scholars from all over the world. The collaborative research and cooperation between African and foreign scholars has been a major development in African archaeology.

Looking back, there is strong temptation to argue that the basic framework for modern inquiry in archaeological research on the African continent began when New Archaeology became established as a paradigm (Binford 1962; Clarke 1968). Archaeologists saw the potential of this new approach to interpreting the human past and vigorously pursued it in their quest to study past human behavior. Lewis Binford and his colleagues articulated evolutionary views and proposed a number of changes that laid a new foundation for how archaeologists interpret and perceive the archaeological record. The new approach stressed the use of generalizations about the causative factors of cultural variability such as the environment (Binford 1972b). These views were in reaction to the culture-historical approach or traditional archaeology that had characterized the late and early 20th centuries (Trigger 1990, 1996). This paradigm shift was an interesting process that resulted not only in new developments in archaeological thought but laid the theoretical framework for developments that followed in the study of culture systems (Binford 1972b).

It is interesting to note that the approaches proposed by Binford and his colleagues to the study of human cultures created an unstoppable stimulus to the development of new methods and techniques in the discipline. As Sahlins and Service (1960) observed, the new approaches were accompanied by a serious consideration of correlating cultural evolution with biological evolution which subsequently made it possible to correlate cultural industries such as Oldowan and Acheulian with hominids. It was through their concept of universal evolution that archaeologists began to link the stages of evolution to human culture and recognize that fundamental principle of human development was indeed

associated with the different stages of cultural development. These new theoretical constructs of the 1960s and 1970s led to an increase in the level of influence on the interpretation of stone tool assemblages and development of cultural evolutionary ideas. Paleoanthropological research in Africa undoubtedly benefitted from these developments as Glynn Isaac and his colleagues set out to investigate early stages of proto-human behavior and ecology east of Lake Turkana in the latter part of the 1970s. Thus it can be argued that some interpretations of the archaeological record at major sites in Africa such as Olduvai Gorge (Leakey 1971), Kalambo Falls (Clark 1974), and Olorgesailie (Isaac 1977) and many others benefitted tremendously as understanding evolutionary processes became a foundation for explaining the complexity of early stages of cultural systems.

Lewis Binford and his colleagues not only rebelled against the paradigm of cultural history but they emphasized the importance of hypothesis testing and the use of scientific methods in archaeological research (Binford 1972b). Isaac (1972) was also a strong, early advocate of hypothesis testing and innovative methodologies in empirical research. Today, no one can ignore the influence of hypothesis testing, emphasis on the study of human societies as adaptive systems, use of computers for fieldwork and analysis and application of scientific reasoning in archaeological research as these played a significant role in the transformation of archaeology from its cultural historical approach to what became known as contemporary archaeology in the 1980s (Leone 1972).

The developments that occurred in the areas of method and theory made archaeology a solid discipline of scientific knowledge that occupies a special place among other disciplines. It was during this period when important publications in archaeology emerged that were exclusively devoted to elucidating some of the latest approaches in method and theory in the discipline. Some of these include Schiffer's *Behavioral Archaeology* (1976) and *Advances in Archaeological Method and Theory* (1978) and Binford's *For Theory Building in Archaeology* (1977) and *Nunamiut Ethnoarchaeology* (1978). The influence of Glynn Isaac on the discipline which came in the form of problem-oriented research was a catalyst for growth and development in method and theory and had tremendous impact on his students as could be seen in the kinds of questions that were being asked. Glynn Isaac and his students set out to investigate hominid activities dating to between one and two million years east Lake Turkana and these investigations were of particular significance to the developments that were taking place in the discipline of archaeology in several ways. First his investigations were problem-oriented. Second, he focused his attention on building on the multidisciplinary research efforts that began in the early 1970s, which resulted in the publication of important monographs (Coppens et al. 1976; M. G. and R. E. Leakey 1978). Third, he paid considerable attention to the formulation of specific problems and hypotheses in the study of proto-human behavior and ecology. His research had the characteristic of developing specific problems and hypotheses that enabled his research teams to focus on those aspects that were deemed essential to the overall investigation. For example, his research east of Lake Turkana, in which I had the privilege to participate in 1979, considered stratigraphy and paleogeography, site formation processes, distribution patterns of artifacts and bones, interpretation of stone artifacts and intra-site spatial patterning and past vegetation patterns. He developed the idea of home base or central place as a

temporary dwelling place where animal and plant foods were shared by members of a social group (Isaac 1978). This idea resulted in the rise of variants that emphasize several aspects of hominid study that were hitherto either unknown or less considered. Archaeologists began to look at hominid behaviors such as butchery practices, meat-sharing, and carcass processing that was ancillary to hunting and closely associated behaviors as important components in the study of the human past.

The Koobi Fora Research Project (KFRP), which began in 1968 and was co-directed by Richard Leakey and Glynn Isaac, generated exciting ideas and research avenues in the study of hominid behavior. Glynn Isaac and his colleagues used hypothesis testing as an important tool in their quest to gain broader understanding of early hominid life and the adaptation of early hominids to their environments. There are important reconstructions using hypothesis testing that have emerged from the KFRP that have considered aspects such as food sharing (Isaac 1978, 1983), meat eating, hunting and scavenging (Bunn 1981), plant exploitation (Sept 1984), stone tool use and lithic microwear analysis (Keeley and Toth 1981; Toth 1985), taphonomy (Behrensmeyer 1975; Schick 1984; Gifford 1980) and spatial configurations of artifacts and bones (Kroll 1984; Kroll and Isaac 1984) and many others. These are important studies leading to our present understanding of early hominid life. They heralded a shift from the way archaeology was practiced from making implicit interpretations based on recovered archaeological materials and the nature of sites to an emphasis on hypothesis testing. Let me cite a few examples that highlight some of the emergent issues associated with the KFRP. For example, excavations at Olduvai Gorge (Leakey 1971) had revealed concentrated areas where stone artifacts and broken up bones were found asso-ciated together and these were widely interpret-ed to represent living floors. However, when excavations at the FXJj50 site at Koobi Fora revealed similar patterning, Isaac and his team decided to find out how these concentrations of stone artifacts and broken bones were formed (Bunn et al. 1980). They began to ask questions in an attempt to seek solutions to a bewildering array of propositions. They considered an earlier model presented by anthropologists in the 1960s that proposed big game hunting as the primary catalyst for human evolution based on the subsistence practices of contemporary hunter-gatherers (Lee and Devore 1968). The idea of large-game hunting had indeed gained a lot of popularity leading to coining "the hunting hypothesis" by scholars who saw the association of humans with big game hunting as offering the most plausible explanation for tool use, lan-guage, sexual division of labor, food sharing, intelligence, and development of the brain (Ardrey 1976; Washburn and Lancaster 1968). The popular view was that the association of stone artifacts and broken pieces of animal bones suggested that ancient man possessed cognitive abilities and sophisticated cooperation that enhanced hunting (Washburn and Lancaster 1968). The work of Bunn (1981) can therefore be considered as re-evaluation of the hunting hypothesis and earlier explanations of the presence of animal bones at archaeological sites and their meaning to hominid studies. He focused his attention on animal bones at two localities in east Africa, Olduvai Gorge and Koobi Fora in an attempt to elucidate when in the course of human evolution meat became an important component of human diet. He con-cluded that there was indeed direct evidence between man's activities and the bones left behind between two and 1.5 million years ago.

However, there have been other important stud-ies looking at competing hypotheses. Shipman (1986) has articulated the scavenging hypothe-sis based on evidence from Olduvai Gorge Bed I sites and is strongly of the opinion that scaveng-ing and not hunting was the major meat pro-curement strategy of early hominids and that it was an important behavior at Olduvai Sites. This aspect of study has witnessed several advances leading to expansion of the focus of hominid subsistence from hunting to scavenging and other forms of meat acquisition and distribution. The use of ethnoarchaeological data in the inter-pretation of archaeological data has been anoth-er significant development in hominid interpre-tation (Agorsah 1990)

Similarly, the importance of plant foods in the diet of early hominid has entered discussions among archaeologists since scholars such as Dart (1949) and Ardrey (1961, 1976) began to draw relationships between diet and the devel-opment of early humans. These early discus-sions on plant foods persuaded archaeologists such as L. S. B. Leakey (1963), M. D. Leakey (1971), Clark (1970, 1972), and Isaac (1971) to begin to examine archaeological evidence in relation to human diet. Isaac (1982:164) has even argued that while hunting and scavenging led to a critical adaptive shift towards social organization that involved some co-operative economic effort and partial division of labor, gathering of plant foods was an important com-ponent of early man's foraging strategies. Subsequently, Sept (1984) reinforced the impor-tance of plants to early man's diet.

The significance of hypothesis testing to inter-pretation of archaeological records is best exem-plified in its ability to provide broad understand-ing of early hominid life and the adaptation of early hominids to their environments. There are exciting results from eastern Africa where hypothesis testing has contributed to a wide range of complex studies including the interpretation of hominid footprints at Laetoli, Tanzania. When White and Suwa (1987) examined hypotheses advanced by Tuttle (1984, 1985) and Stern and Susman (1983) which contradicted earlier interpretations, they used hypothetical testing to conclude that the Laetoli footprints were made by *Australopithecus afarensis* and not a different hominid species. My Berkeley training enabled me to use hypothesis testing during my archaeological investigations in the Muchinga escarpment of Zambia (Musonda 1983) and to begin to place more emphasis on post-Pleistocene subsistence systems and social organization in terms of adaptive systems and less on typological considerations. In this study, I was not primarily concerned with the physical activities of the ancient inhabitants of the area but to sought to understand the co-existence of the prehistoric communities that practiced differing and varied economies (Musonda 1987:155). It seems to me that the presence of potsherds in the upper layers of Later Stone Age sequences at several sites in the Muchinga escarpment and many other sites in similar situations in the region may be a reflection of a complex food-procurement strategy that involved collection of potsherds by hunter-gatherers in a random fashion with little or no contact with Iron Age Communities. This line of inquiry presupposes that there was an element of attraction of those elements that were encountered on the landscape and subsequently transported to the home bases and that the differences in the levels of adaptation to the environment did little to bring them together. These were behavior patterns that constituted a continuation of the old mode of subsistence that depended on hunting and gathering even after the arrival of food producers in the area. The importance of this approach enabled us not only

to analyze artifacts as reflections of human behavior, but as tests for hypothetical reasoning. With Glynn as my mentor, I began to view evidence differently and was able to generate a set of predictions about the nature of contact between the different communities and how the new populations occupied and utilized ecological zones without coming in direct conflict or competition with the hunter-gatherers.

It is through hindsight that one sees the positive link between the KFRP which enabled Glynn Isaac to develop some of his earlier thoughts regarding concentrations of stone artifacts and bones and to attempt and understand what principal agents may have been responsible in their transportation and deposition and the subsequent studies that have added a new dimension to interpretations of past hominid activities. Undoubtedly, the work conducted east of Lake Turkana generated a lot of hypotheses on aspects of technology and social-economic organization of early hominids that opened up new insights into the study of ancient humans.

Conclusion

The paper has highlighted some of the major developments in African archaeology that were largely influenced by Glynn Isaac's approach to the study of the past since the beginning of the 1960s. Isaac was at the core of important studies conducted at Koobi Fora on a wide range of topics including stone tool assemblages, bone assemblages, and processes of bone accumulations, which have undoubtedly shed light on aspects of socioeconomic and technological organization of early hominids that are important to our understanding of human development. These new pursuits have created an interesting insight into new debates that have characterized the last 50 years. While Lewis Binford and his colleagues may be credited for initiating a change

in the approaches and methods of doing archaeology, it was field investigations such as those
conducted east of Lake Turkana that have pro-
pelled our understanding of the technology and
socio-economic organization of early hominids.
It is on the basis of work done at Koobi Fora and
the conclusions drawn from the myriad studies
conducted by Isaac and his colleagues that one
can conclude that the methods and theory of the
last fifty years enabled archeologists to acquire
objective knowledge of the past and contribute
to interpretations of hominid activities.

Investigations east of Lake Turkana have led
to a dramatic shift in the way artifacts are
viewed as sources of information about prehistoric human behavior leading to development of
new classification systems. While artifacts have
remained central to the study of early humans,
emphasis on how they are studied has been
steadily changing. Isaac (1982:158) has
observed that artifacts remain central ingredi-
ents of evidence because "it is through the presence of artifacts that one recognizes the places
where early humans were particularly active
and it is through the presence of artifacts that
archaeologists can distinguish prehistoric food-
refuse from natural materials". He further
argues that "artifacts preserve in their form evidence regarding their practical function and
regarding the culturally determined habits and
preferences of their makers". The underlying
principle here is that artifacts did not only play
a role as sources of information on human and
cultural development but can also be used to
understand archaeological contexts, patterning
of stone artifacts and human behavioral activities. His emphasis on excavated material has
contributed to the development of African

archaeology and has made scholars revisit
archaeological sites in order to study them using
new techniques (Potts et al. 1999).

However, despite all the positive developments that have been made in archaeological
method and theory, interpretations, and data
analysis, not every archaeologist working in
Africa has discarded mere descriptions of
archaeological material and other aspects of traditional study such as ecological and environmental studies. This remains a well appreciated
approach as evidenced in archaeological reports
in journals like *South African Archaeological
Bulletin*, *AZANIA*, and *African Archaeological
Review*. But an important outcome of developments during Glynn Isaac's career is strong
emphasis on inter-disciplinary approaches to
seeking solutions to problems. We have seen
scholars from a wide range of disciplines sharing
knowledge with archaeologists to illuminate the
solutions for the African past. The researches
undertaken in east Africa have revealed the
importance of sharing information which has
the potential of exploiting a wide range of data in
a coordinated fashion. Such an approach has
also enabled individuals to pursue specialization
in areas of particular interest since very few people can claim a high level of competence in all
branches of archaeology. There is no doubt that
Glynn Isaac's approach to doing archaeology has
enabled them to develop good working methods,
analysis, and interpretation and an excellent
grasp of the subject matter. Looking back to the
achievements made since the time Glynn Isaac
stepped on the archaeology scene, one is tempt-
ed to delight at the emergent issues in African
archaeology that continue to make it a fascinating and challenging discipline.

References

Ardrey, R.
1961 *African Genesis.* London, Atheneum.

1976 *The Hunting Hypothesis.* New York.

Agorsah, E. K.
1990 Ethnoarchaeology: The search for a self-corrective approach to the study of past human behaviour. *The African Archaeological Review* 8:189-208.

Behrensmeyer, A. K.
1975 The taphonomy and paleo-ecology of Pliopleistocene vertebrate assemblages east of Lake Rudolf, Kenya. *Bulletin of the Museum of Comparative Zoology at Harvard University* 146(10):473-578.

Binford, L. R.
1962 Archaeology as anthropology. *American Antiquity* 28(2):217-225.

1972a Contemporary model building: Paradigms and the current state of Paleolithic research. In *Models in Archaeology*, edited by D. L. Clarke, pp. 109-166. Methuen, London.

1972b *An Archaeological Perspective.* Seminar Press, New York.

Binford, L. R. (Editor)
1977 *For Theory Building in Archaeology: Essay on Faunal Remains, Aquatic Resources, Spatial Analysis, and Systemic Building.* Academic Press, New York.

1978 *Nunamiut Ethnoarchaeology.* Academic Press, New York.

Binford, S. R., and L. R. Binford (Editors)
1968 *New Perspectives in Archaeology.* Aldine Publishing Company, Chicago.

Bishop, W. W., and J. D. Clark (Editors)
1967 *Background to Evolution in Africa.* University of Chicago Press, Chicago.

Bunn, H. T.
1981 Archaeological evidence for meat-eating by Plio-Pleistocene hominids from Koobi Fora and Olduvai Gorge. *Nature* 291:544-547.

Bunn, H., J. W. K. Harris, G. Isaac, Z. Kaufulu, E. Kroll, K. Schick, N. Toth, and A. K. Behrensmeyer
1980 FXJj50: An early Pleistocene site in northern Kenya. *World Archaeology* 12(2):109-136.

Caton Thompson, G.

1931 *The Zimbabwe Culture. Ruins and Reactions.* Clarendon Press, Oxford.

Clark, J. D.

1970 *The Prehistory of Africa.* Praeger, New York.

1972 Palaeolithic butchery practices. In *Man, Settlement and Urbanism,* edited by P. J. Ucko, R. Tringham, and G. W. Dimbleby, pp. 149-156,

1974 *The Kalambo Falls Prehistoric Site, Volume II.* Cambridge University Press, Cambridge.

1986 Archaeological retrospect. *Antiquity* LX:179-188

1990 A personal memoir. In *A History of African Archaeology,* edited by P. Robertshaw, pp. 189-204. James Currey, London.

Clarke, D. L.

1968 *Analytical Archaeology.* Methuen, London.

Cooke, H. B. S.

1958 Observations relating to Quaternary environments in East and Southern Africa. *Transactions of the Geological Society of South Africa (Annex)* 61:1-73.

Coppens, Y., F. C. Howell, G. L. Isaac, and R. E. F. Leakey (Editors)

1976 *Earliest Man and Environments in the Lake Rudolf Basin.* University of Chicago Press, Chicago.

Dart, R. A.

1949 The predatory implemental technique of Australopithecus. *American Journal of Physical Anthropology* 7:1-38.

Flint, R. F.

1959 Pleistocene climates in Eastern and Southern Africa. *Bulletin of the Geological Society of America* 70:343-374.

Garlake, P

1973 *Great Zimbabwe.* Thames and Hudson, London

1982 *Great Zimbabwe Described and Explained.* Zimbabwe Publishing House, Harare.

1983 Prehistory and Ideology in Zimbabwe. In *Past and Present in Zimbabwe,* edited by J. D. Peel and T. O. Ranger pp. 1-19, Manchester University Press, Manchester.

1984 Ken Mufuka and Great Zimbabwe. *Antiquity* 58:121-123.

Gifford, D. P.

1980 Ethnoarchaeological contributions to the taphonomy of human sites. In *Fossils in the Making, Vertebrate Taphonomy and Paleoecology*, edited by A. K. Behrensmeyer and A. P. Hill, pp. 93-106. University of Chicago Press, Chicago.

Goodwin, A. J. H., and C. Van Riet Lowe.

1929 The Stone Age cultures of South Africa. *Annals of the South African Museum* 27:1-289.

Holl, A.

1990 West African archaeology: Colonialism and nationalism. In *A History of African Archaeology*, edited by P. Robertshaw, pp. 296-308. James Currey, London.

Isaac, G. Ll.

1971 The diet of early man: Aspects of archaeological evidence from lower and middle Pleistocene sites in Africa. *World Archaeology* 2:278-299.

1972 Early phases of human behaviour: Models in Lower Palaeolithic archaeology. In *Models in Archaeology*, edited by D. L. Clarke, pp. 167-199, London Methuen.

1977 *Olorgesailie: The archaeology of a Middle Pleistocene lake basin in Kenya*. University of Chicago Press, Chicago.

1978 The Food-sharing behaviour of protohuman hominids. *Scientific American* 238:90-108

1982 The earliest archaeological traces. In *The Cambridge History of Africa, Volume I From the Earliest Times to C. 500 B.C.*, edited by J. Desmond Clark, pp. 157-247. Cambridge University Press, Cambridge.

1983 Bones in contention: Competing explanations for the juxtaposition of early Pleistocene artifacts and faunal remains. In *Animals and Archaeology: Hunters and their Prey*, edited by J. Clutton-Brock and C. Grigson, pp. 3-19. British Archaeological Reports.

Keeley, L., and N. Toth

1981 Microwear polishes on early stone tools from Koobi Fora. *Nature* 293:464-465

Kleindienst, M. R.

1961a Variability within the Late Acheulian assemblage in Eastern Africa. *South African Archaeological Bulletin* 16:35-52.

1961b Components of the east African Acheulian assemblage: An analytical approach. *Acts of the 4th Pan African Congress of Prehistory*, pp. 81-111.

Kroll, E. M.

1984 The anthropological meaning of spatial configurations at Plio-pleistocene archaeological sites in East Africa. Ph.D. thesis, Department of Anthropology, University of California, Berkeley.

Kroll, E. M., and G. L. Isaac

1984 Configurations of artifacts and bones at Early Pleistocene sites in East Africa. In *Intrasite Spatial Analysis in Archaeology*, edited by H. J. Hietala, pp. 4-30. Cambridge University Press, Cambridge.

Leakey, L. S. B.

1931 *The Stone Age Cultures of Kenya Colony*. The University Press, Cambridge.

Leakey, M. D.

1971 *Olduvai Gorge, Volume III: Excavations in Beds I and II, 1960-1963*. Cambridge University Press, Cambridge.

Leakey, M. G., and R. E Leakey (Editors)

1978 *Koobi Fora Research Project, Volume I: The Fossil Hominids and an Introduction to their context, 1968-1974*. Clarendon Press, Oxford.

Leakey, L. S. B.

1963 Very early East African hominidae, and their ecological setting. In *African Ecology and Human Evolution*, edited by F. C. Howell and F. Bourliere, pp. 448-457. Chicago University Press, Chicago.

Lee, R. B., and I. De Vore (Editors)

1968 *Man the Hunter*. Aldine, Atherton, Chicago.

Leone, M. P. (Editor)

1972 *Contemporary Archaeology. A Guide to Theory and Contributions*. Southern Illinois University Press.

Mufuka, K.

1983 *Dzimbabwe: Life and Politics in the Golden Age 1100-1500 AD*. Harare.

Musonda, F. B.

1983 Aspects of the Prehistory of the Lunsemfwa Drainage Basin, Zambia, during the last 20,000 years. Ph.D. Dissertation, University of California, Berkeley.

1987 The significance of pottery in Zambian Later Stone Age contexts. *The African Archaeological Review* 5:147-158.

1990 African Archaeology: Looking forward. *The African Archaeological Review* 8:3-22.

Ndoro, W.

1997 Great Zimbabwe. *Scientific America* November:94-99.

Posnansky, M.

1982 African Archaeology comes of age. *World Archaeology* 13:345-358.

Potts, R., A. K. Behrensmeyer, and P. Ditchfield

1999 Paleolandscape variation and Early Pleistocene hominid activities: Members 1 and 7, Olorgesailie Formation, Kenya. *Journal of Human Evolution* 37:747-788.

Robertshaw, P. (Editor)

1990 *A History of African Archaeology*. James Currey, London.

Sahlins, M. D., and E. R. Service (Editors)

1960 *Evolution and Culture*. University of Michigan Press, Ann Arbor, Michigan.

Schick, K.

1984 Towards the interpretation of early archaeological sites: Experiments in site formation processes. Ph.D. Dissertation, University of California, Berkeley.

Schiffer, M. B.

1976 *Behavioral Archaeology*. Academic Press, New York.

Schiffer, M. B. (Editor)

1978 *Advance in Archaeological Method and Theory, Vol. I*. Academic Press, New York.

Sept, J.

1984 Plants and early hominids in East Africa: A study of vegetation in situations comparable to early archaeological locations. Ph.D. Dissertation, University of California, Berkeley.

Shaw, T.

1990 A personal memoir. In *A History of African Archaeology*, edited by P. Robertshaw, pp. 205-220. James Currey, London.

Shinnie, P.

1990 A personal memoir. In *A History of African Archaeology*, edited by P. Robertshaw, pp. 221-235. James Currey, London.

Shipman, P.

1986 Studies of hominid-faunal interactions at Olduvai Gorge. *Journal of Human Evolution* 15:691-706.

Sinchair, P. J. J.
1988 *The Mombasa Specialist Workshop 1st-6th August, 1988.* Swedish Central Board of National Antiquities, Stockholm.

Stern, J. T., and Susman, R. L.
1983 The locomotor anatomy of *Australopithecus afarensis. American Journal of Physical Anthropology* 60:279-317.

Toth, N.
1985 The Oldowan reassessed: A close look at early stone artifacts. *Journal of Archaeological Science* 12:101-120.

Trevor-Roper, H. R.
1966 *The Rise of Christian Europe.* Second Edition. Thames and Hudson, London.

Trigger, B. G.
1990 The history of African archaeology in world perspective. In *A History of African Archaeology,* edited by P. Robertshaw, pp. 309-319.

1996 *A History of Archaeological Thought. Second Edition.* Cambridge University Press, Cambridge, pp. 211-313.

Tuttle, R. H.
1984 Bear facts and Laetoli impressions. *American Journal of Physical Anthropology* 63:230.

1985 Ape footprints and Laetoli impressions: A response to the SUNY Claims. In *Hominid Evolution: Past, Present and Future,* edited by P. V. Tobias, pp. 129-133. Alan R. Liss, New York.

Ucko, P. J.
1993 Foreword. In *The Archaeology of Africa; Food, Metals and Town* edited by T. Shaw, P. Sinclaire, B. Andah and A. Okpoko. Routledge, London.

Washburn, S. L., and C. S. Lancaster
1968 The evolution of hunting. In *Man the Hunter,* edited by R. B. Lee and I. De Vore, pp. 293-303. Aldine. Atherton, Chicago.

White, T. D., and G. Suwa
1987 Hominid footprints at Laetoli: Facts and interpretations. *American Journal of Physical Anthropology* 72:485-514.

15

The Tortoise and the Ostrich Egg: Projecting the Home Base Hypothesis into the 21st Century

Brian A. Stewart, John Parkington, and John W. Fisher, Jr.

Introduction

In the late 1970s Glynn Isaac was promoting a home base with food-sharing model to understand the residual sites of the Plio-Pleistocene of east Africa. The debate was around the competencies of our early ancestors and how these could be accessed through imaginative manipulation of the archaeological record (Binford 1981, 1985; Blumenschine 1986; Bunn 1981; Isaac 1978a, 1978b, 1984; Potts 1984, 1988; Sept 1982; Shipman 1983, 1984). The contemporary florescence of ethnographic documents on hunter-gatherer sites and spatial layouts was an important inspirational platform (Binford 1978, 1983; Gould 1978, 1980; O'Connell 1977, 1987; Yellen 1976, 1977). Many felt that it was a brave, not to say dangerous, leap backward from the 20th century Kalahari to the 2 million-year-old Turkana Basin. But we share with Glynn and his students the belief that it is the implementation, not the source, of an idea that matters. We support the view that 'eureka' can happen anywhere but it is the excavated record that has the final say. We applaud his determination to integrate insights from experiments, from ethnographies, from excavations and from innovative thinking into reconstructions of what might have happened in the past.

Our feeling is that an important strut in this kind of imagining of the distant past comes from an application in the recent past where preservation is usually very good, where analogous behaviors are more reliably postulated and where observations, including aspects of site structure, are more densely packed. Kroll and Isaac (1984) explored this possibility in more detailed mapping of the spatial patterning of artifacts, features and 'site furniture' (Binford 1983), showing the potential for reading the social from the spatial. Dunefield Midden or DFM, a very late pre-colonial Stone Age site in the Western Cape Province of South Africa provides just such an opportunity. Because the most persuasive and pervasive models are derived from Kalahari hunter-gatherer observations, this southern African forager behavioral residue should be amenable to the kind of inspired reconstruction Isaac favored. Here we look at the recognition of home base structure and sharing through the lens of a large, carefully excavated site with the preservation of a wide range of material evidence.

We have chosen to focus on tortoise bones and ostrich eggshell fragments because, in addition to being very common and almost always precisely mapped at DFM, they reflect interesting trajectories of objects brought to the site as food items, partly discarded and partly recycled as artifacts, curated, likely shared among families and ultimately abandoned. As we have shown elsewhere (Fisher and Strickland 1989, 1991), even briefly occupied hunter-gatherer camps have a life history that compounds the difficulties of reading behavior from spatially mapped residues. We can try to follow these use-trajectories through the brief time-span of the DFM occupation(s). We emphasize that DFM differs in significant ways from ethnographic Kalahari San sites: DFM's coastal setting and abundance of shellfish, a greater variety of thermal feature types (hearths, roasting pits, stone-packed hearths, etc.) than seems to be the case at Kalahari sites, a profusion of seal bones, and so forth. We look to the Kalahari San for hypotheses but not for direct analogues.

Below, we present the spatial patterns of tortoise and ostrich eggshell remains at DFM, and investigate the diverse use-histories and depositional trajectories that underpinned them. We then explore the fluid boundary between subsistence remains and technological remains by drawing comparisons between tortoises, ostrich eggshells and ceramic vessels at the site. We hope to showcase the importance of using relatively recent, exemplar home bases such as DFM to inform spatial interpretations of more ancient home bases like those which Glynn worked so energetically to understand.

Dunefield Midden: Site Overview

DFM is a series of very briefly occupied and marginally overlapping campsites located on the Atlantic coast at Elands Bay in the Western Cape Province, South Africa. It lies about 2 km north of the mouth of the Verlorenvlei River and some 600 m inland from the shoreline (Figure 15.1). The rich archaeological deposits contain just under 2,500 kg of shellfish remains, some 18,287 stone artifacts, in excess of 10,000 identified bones and teeth of a variety of mammals and other vertebrates, 796 ostrich eggshell fragments, 408 ostrich eggshell beads, 1,011 ceramic sherds, and other materials (Orton 2002; Parkington et al. 1992, 2009; Parkington and Fisher 2006; Stewart 2005a, 2005b, 2008, 2010, 2011; Stynder 2008; Tonner 2005). The majority of the occupation of DFM took place during a span of about a century between AD

1300-1400. A set of 27 radiocarbon dates from DFM on charcoal and shell has a narrow temporal distribution, with an average of the 27 calibrated dates being AD 1354 (Parkington et al. 2009; Tonner 2005).

Our excavations exposed 859 m² of shallow living debris with many different ashy features and highly variable distributions of both artifacts and foodwaste. The location of all materials is known to at least its 1 m², and many items were piece plotted. Various lines of evidence support our belief that the spatial distribution of materials retains a high degree of integrity (Parkington et al. 2009). We have previously defined hearths, roasting pits, processing fires and other facilities (Figure 15.2a), demarcated domestic zones and dump zones and suggested different roles for different hearths (Parkington 2006; Parkington et al. 2009; Stewart 2008; Tonner 2005). Much of the site structure has survived and provides the spatial framework for looking at activities such as food consumption, sharing and waste disposal (Jakavula 1995; Nilsson 1989; Reeler 1992; Stewart 2008; Tonner 2002, 2005).

We discern a probable domestic area and a 'main dump' at DFM, which lie roughly parallel to one another on a northwest-southeast axis (Figure 15.2b). The domestic area is delineated by a linear, arcuate arrangement of hearths that extends from the site's northern extremity southeastward toward the southeastern extremity. The main dump is situated to the west of the domestic area and consists of a dense concentration of shell, animal bones and artifacts. A number of hearths and many other ashy features also occur in the dump area; some of these might have been created early in the site occupation and subsequently were covered with dumped materials. Several small concentrations of shell and other materials – 'satellite dumps' – are situated near or east of some domestic hearths in the northern

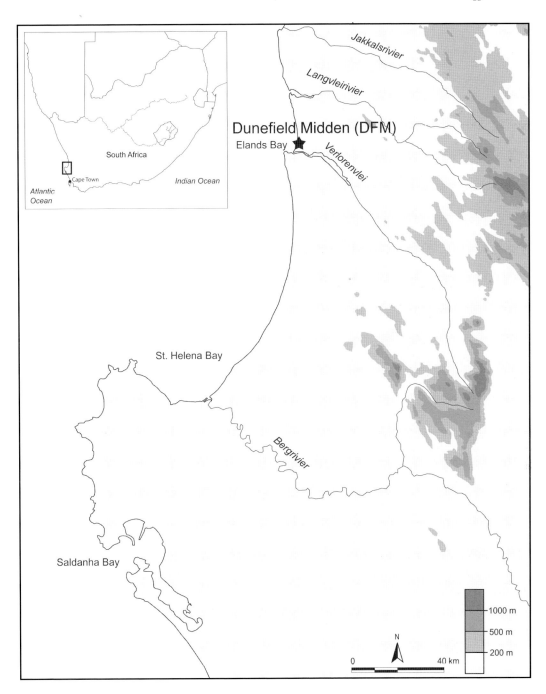

Figure 15.1 Map of South Africa's west coast with Elands Bay and location of Dunefield Midden (DFM).

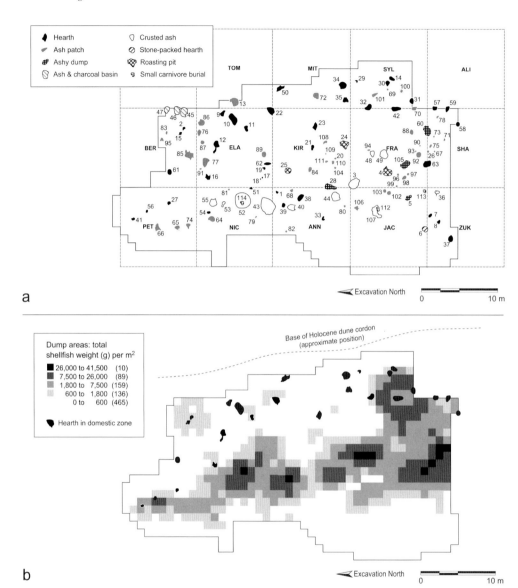

a

b

Figure 15.2 Maps of DFM showing (a) the positions of the various feature types, and
(b) the position of the eastern domestic zone of hearths in relation to the main dump to the west.

half of the site (Parkington et al. 2009; Stewart 2008; Tonner 2005; Figure 15.2b).

Tortoises

More than 7,500 dispersed fragmentary and complete bones of angulate tortoise (*Chersina angulata*) have been recorded at DFM. The assemblage includes some 5,140 carapace specimens, 1,500 plastron specimens and some 920 non-shell skeletal elements (limbs, axial elements and crania). This species is extremely common in the near coastal sandveld surrounding the site. From indications of charred plates on both plastron and carapace bones, it is clear that tortoises were picked up, brought back to camp and cooked by simply placing the living animal on the coals of a fire (see Silberbauer 1981:216; Yellen and Lee 1976:41). The tortoise effectively cooks in its own juices. From the size of the bones all animals were fairly small, offering much less than 1 kg of edible material.

Carapaces often were made into bowls by simply removing the marginal plates and smoothing the lateral sutures of the costal plates. Just over 600 carapace bowl parts were recovered from DFM, which refit to form a minimum of 30 bowls and probably closer to 50. Adult tortoise carapaces were typically chosen for bowls, but some were made from (occasionally very young) juvenile individuals, raising the possibility that some functioned as spoons or compact bowls for special tasks such as preparing arrow poison (Schapera 1930:94, 132, 144). Adult or juvenile specimens may have also been used as ladles (Silberbauer 1981:72). The anteriorly furrowed nuchal plate was always left intact and, whether these bowls were used in food consumption, preparation or both, may have acted as a spout for drinking or pouring. Unmodified carapaces may sometimes also have been used as bowls (Lee 1979:152).

All tortoise bones, including carapace, plastron and non-shell skeletal elements, are more numerous in the high density shell zone that we regard as a secondary dump area (the 'main dump'; carapace: 50.2%, n = 2,249; plastron: 69.1% n = 1,106; non-shell skeletal: 59.7%, n = 610) than in the hearth-dominated eastern area we interpret as domestic space. All also show relatively homogenous distributions throughout the main dump, occurring more frequently near dump features than other feature types. In the domestic area, however, the relative frequency and spatial configurations of each anatomical category varies tremendously. We begin with non-shell skeletal elements then move on to plastron and carapace parts. Skeletal elements in the domestic zone are predominantly distributed in and around hearths: 49% (n = 202) are within 1 m of hearths. Limb bones and other non-shell bones of angulate tortoise are small and after consumption would have become easily buried through trampling or other vertical displacement in DFM's unconsolidated sand substrate (Gifford-Gonzalez et al. 1985). Their hearth-centered domestic distribution thus presumably reflects primary discard (Schiffer 1987) - a drop or drool zone in Binford's (1983) sense (Figure 15.3a).

Plastron parts occur in the domestic area, but they are less densely concentrated than non-shell skeletal elements in and near hearths. Rather, plastrons show a more patchy distribution in this area, with tight clusters near ashy features we interpret as components of satellite dumps that serviced the domestic hearths (Figure 15.3b). Plastrons have no further use after consumption and thus, like skeletal bones, their distribution presumably also strictly reflects discard. However, plastron plates are generally much bulkier than non-shell bones and their upright hypoplastron bridges could

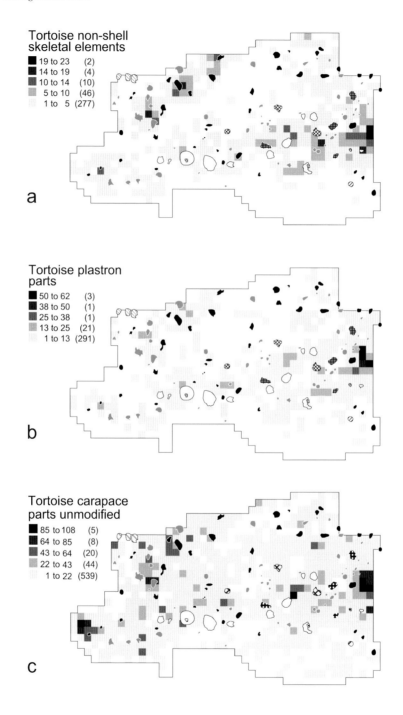

Figure 15.3 Spatial distributions at DFM of different categories of (a-c) tortoise foodwaste, (d) carapace bowl parts, and (e) carapace bowl refits in relation to the various feature types.

Tortoise carapace
bowl parts

■ 23 to 28   (4)
■ 17 to 23   (2)
■ 12 to 17   (8)
▦ 6 to 12 (20)
▫ 1 to 6 (87)

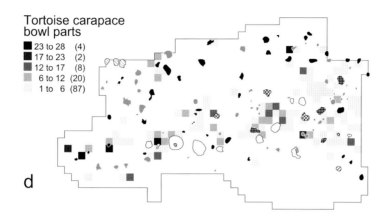

d

Tortoise carapace
bowl refits

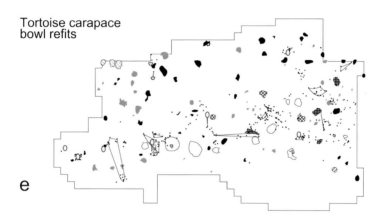

e

◗ Hearth	⊘ Stone-packed hearth
🝆 Ash patch	⬡ Roasting pit
▦ Ashy dump	⬡ Small carnivore burial
⬦ Ash & charcoal basin	○ Carapace bowl (>90% whole)
◔ Crusted ash	▽ Carapace bowl (>50% whole)

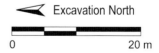

Excavation North

0   20 m

potentially cause injury if left dorsal side-up in zones with high foot traffic. Their close association with ashy patches in satellite dumps therefore likely results from efforts to clear these plates from immediate domestic spaces where tortoises were typically cooked and consumed, indicating either a toss zone (cf. Binford 1983) or a localized secondary refuse aggregate (Murray 1980; Wilson 1994).

Carapace parts are far more widespread in both the domestic and dump areas than either non-shell skeletal elements or plastron parts, occurring at some frequency in most squares at the site. In the domestic area, some very high density concentrations of carapace (>50 specimens/m^2) exist in/near both hearths and ashy features of satellite dumps (Figure 15.3c). Indeed, comparable frequencies of carapace parts located outside of the main dump are situated 1 m from hearths and ashy patches. Carapace plates are more brittle than non-shell bones and plastron plates as shown by their high rate of fragmentation. This, along with the possibility that fragmented carapace plates were often trampled into the sand substrate (like non-shell bones), may help explain their ubiquity across the site. The most localized distribution is that of carapace bowl parts (Figure 15.3d). Bowls are skewed heavily towards the main dump (65.4%, n = 398) in proportions similar to that of plastron parts. The few bowls in the domestic area exhibit spatial patterns that diverge markedly from those of each food-waste category. Rather than diffuse distributions with localized patches of higher density in/near ashy patches and/or hearths, bowl parts in the domestic zone almost exclusively form several dense (>20 specimens/m^2) clusters concentrated next to several hearths (Features 9, 27, 32, 41 and 56). Indeed, 70.6% (n = 149) of bowls in the domestic area were

recovered within 1 m of hearths, whereas only 0.92% (n = 2) were situated within the same range of ashy patches. These domestic hearth-centered specimens show some of the tightest spatial clustering of any carapace bowl refits, and include the only complete bowl and half of the nearly-complete bowls at DFM (Figure 15.3e). Such low dispersion of carapace bowl parts supports primary discard, perhaps at a late enough occupational stage to preclude secondary clean-up and disposal onto the dump, or perhaps the tight clustering reflects caching in anticipation of future visits.

The overall distribution of carapace bowls is especially striking relative to unmodified carapace plates and non-shell skeletal elements, both of which are far more diffusely scattered throughout DFM (compare Figures 15.3a, c–d). This is particularly true of the unmodified carapace assemblage. Non-shell skeletal elements, although not much more abundant than bowl carapace plates, also show a wider distribution across the site and occur, at some frequency, in and near most hearths in the domestic area. By contrast, bowls are concentrated in isolated, high density clusters that in the domestic zone take the form of tightly refitting 'puddles' near a very limited number of hearths. We suggest that this is a direct reflection of differences in the life histories of various tortoise body parts, with small disposable non-shell skeletal elements casually discarded in and around domestic hearths, bulkier carapace and plastron plates typically taken further afield to satellite or main dumps, and useful carapaces modified into bowls, curated, cached and more formally discarded after use.

Ostrich Eggshells

There are just under 800 mapped fragments of ostrich eggs at DFM, of which at least 70% have so far been refitted, allowing us to estimate the existence of four to seven once-complete eggs. The refitted eggs had all been used as water flasks, recognizable from the perforations and rounded rim pieces. It is almost certain that more flasks were in use than we can detect (they simply did not break), and that flasks were transported and cached across the landscape for future use. Only 2 km away we have excavated a small portion of a site remarkably similar to DFM with a cache of five unbroken eggs (Jerardino et al. 2009; Parkington 2006). It is possible that ostriches were not common in this part of the sandveld and that eggs were systematically brought in as flasks and regularly buried as a source of fresh water. These caches are nearly impossible to detect in the sandy landscape and are probably far more common than we realize.

Eggshell fragments are, by comparison with other artifacts and foodwaste, rather restricted in their distribution. They are far more common in the northern half of the site than in the south (Figure 15.4a); in the latter they form small patches that often refit with larger clusters in the north (the exception is one much larger, diffuse concentration near the southern hearth Feature 63). Most ostrich eggshell fragments were recovered from the main dump (71.3%, n = 561). This proportion is higher than any of the tortoise categories discussed above, but not far off from plastron (69.1%, n = 1,106) and bowl parts (65.4%, n = 398). In the main dump, ostrich eggshell fragments occur in three major concentrations in the north part of the site and a fourth in the south (near Feature 63). They associate with highest frequency near dump features, followed by processing features and then hearths. The relatively high frequency near processing features is something eggshell fragments share with tortoise carapace bowls. The behaviors responsible for these associations are obscure

since it is difficult to envisage ostrich eggshells being used in food processing, unless, of course, it was the eggs themselves being prepared. Among Kalahari groups, however, ostrich eggshells are never broken during consumption, but rather the egg contents are accessed through the same perforation that later serves as the flask spout (Lee 1979; Silberbauer 1981).

In the domestic area, the distribution of ostrich eggshell fragments recalls tortoise carapace bowls in that relatively large concentrations cluster around a few specific hearths (Features 10, 11 and 22), with other hearths hosting one or two isolated fragments (Figure 15.4a). Thirty one percent (n = 70) of eggshell fragments outside the main dump are within 1 m of hearths, whereas 17.3% (n = 39) occur within 1 m of ashy patches. But unlike the bowls, the specific hearths around which eggshell fragments concentrate are not overlain by subsequent dump deposits and so can be more confidently associated with their adjacent eggshell scatters. We thus interpret the domestic distribution of eggshell fragments as indicating that flasks were in use at these hearths and subsequently discarded when broken. Fragments recovered near hearths presumably reflect incomplete hearthside maintenance or fragment reuse. Fragments deemed waste were discarded either in nearby satellite dumps or in the main dump.

A detailed picture of ostrich eggshell flasks use, reuse and disposal can be reconstructed from the flask refit patterns, which are some of the most interesting of any class of archaeological material at DFM (Jakavula 1995; Figure 15.4b). Numerous connections exist within and between different eggshell clusters, features and areas of the site, showing that the flasks underwent remarkably dynamic depositional trajectories often involving long-distance movements. We recognize three broad types of connection:

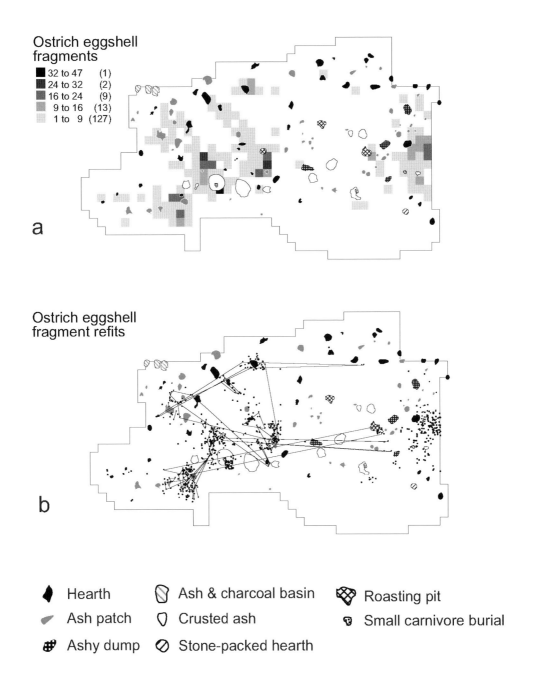

Figure 15.4 Spatial distributions at DFM of (a) ostrich eggshell flask fragments, (b) ostrich eggshell flask refits, and (c–d) ostrich eggshell beads in relation to the various feature types.

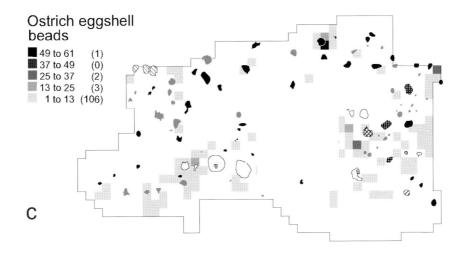

Ostrich eggshell
beads
■ 49 to 61   (1)
■ 37 to 49   (0)
■ 25 to 37   (2)
■ 13 to 25   (3)
□ 1 to 13 (106)

c

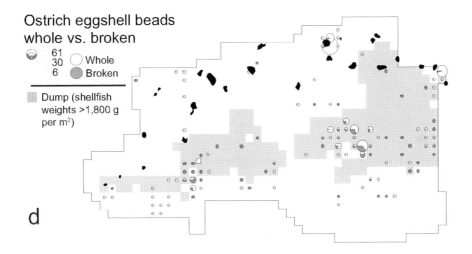

Ostrich eggshell beads
whole vs. broken
  61
  30   ○ Whole
  6   ● Broken

  Dump (shellfish
  weights >1,800 g
  per m²)

d

Excavation North

0   20 m

1) refits within the domestic area; 2) refits between the domestic area and the main dump; and, 3) refits within the main dump.

Within the domestic zone a web of refits links five individual domestic hearths (Features 11, 12, 22, 33, and 35; Figure 15.4b). We suggest that this reflects the daily movements of people between individual domestic spaces organized around fireplaces. As with broken pots (see below), large fragments of broken ostrich eggshell flasks still would have been useful as plates, bowls, scoops, spoons, bead blanks and so forth – particularly if, as we suspect, ostrich eggs were difficult to come by in the local sandveld landscape. These hearth-to-hearth connections probably thus reflect the extended use-lives of these valuable objects as people transported and broke flasks, then shared and discarded the fragments. That the web knits together three neighboring hearths (Features 11, 12, and 22) raises the possibility that these features belonged to people who were close, whether kin relations or friends (cf. Yellen 1977). In addition to hearth-to-hearth connections, refits within the domestic zone also imply discard processes. Two of the domestic hearths linked by the web discussed above (Features 22 and 12) connect to the satellite dump directly north of Feature 12 (associated with two ash patches, Features 85 and 87), which suggests that different domestic units (families?) were using the same satellite dump, further strengthening their spatial (and social?) connectedness.

A number of the eggshell clusters involved in these hearth-to-hearth and hearth-to-satellite dump relationships link into the larger eggshell concentrations in the main dump (Figure 15.4b). These domestic-to-dump refits provide fairly unambiguous evidence for secondary discard processes; exhausted flask portions were collected during the cleaning of hearthside domestic

spaces or satellite dumps and re-deposited in the site's primary waste accumulation, the main dump. But they also give a sense of how the main dump formed, with (what we presume to have been) individual domestic units consistently depositing their waste in specific (prescribed?) parts of the dump, creating initially separate middens that coalesced with time (Parkington et al. 2009:116). The refit patterns show that these individual dumps were located in areas of the main dump nearest to the living spaces of the given domestic units that produced them.

Finally, there are refits between ostrich eggshell fragments within the main dump. The main dump includes four major concentrations of eggshell fragments, each exhibiting considerable internal connectivity in refits (Figure 15.4b). However, occasional refits also link the concentrations to one another, which indicates they are contemporary, and again supports the notions that the main dump served interlinked domestic units and consists of several separate dumps that eventually fused together. The distances over which eggshell fragments refit in the main dump are extremely variable, ranging from a few centimeters to over 18 m (Figure 15.4b). These attenuated conjoins show that although discarded eggshells at DFM are intimately bound up with domestic processes in the north, occasional activities brought fragments further south, echoing the north-south connectivity exhibited by the domestic hearth-to-hearth refit patterns.

The above distributional and refit patterning gives an idea of ostrich eggshell flask use, reuse, sharing and discard at DFM. Also present at the site are ostrich eggshell beads which are, of course, another form of flask fragment reuse. Beads represent the far 'distal end' of an ostrich egg's use-life. Interestingly, broken but finished beads are more common in the dump, and whole finished ones are more common near hearths (Figure 15.4d). But despite the evidence presented above for hearth-to-hearth reuse of broken flasks, little attempt appears to have been made at DFM to turn ostrich eggshell fragments into beads. Out of a total of 408 ostrich eggshell beads at DFM, only three are unfinished and none were abandoned or broken in the course of manufacture (Kandel and Conard 2005; Orton 2008).

Spatially, beads and flask fragments consistently occur in different areas of both the domestic zone and especially the main dump (Figure 15.4a, c). In the domestic zone, the bead distribution is heavily skewed toward the south where several large clusters center directly on hearths. This pattern is the reverse of that exhibited by eggshell flask fragments; whereas the southern domestic area is virtually devoid of flask fragments, the north contains several sizeable concentrations. In the main dump these two artifact types coincide more frequently, but again their distributions diverge. From Figure 15.4a and 15.4c it is clear that squares with beads often occupy the spaces 'between' those with flasks fragments. As with the domestic zone, most beads in the main dump are situated in the south. This, again, contrasts sharply with the distributions of flask fragments, the vast bulk of which comprise one of three major main dump concentrations in the north. This spatial mismatch of ostrich eggshell beads and flask fragments suggests that these artifacts experienced independent depositional trajectories. This is consistent with relatively minimal evidence for on-site bead production, indicating that although the use-lives of the ostrich eggshell flasks may have extended beyond breakage, they were not attenuated enough, at least at DFM, for fragments to be recycled into wholly new artifact forms.

Comparing Containers: Carapace Bowls, Eggshell Flasks and Ceramic Vessels

Ceramic vessels constitute a third container type employed by the DFM inhabitants in addition to tortoise carapace bowls and ostrich eggshell flasks (see Stewart 2005a, 2005b). Unlike the latter two, however, pots never underwent a comparable phase of being a food resource prior to becoming an artifact, and thus serve as a useful 'yardstick' against which to compare the spatial distributions of bowls and flasks. Here, we expand our view to consider the life histories of these three container types, and contemplate how much of that life history transpired at DFM and was captured in this site's archaeological record. Because they differed sharply from one another in raw material availability, manufacturing effort, size, robusticity and, very probably, perceived value, our expectation is that the life histories of flasks, bowls and pots were diverse. Among our considerations is the notion that the number of specimens deposited at DFM and their spatial distribution might have been conditioned by the 'turnover rate' of each artifact type.

Tortoise carapace bowls began their artifact life history in the landscape outside of DFM when they were collected, most likely as live tortoises, and subsequently brought to DFM. At DFM the tortoises presumably were cooked and consumed, and afterward some carapaces were made into bowls. We suggest that carapace bowls typically spent most or all of their use-life at DFM. They could be replaced easily, so long as the availability of living tortoises remained adequate. Thus the turnover rate of carapace bowls might have been relatively rapid. If the DFM occupants used carapace bowls for serving food and eating, as in the Kalahari, they were likely to enter the archaeological record at DFM rather than at an offsite location. A high turnover rate coupled with discard at DFM rather than offsite

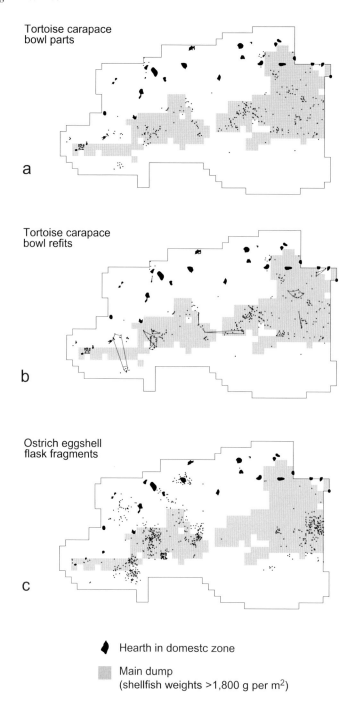

Figure 15.5 Spatial distributions at DFM of (a-b) tortoise carapace bowl parts and refits, (c-d) ostrich eggshell flask fragments and refits, and (e-f) ceramic sherds and refits in relation to the domestic zone of hearths and the main dump.

Ostrich eggshell
flask refits

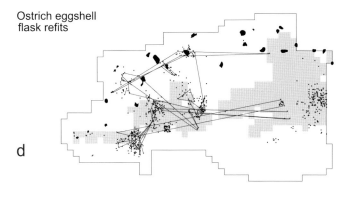

d

Ceramic sherds

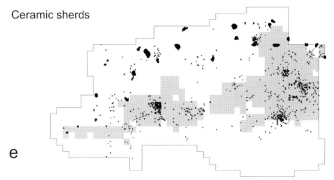

e

Ceramic sherd
refits

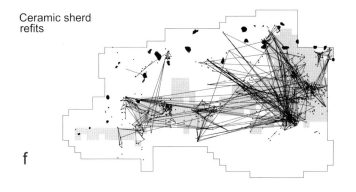

f

Excavation North

0   20 m

would result in a relatively high frequency of specimens; this is indeed the case, with 30 bowls a minimum estimate for DFM and upwards of 50 more likely.

Carapace bowl refits occur almost exclusively within the dump (Figure 15.5b). The few that occur in the domestic zone never connect two or more domestic hearths but instead form tight 'puddles' of near-complete bowls near several individual domestic hearths, and only one bowl connects the domestic area with the main dump. Moreover, the refit distances of carapace bowls within the dump span much shorter distances than is the case for ostrich eggshell and ceramic fragments. The combined evidence suggests that carapace bowls were rendered useless quickly after breakage, perhaps once one or two plates had fallen out, and disposed of in the dump while still relatively intact rather than after fragments had become dispersed. This is consistent with our notion that the turnover rate of these artifacts was relatively rapid.

We suggest that the life history of ostrich eggshell as an artifact differed significantly from tortoise carapace bowls in that ostrich eggs typically had lower turnover rates and generally 'passed through' DFM, whereas bowls were something more akin to 'permanent residents' at the site. Ostrich eggs also began their artifact life history away from DFM when they were obtained somewhere in the landscape. If ostriches and their eggs were rare in the sandveld, efforts would have been made to prolong their use-lives (Lee 1979:274). Also, if use as water flasks by the DFM occupants was the principal function of ostrich eggs, we suggest that flasks spent much of their life away from DFM. Flasks presumably would have accompanied women and men on gathering and hunting expeditions, respectively (Lee 1979; Silberbauer 1981). Further, the absence at DFM of whole flasks or even large fragments suggests

those that remained intact when the inhabitants moved on were taken from DFM. The upshot of all this is that most flasks whose artifact life ended in breakage probably came to this end away from DFM. DFM was just a "stop" in what was a geographically broad network of use, and few flasks entered the archaeological record there. This scenario is consistent with the small number of flasks at the site - five to seven specimens.

Flask refits, unlike those of tortoise carapace bowls, connect between different domestic hearths with some regularity (Figure 15.5d). Perhaps broken flask fragments had useful functions as implements of some kind (e.g., plates, spoons, scoops), and once a flask broke the pieces were shared among multiple hearths (households?). We submit that the most 'intuitive' reason for sharing broken ostrich eggshell fragments is that they provide raw material for ostrich eggshell bead manufacture. But there is considerable evidence to suggest that the site occupants engaged only minimally in bead production at DFM. For now we lack a compelling explanation for this pattern, but the limited number of water flasks at DFM combined with their long refit distances is consistent with the idea that ostrich eggshells represented rare and thus relatively precious artifacts to be handled carefully, curated, perhaps reused as implements (if not recycled as beads) or, if still whole, transported away from DFM.

Ceramic vessels, of which there are at least 20 at DFM, also started their life history offsite where they were manufactured or acquired through trade with neighboring ceramic-producers (Stewart 2005b, 2008). Since the DFM pots were clearly employed in cooking (Stewart 2005a), their life history, like that of carapace bowls, was probably more strongly centered than ostrich eggshell flasks at DFM than at offsite locations. But similar to flasks (and unlike carapace bowls), complete pots or serviceable pot portions would have been transported offsite when the occupants moved. Whether the pots were manufactured by the DFM occupants or traded in, the ceramic assemblage represents a substantial investment in time and/or resources. We suggest that efforts to protect, repair and reuse pots (Stewart 2008), along with their relatively high robusticity, would have resulted in use-lives that were certainly longer than carapace bowls and perhaps also ostrich egg flasks. But we also expect that the greater amount of time pots spent at DFM relative to flasks would have offered more opportunities for breakage on-site, resulting in more vessels than flasks but perhaps fewer than carapace bowls (which, though also 'site-based', were clearly more expendable). This is indeed what we see; DFM contains at most 25 ceramic vessels, more than the maximum number of ostrich egg flasks (n ≈ 7) and less than carapace bowls (n ≈ 50).

The distributions of ceramic sherds (Figure 15.5e) echo those of ostrich egg flasks and tortoise carapace bowls in that each of these classes of artifact shows spatially tight concentrations of small numbers of specimens adjacent to and clearly associated with domestic hearths. These clusters might represent specimens that were broken shortly before the occupants departed DFM and thus were not cleaned up onto the dump, or they could be cached vessels (Nilssen 1989; Stewart 2008). Relative to ostrich eggshell flasks, the ceramic refits infrequently connect separate domestic hearths, perhaps implying that ceramic vessels rarely moved between domestic households. But the ceramics exhibit a much more expansive network of refits than either ostrich egg flasks or carapace bowls, with frequent connections between the domestic area and the main dump and within the dump (Figure 15.5f). We suggest that the relatively long use-lives and the

site-based life histories of the DFM pots are reflected in these refit patterns. The use-life trajectory of ceramic vessels harbors potentially substantial complexity, and broken specimens could still be functional, albeit for uses different from their original function (Stewart 2005b, 2008). A range of stages in the use-life of vessels is represented by specimens of varying completeness, and different pieces of the same vessel are sometimes distributed among various locations at DFM, probably reflecting a "sequence" through time of a vessel being reduced by breakage and pieces re-used then eventually discarded (Stewart 2008).

Discussion: From Foodwaste to Artifact
Making sense of the spatial distributions of tortoise remains and ostrich egg fragments at DFM involves a consideration of seemingly complex life history trajectories, inasmuch as these two items offered the DFM inhabitants both a meal and a raw material from which to manufacture artifacts ranging from the practical to the ornamental. We find ourselves striving to integrate a diverse set of factors as we attempt to disentangle these differing but interconnected pathways. The factors involved range from cooking, consumption, and discard patterns conditioned by the size and structure of the food "package" to the trajectory and location of artifact manufacture, use, and discard to the "nuisance value" (O'Connell 1995) of objects and how that influenced the manner and location of disposal. Here we discuss several implications of these complex foodwaste-to-artifact trajectories for interpreting archaeological spatial patterning both at DFM and at older, less well preserved archaeological home bases.

A first observation is that the difference in availability of a food resource can directly affect the value of an artifact made from its waste byproducts and, by extension, the artifact's spatial distribution at home bases in the archaeological record. Tortoises abound in DFM's sandveld landscape. Ample access to these animals reduced the value of carapace bowls. Even if a single carapace plate of a bowl fell out or broke, another bowl could be rapidly produced from an almost inexhaustible supply of live tortoises in the landscape or from discarded carapaces at DFM. Thus, carapace bowls at DFM probably had high turnover rates and negligible reuse of fragments upon breakage. The result is that although carapace bowls are the most abundant container type at DFM, they exhibit the shortest refit distances. This foodwaste-to-artifact pathway contrasts sharply to that of ostrich eggs, which are far less frequent in the sandveld landscape than tortoises. The limited accessibility of ostrich eggs and their importance as water flasks lowered their turnover rates and raised frequencies of flask fragment reuse. Consequently, there are far fewer ostrich egg flasks relative to carapace bowls at DFM, but individual flasks are widely dispersed with webs of long-distance refits linking domestic hearths with one another and with the main dump.

The analysis of spatial patterns of items that could be both foodwaste and artifact also has methodological implications for transcending the distinction between subsistence and technology at archaeological home bases. Striking differences in the spatial distribution of tortoise carapace bowls compared to unmodified carapace at DFM offers perhaps the strongest signal separating foodwaste and artifact. Carapace bowls occur largely in the dump whereas unmodified carapace occurs all throughout the dump and the domestic area, where they often associate with hearths. This dissimilarity suggests a significant difference in disposal between artifact and foodwaste - when it came time to dispose of carapace bowls they usually

went to the dump, while disposal of unmodified carapace was more widespread across the site. The progressive disintegration and dispersal through time of a discarded, unmodified carapace or plastron into individual plates in the domestic area could result in rather widespread distribution of fragments. But could the differences in carapace distribution be detected at archaeological sites of people, such as the Ju/'hoansi, who did not modify carapace bowls before using them? If not, these important artifacts and their use-contexts would be misinterpreted as strictly subsistence-related. Archaeologists must be able to distinguish the spatial patterning of artifact and foodwaste categories, even when they are manifested by specimens made of the same raw material. We have presented steps in this direction in our analyses of tortoise and ostrich egg at DFM.

Finally, the foodwaste/artifact dichotomy holds implications for discerning social relationships and organization at archaeological home bases. Most efforts to reconstruct social connectivity at prehistoric campsites have focused on subsistence remains, particularly those of large animals that would have been shared between domestic units. The success of this approach depends directly on these bones being in primary context, such as near hearths around which consumption took place, which is most likely to happen at sites where occupation spans were short. But ethnoarchaeological research demonstrates that prehistoric home bases occupied for a longer period of time are more likely to have survived for archaeologists to detect them and, unfortunately, longer occupational durations mean more intensive secondary discard processes. DFM is a case in point. Here, larger animals such as eland and steenbok are more likely to have been shared between social units such as nuclear families, but also are more likely to have ended up on the dump where the social connections are harder to trace (Stewart 2008). Smaller animals such as tortoise are more likely to have remained in primary context, but are less likely to have been shared between social units. By contrast, although some artifact types at DFM, notably carapace bowls, generally ended up on the dump, ceramics and ostrich egg flasks occur in the domestic area, and refits connect separate hearths. It might be, therefore, that artifacts - and especially those that were curated and shared - hold greater potential than subsistence remains for recognizing social connections at many archaeological home bases since efforts to remove them from domestic contexts were less rigorous.

Final Reflections

The objective of Glynn Isaac's pursuit of a home base model in the 1970s was to demonstrate that early Pleistocene sites could plausibly be interpreted as home bases - places to which ancient men and women returned with food items separately gathered. The 'home base' concept implied the existence of a number of behavioral associations, including a division of labor, the capacity to agree on reuse of specific places as central locations from which to forage, and food sharing. In Isaac's view, early Pleistocene foragers could be shown already to exhibit behavioral patterns well documented in the ethnographic record. Other archaeologists, of course, preferred to see a much later appearance of such 'modernity' (and Isaac himself modified his views). Despite this, the objective was rather simple. Positing the existence of a home base needed, according to Isaac, only the demonstration that faunal assemblages documented the residues of multiple episodes of animal food acquisition.

Since then ethnoarchaeologists and archaeologists have adopted much more ambitious

agendas for the interpretation of scatters and patches of apparently associated faunal and artifactual remains. At DFM, for example, we recognize that a primary obligation is to persuasively argue for brevity of occupation and reliability of association as the underpinnings for a number of further inferences and conclusions. A large, reliably associated assemblage that includes food debris and artifactual wastes deriving from many sources and materials provides invaluable evidence as to the identity of occupants, the seasonality of visits, the exploitation of resources and the domestic organization of groups. Included in the behavioral objectives are the recognition and documentation of food and artifact sharing, waste disposal, artifact manufacture and gendered uses of features and spaces. Comparing the mapped distributions of similar and different sets of items, and tracing their variable trajectories using refits, lets us hypothesize about the behavioral dynamics that suffused and structured home bases in prehistory. Because even a hunter gatherer camp has a life history, it becomes possible to try to trace the passage of objects into, through and out of briefly occupied sites.

From our discussion above of tortoise and ostrich eggshell remains, it is obvious that there is a deep level of connectivity across the 859 m^2 we have excavated. Links between features that reveal contemporaneity of occupation across the whole excavated space boost our confidence in the brevity of occupation, and increasingly support the tight radiocarbon chronology. Dissecting this connectivity, we have argued for some specific uses of features within a broad distinction between domestic and waste disposal zones, with some features associated with particular types of artifacts and/or non-artifactual animal remains and others not, some quite focused, others less so. We have examined in some detail two such remains - tortoise and ostrich eggshell - that are particularly revealing since both often transcended the porous boundary between subsistence and technology. We have argued that: 1) the availability of a food resource directly influences the abundance and spatial patterning of the artifactual byproduct; 2) interpretations of spatial patterning depend on the ability to distinguish between foodwaste and artifacts; and, 3) artifacts might yield better information than foodwaste about social connectivity at many home bases. We feel this last point to be particularly salient.

The success of spatial research at archaeological home bases in the 21st century depends on our capacity to bridge the yawning temporal and inferential gap between 20th century hunter-gatherers and archaeological traces of far more ancient foragers. The analytical power of relatively recent sites like DFM lies in the richness, diversity and preservation of remains, which greatly enhances the strictly 'stones and bones'-based archaeology that typifies most Paleolithic enquiries. But even at sites like DFM interpretive challenges loom. Efforts must be made to recognize the early, patchy accumulation of dumps that subsequently expanded and coalesced, swamping some hearths with their associated debris. More detailed work on tortoise limb bone distributions will likely firm up the recognition of consumption zones. That minimal bead making took place may have to do with the vagaries of behaviors on briefly occupied sites, but accurate interpretation must wait until other local and regional comparisons become available. These future directions should further strengthen our inferences at DFM, and with them our confidence that recent, well preserved and extensively excavated home bases like DFM can serve as stepping stones back into earlier times of the Paleolithic.

Acknowledgements

We thank Dolores Jacobs, whose indefatigable work in processing and refitting archaeological materials from DFM underlies much of what we have accomplished here. A good deal of what we have presented in this paper builds on earlier refitting and/or recording work by Zukisani Jakavula, Peter Nilssen and Tobias Tonner, all of whom we gratefully acknowledge. We are also grateful to Stuart Challender for providing GIS expertise. B. A. S thanks Oxford University Press and St. Hugh's College, Oxford, for doctoral funding. J. P. thanks the University of Cape Town for sustained logistical and financial support. J. W. F. thanks Montana State University for a Scholarship and Creativity grant that supported his research on DFM materials. We thank Jeanne Sept and David Pilbeam very deeply for inviting us to contribute to this memorial for Glynn. Above all we acknowledge Glynn Isaac, who with warm and cheerful countenance oriented Stone Age archaeology in directions that continue to remain so relevant today.

References

Binford L. R.

1978 *Nunamiut Ethnoarchaeology*. Academic Press, New York.

1981 *Bones: Ancient Men and Modern Myths*. Academic Press, New York.

1983 *In Pursuit of the Past: Decoding the Archaeological Record*. Thames and Hudson, New York.

1985 Human ancestors: Changing views of their behavior. *Journal of Anthropological Archaeology* 4:292-327.

Blumenschine, R. J.
1986 Carcass consumption sequences and the archaeological distinction of scavenging and hunting. *Journal of Human Evolution* 15:639-659.

Bunn, H. T.
1981 Archaeological evidence for meat-eating by Plio-Pleistocene hominids from Koobi Fora and Olduvai Gorge. *Nature* 291:574-577.

Fisher, Jr., J. W., and H. C. Strickland
1989 Ethnoarchaeology among the Efe Pygmies, Zaire: Spatial organization of campsites. *American Journal of Physical Anthropology* 78:473-484.

1991 Dwellings and fireplaces: Keys to Efe Pygmy campsite structure. In *Ethnoarchaeological Approaches to Mobile Campsites*, edited by C. S. Gamble and W. A. Boismier, pp. 215-236. International Monographs in Prehistory, Ann Arbor.

Gifford-Gonzalez, D., D. B. Damrosch, J. Pryor, and R. L. Thunen
1985 The third dimension in site structure: An experiment in trampling and vertical dispersion. *American Antiquity* 50:803-818.

Gould, R. A. (Editor)
1978 *Explorations in Ethnoarchaeology.* University of New Mexico Press, Albuquerque.

Gould, R. A.
1980 *Living Archaeology.* Cambridge University Press, Cambridge.

Isaac, G. L.
1978a The food sharing behaviour of protohuman hominids. *Scientific American* 238:90-106.

1978b Food sharing and human evolution: Archaeological evidence from the Plio-Pleistocene of East Africa. *Journal of Anthropological Research* 34:311-325.

1984 The archaeology of human origins: Studies of the Lower Pleistocene in East Africa 1971-1981. *Advances in World Archaeology* 3:1-97.

Jakavula, Z.
1995 More Than Lines on a Map?: Refitting Dunefield Midden. Unpublished Honours thesis, Department of Archaeology, University of Cape Town.

Jerardino, A., L. K. Horwitz, A. Mazel, and R. Navarro
2009 Just before Van Riebeck: Glimpses into terminal LSA lifestyle at Connies Limpet Bar, West Coast of South Africa. *South African Archaeological Bulletin* 64:75-86.

Kandel A. W., and N. J. Conard
2005 Production sequence of ostrich eggshell beads and settlement dynamics in the Geelbek Dunes of the Western Cape, South Africa. *Journal of Archaeological Science* 32:1711-1721.

Kroll, E. M., and G. L. Isaac
1984 Configurations of artifacts and bones at Early Pleistocene sites in East Africa. In *Intrasite Spatial Analysis in Archaeology*, edited by H. Hietala, pp. 4-31. Cambridge University Press, Cambridge.

Lee, R. B.
1979 *The !Kung San: Men, Women and Work in a Foraging Society*. University of Cambridge Press, Cambridge.

Murray, P.
1980 Discard location: The ethnographic data. *American Antiquity* 45:490-502.

Nilssen, P. J.
1989 Refitting Pottery and Eland Body Parts as a Way of Reconstructing Hunter-Gatherer Behavior: An Example from the Later Stone Age at Verlorenvlei. Unpublished Honours thesis, Department of Archaeology, University of Cape Town.

O'Connell, J. F.
1977 Ethnoarchaeology of the Alyawarra: A report. *AIAS Newsletter* 7:47-49.

1987 Alyawara site structure and its archaeological implications. *American Antiquity* 52:74-108.

1995 Ethnoarchaeology needs a general theory of behavior. *Journal of Archaeological Research* 3:205-255.

Orton J.
2002 Patterns in stone: The lithic assemblage from Dunefield Midden, Western Cape, South Africa. *South African Archaeological Bulletin* 57:31-37.

2008 Later Stone Age ostrich eggshell bead manufacture in the Northern Cape, South Africa. *Journal of Archaeological Science* 35:1765-1775.

Parkington, J. E.
2006 *Shorelines, Strandlopers and Shell Middens*. Krakadouw Trust, Cape Town.

Parkington, J. E., and J. W. Fisher, Jr.
2006 Small mammal bones on Later Stone Age sites from the Cape (South Africa): Consumption and ritual events. In *Integrating the Diversity of Twenty-First-Century Anthropology: The Life and Intellectual Legacies of Susan Kent*, edited by W. Ashmore, M.-A. Dobres, S. M. Nelson, and A. Rosen, pp. 71-79. Archaeological Papers of the American Anthropological Association, Vol. 16.

Parkington, J. E., J. W. Fisher, Jr., and T. W. W. Tonner
2009 'The fires are constant, the shelters are whims': A feature map of Later Stone Age campsites at the Dunefield Midden site, Western Cape Province, South Africa. *South African Archaeological Bulletin* 64:104-121.

Parkington, J. E., P. Nilssen, C. Reeler, and C. Henshilwood
1992 Making sense of space at Dunefield Midden campsite, western Cape, South Africa. *Southern African Field Archaeology* 1:63-70.

Potts, R.
1984 Home bases and early hominids. *American Scientist* 72:338-347.

1988 *Early Hominid Activities at Olduvai*. Aldine de Gruyter, New York.

Reeler, C.
1992 Spatial Patterns and Behaviour at Dunefield Midden. Unpublished M.A. thesis, Department of Archaeology, University of Cape Town.

Schapera, I.
1930 *The Khoisan Peoples of South Africa: Bushmen and Hottentots*. Routledge and Kegan Paul, London.

Schiffer, M. B.
1987 *Formation Processes of the Archaeological Record*. University of New Mexico Press, Albuquerque.

Sept, J.
1982 Was there no place like home? *Current Anthropology* 33:187-207.

Shipman, P.
1983 Early hominid lifestyle: Hunting and gathering or foraging and scavenging? In *Animals and Archaeology: Volume I, Hunters and Their Prey*, edited by J. Clutton-Brock and C. Grigson, pp. 31-49. British Archaeological Reports, Oxford.

1984 Ancestors: Scavenger hunt. *Natural History* 93:20-27.

Silberbauer, G. B.
1981 *Hunter and Habitat in the Central Kalahari Desert*. Cambridge University Press, Cambridge.

Stewart, B. A.

2005a Charring patterns on reconstructed ceramics from Dunefield Midden: Implications for Khoekhoe vessel form and function. *Before Farming* [online version] 2005/1 article 1.

2005b The Dunefield Midden ceramics: Technical analysis and placement in the Western Cape sequence. *South African Archaeological Bulletin* 60:103-111.

2008 Refitting Repasts: A Spatial Exploration of Food Processing, Sharing and Disposal at the Dunefield Midden campsite, South Africa. Unpublished D.Phil. Dissertation, Institute of Archaeology, University of Oxford.

2010 Modifications on the bovid bone assemblage from Dunefield Midden, South Africa: Stage one of a multivariate taphonomic analysis. *Azania: Archaeological Research in Africa* 45:238-275.

2011 'Residues of parts unchewable': Stages two and three of a multivariate taphonomic analysis of the Dunefield Midden bovid bones. *Azania: Archaeological Research in Africa* 46:141-168.

Stynder, D. D.

2008 The impact of medium-sized canids on a seal bone assemblage from Dunefield Midden, West coast, South Africa. *South African Archaeological Bulletin* 63:159-163.

Tonner, T. W. W.

2002 A spatial database for the Later Stone Age site 'Dunefield Midden' (Western Cape, South Africa). Unpublished M.A. thesis, Department of Archaeology, University of Cape Town.

2005 Later Stone Age shellfishing behaviour at Dunefield Midden (Western Cape, South Africa). *Journal of Archaeological Science* 32:1390-1407.

Wilson, D. C.

1994 Identification and assessment of secondary refuse aggregates. *Journal of Archaeological Method and Theory* 1:41-68.

Yellen, J. E.

1976 Settlement patterns of the !Kung: An archaeological perspective. In *Kalahari Hunter-Gatherers: Studies of the !Kung San and their Neighbors*, edited by R. B. Lee and I. DeVore, pp. 47-72. Harvard University Press, Cambridge (MA).

1977 *Archaeological Approaches to the Present: Models for Reconstructing the Past.* Academic Press, New York.

Yellen, J. E., and R. B. Lee

1976 The Dobe-/Du/da environment: Background to a hunting and gathering way of life. In *Kalahari Hunter-Gatherers: Studies of the !Kung San and their Neighbors*, edited by R. B. Lee and I. DeVore, pp. 27-46. Harvard University Press, Cambridge (MA).

16

A VERY PARTICULAR KIND OF ARCHAEOLOGIST

Bernard Wood

"... if Watson and Crick had not discovered the structure of DNA, one can be virtually certain that other scientists would eventually have determined it. With art -whether painting, music, or literature - it is quite different. If Shakespeare had not written Hamlet, no other playwright would have done so".

- Lewis Wolpert

It is exactly 25 years, and thus a generation ago, since the posthumous publication of the last of Glynn Isaac's single author papers. Thus, for today's graduate students and for younger researchers the only contact they have with Glynn is his publications. Fortunately many of these have been reproduced in *The Archaeology of Human Origins* edited by Barbara Isaac (1989). In the Foreword to that edited volume no less an archeologist than Mary Leakey paid tribute to the way Glynn helped sweep away her "fallacious beliefs" leaving her with "a greater clarity of mind" (Isaac 1989:xv) and in his excellent and informative introduction John Gowlett refers to Glynn contributing "many of the most penetrating thoughts conceived within the domain of human evolution" (Isaac 1989:1). To those who did not know Glynn this tribute might seem hyperbolic but in this short essay I hope to demonstrate that for those of us who had the good fortune to be his colleagues it is a realistic and even-handed assessment of Glynn's contribution to human origins research.

My initial contact with Glynn came through our participation in the East Rudolf Research Project (ERRP). I had been to East Rudolf in 1968 as a medical student, before Glynn's involvement, but by the time I returned in 1971

Glynn had joined the ERRP as its co-leader and I was still a very junior participant. Many of my friends know what my critics have long suspected; I have no formal education in archaeology. At the time this was both "bad news" and "good news". The bad news is obvious; I did not know a flake from a hand-axe. The good news is that it meant I needed to learn "on the job" and Glynn was a willing and generous tutor.

Although there was a central permanent camp at Koobi Fora in the 1970s as far as field camps were concerned the archeologists and the paleontologists used to operate more or less independently, with Glynn in charge of the archaeology camp and Kamoya Kimeu the *de facto* leader (between visits from Richard Leakey who was by that time running the National Museums of Kenya as well as co-leading the ERRP) of the paleontology field effort. My prowess as a fossil finder was on a par with my archeological knowledge, so on several occasions I was "released" from being an ineffective field paleontologist to go and work as a field assistant with Glynn and/or Kay Behrensmeyer.

My main memory is of the "scatter between the patches" project, which was a quintessential Glynn Isaac-type question-driven endeavor. It was obvious that there were places called sites where stone tools and other evidence of hominin activity were concentrated, but did that mean that stone tools were confined to such places, or were they more widely distributed across the landscape, but in such low numbers and thus at such a low density that they were effectively invisible? My memory is that 5-m-wide squares were randomly chosen across the landscape and these were to be crawled over to make sure we picked up each broken bone and small flake. I was used to random trials in medicine, but random squares were something else. Sometimes a random square would coincide with a thicket of thorn bushes or with a steep scree-like slope. No amount of special pleading on my part could persuade Glynn to transfer to a more convenient adjacent square, so off you went crawling under "wait-a-bit" (*Acacia mellifera*) thorns, your nose within inches of the ground. You learn a lot about someone when you share that type of experience, and if you have the good fortune to share it with Glynn Isaac you learn a lot else as well.

The first thing I learned from that project was that Glynn was not just an archeologist. He was an evolutionary biologist with a special interest in the evolution of human behavior *who happened to use archeological methods*. Glynn used hypotheses in archaeology the way biologists do. He thought in terms of testable models, but whereas biologists might be able to perturb nature with a well-designed experiment, Glynn would test the hypothesis with empirical observations. He did not merely "talk the talk" of a Popperian, he really "walked the walk"; he was his own fiercest critic. But he was not averse to experimentation and together with his students and geological colleagues he conducted several important taphonomic experiments involving recently killed or scavenged crocodiles and large mammals.

The second lesson I should have learned from Glynn is intellectual humility. Despite his towering contributions to human origins research one never had the sense that any of these contributions were "his" property. He never sought recognition for them. To him they were like a contribution to a communal meal; it was no more or no less than was expected of any participant. His assessment was that his idea or hypothesis was just one among many, *primus inter pares*, but the rest of us knew that was not true.

The third lesson concerned the manner and tone of scientific discourse. Glynn was adamant that the aim was hypothesis building and then hypothesis testing. In an ideal world you should test your own hypotheses, but in the real world it is inevitable that on occasion others will test them for you and find them wanting. Glynn refused to take such attempts personally and I have a vivid recollection during the time of the "KBS tuff controversy" that Glynn worked hard to maintain his friendships with those on the other side of the debate. Glynn did not worry about who "won" or "lost"; he was content as long as our collective ignorance was being reduced.

So how does the quote from Lewis Wolpert's book relate to Glynn Isaac? In an important sense the quotation is correct. If Glynn had not excavated at Olorgesailie or at Koobi Fora, someone else would have done and perhaps they would have made as good a technical job of it as Glynn did. But in another sense the quotation fails to capture the intangible and arguably the more important contributions made to science by special human beings such as Glynn. He set a standard that made other scientists perform above their potential and his demeanor and behavior stimulated others to think in ways they would not have done without Glynn's benign but powerful influence.

If all this sounds hopelessly airy-fairy then I urge you to read a paper of Glynn's that is not included in the volume of his collected papers. It is a review contributed to a Nobel Symposium held in 1978, but not published until 1981 (Isaac 1980). Entitled, "Casting the net wide: A review of archaeological evidence for early hominid land-use and ecological relations", it demonstrates both the breadth of Glynn's interests and expertise and the depth of his wisdom. In it he advocated that "behavior, adaptation, phylogeny, and taxonomy cannot be seen as matters for separate investigation" (Isaac 1980:245) and he wrote of his own efforts to "try to see if the archaeological record of the development of human behavior helps us in any way to formulate explicit, testable hypotheses that can serve as the *first approximations of answers* (my italics)" (Isaac 1980:230). He also wrote urging scientists engaged in human evolution "to adopt less gladiatorial poses" and he went on to opine that "whereas the prizes used to go for offering or defending a single plausible version of the evolutionary story, I advocate that from now on the laurels should be awarded for providing a series of alternative models which can be tested and used to guide future research" (Isaac 1980:230). It is my own belief that if Glynn had lived out his life in the way we all hope we will do (i.e., to well beyond "three score years and ten") he would have made a real difference by influencing his colleagues to set aside the confrontational style that unfortunately still persists, and apparently prospers, today.

The last lesson I learned from Glynn is the value of enthusiasm. Not the breathless enthusiasm beloved of announcers in the media, but the type of enthusiasm that never lets a research opportunity or a "teachable moment" slip by. But I also learned that Glynn's enthusiasm could have costs as well as benefits. That learning

experience took place at the "First International Congress of Human Palaeontology" held in 1982 in Nice. No one who attended that meeting will ever forget it. Not, I might add, for the science, but for the sheer scale of the hospitality. Henry de Lumley, who seemed to have had access to half the annual GDP of France, organized it. The final banquet is the stuff of legend. It was held in the gardens of the Matisse Museum at Cimiez, on the outskirts of Nice. The term "garden" is a misnomer, for it consists of olive groves within and around a Roman Amphitheatre. Fleets of buses took us there and we strolled along the pathways enjoying the generous food and the flowing wine. However, this was merely the hors d'oeuvre for at an appointed hour cloths were whipped off tables that positively groaned with roasted meat cut from carcasses on spits and even better wines. It is fair to say that my memories of that evening are vague, but I do distinctly remember both Glynn and Desmond Clark each holding a ribbon and dancing round a Maypole, and police motorcyclists with whistles trying to herd human paleontologists towards the waiting buses in the early hours of the morning. Much earlier in the day, when the early morning sun was shining brightly and Mediterranean was at its bluest, Glynn had decided to take a swim and he persuaded Derek Roe and me to join him; none of us had swimsuits. Those of you who know Derek will not be surprised that he declined the chance of a swim either in his undergarments, or without them, but he offered to watch our clothes. It was a truly beautiful morning, but it was very early and the beach was deserted. After a fine swim we sat down among the pebbles in the watery sun to dry out, trying to look as inconspicuous as possible. But Glynn could not resist the opportunity to explore the fracture properties of the pebbles and he started to make a stone tool. Glynn was in his element, sharp flakes were flying in all directions and the tool was taking shape, but one of the flying flakes cut straight through a vein on the top of my naked foot (with hindsight thank goodness it was the foot!) and dark blue venous blood issued forth in copious quantities. I had learned in surgery that you can stop venous bleeding by elevating the affected part above the heart, so I lifted my naked leg vertically and the bleeding did soon stop. But instead of being inconspicuous the three of us immediately became *very* conspicuous. Mercifully two mostly naked men must have been a common sight on the Nice beach because a passing gendarme merely smiled knowingly. Meanwhile, Glynn continued to fashion the stone tool (which I still have) and whenever I notice the still-visible scar on the dorsum of my foot and I smile outwardly and inwardly and think of Glynn and especially remember the blessing of his uncomplicated and infectious enthusiasm (and the risks connected with the manufacture of stone tools).

A few years ago it was proposed we search for an archaeologist, but in the vain hope that we could attract someone in Glynn's mold I suggested we advertize not for an archaeologist, but for someone with a research interest in the "evolution of human behavior". Not one archaeologist applied, but I feel certain Glynn would have done.

References

Isaac, Barbara (Editor)

1989 The Archaeology of Human Origins. Cambridge University Press, Cambridge.

Isaac, Glynn

1980 Casting the net wide: A review of archaeological evidence for early hominid land-use and ecological relations. In *Current Argument on Early Man*, edited by L.-K. Königsson, pp. 226-251. Pergamon Press, Oxford.

Wolpert, Lewis

1992 *The Unnatural Nature of Science*, p. 86. Faber, London.